Microtechnology and MEMS

Series Editors

Prof. Dr. Hiroyuki Fujita
Institute of Industrial Science, University of Tokyo, Tokyo, Japan
Prof. Dr. Dorian Liepmann
Department of Bioengineering, University of California, Berkeley, USA

For further volumes:
http://www.springer.com/series/4526

The series Microtechnology and MEMS comprises text books, monographs, and state-of-the-art reports in the very active field of microsystems and microtechnology. Written by leading physicists and engineers, the books describe the basic science, device design, and applications. They will appeal to researchers, engineers, and advanced students.

Pierre Lambert
Editor

Surface Tension in Microsystems

Engineering Below the Capillary Length

Springer

Editor
Pierre Lambert
BEAMS Department CP 165/56
Université libre de Bruxelles
Brussels
Belgium

ISSN 1615-8326
ISBN 978-3-642-37551-4 ISBN 978-3-642-37552-1 (eBook)
DOI 10.1007/978-3-642-37552-1
Springer Heidelberg New York Dordrecht London

Library of Congress Control Number: 2013942011

Printed on acid-free paper

Springer is part of Springer Science+Business Media (www.springer.com)

A Virginie. Je demande pardon aux enfants d'avoir dédié ce livre à une grande personne.
To all those whose time I devoted to writing it.

Foreword

Less noticed in the daily life, the surface tension exists everywhere, for example, making almost perfect sphere rain drops on lotus leaves or beautiful soap bubbles. The lotus leaf repels rain drops so well that they roll around the surface collecting all the dusts before falling down; thus the leaf surface is always kept clean. Actually the surface has many microscopic protrusions covered with nano hairs of hydrophobic nature. We can find a real nano tech surface in nature. Recently, artificially made superhydrophobic surfaces based on biomimetic engineering are found many places from table sheets, umbrellas and building painting.

In my own experience, I found the surface tension problematic as a researcher on MEMS (micro electro mechanical systems). The MEMS technology relies on silicon based semiconductor microfabrication technique to make three dimensional micro miniature structures for sensors and actuators. Sometimes, those structures are released from substrate to move freely. Because their typical size is in the order of ten to hundred micrometers, where the surface tension force dominates over inertial or gravitational forces, they were pulled down to the substrate and stuck to it during evaporation of rinsing liquid. Quite a few countermeasures were developed from hydrophobic surface treatment to all dry release processing by gaseous or plasma etching. Almost all those works were, however, conducted intuitively or experimentally, because it was not easy to predict the behavior of surface tension phenomena.

This book edited by Prof. Pierre Lambert gives us the comprehensive understanding of the surface tension effect especially in the microscopic sizes. The book starts with basic physics of the surface tension and gradually proceeds into modeling of static and dynamic behaviors of capillary forces and torques. The models and deducted formulas provide indispensable tools for the scientific understanding of microscopic surface tension phenomena and the engineering design of MEMS devices and fabrication processes. The design includes not only solving problems associated with capillary forces but also utilizing surface tension phenomena in advanced MEMS.

Handling liquid in MEMS, more specifically in microfluidic devices, has become very popular for the applications in chemical, biological and medical fields. The capillary force always plays an important role when filling and flowing liquid in microchannels. In addition to continuous flow based devices, microfluidic devices based on droplet handling are emerging. This book will help researchers designing such devices and analyzing experimental observations.

The last part of the book describes case studies how to utilize surface tension phenomena in small scales, e.g. microactuation by capillary forces and self-assembly of small parts by surface tension. Readers of the book should be stimulated to find novel ways to take advantages of the phenomena. A very new class of micro nano devices will be invented along the way shown in Part IV of the book.

Prof. Lambert and his colleagues successfully explain complicated physics and mechanics of capillary forces in an easy-to-understand sentences and figures. The formulas are useful to predict their behaviors associated with change in parameters such as dimensions and material properties. Also many references are cited to help readers for the further study. I believe this book is a valuable asset in laboratories investigating MEMS and microfluidic systems as well as a good textbook for students.

Tokyo Hiroyuki Fujita

Preface

This book describes how surface tension effects can be used by engineers to provide mechanical functions in miniaturized products (<1 mm). Even if precursors of this field such as Laplace already date back to the eighteenth century, describing surface tension effects from a mechanical perspective is very recent. The originality of this book is to consider the effects of capillary bridges on solids, including forces and torques exerted both statically and dynamically by the liquid along the 6 degree of freedom (DOF). It provides a comprehensive approach to various applications, such as capillary adhesion (axial force), centering force in packaging (lateral force), and recent developments such as a capillary motor (torque). It devises how surface tension effects can be used to provide mechanical functions such as actuation (bubble-actuated compliant table), sealing, and tightness. Different case studies will also be proposed in the fields of energy harvesting, nanodispensing, self-assembly...

Context and Historical Note

The skeleton of this book concerns the modeling, characterization, and application of surface tension effects in precision and micro-engineering.

This project will therefore quickly recall the dominance of surface tension effects at scales smaller than a few millimeters. This limit is known as the *capillary length* L_C, which expresses the ratio between gravity and surface tension effects:

$$L_C = \sqrt{\frac{\gamma}{\rho g}} (\approx 2.7 \text{ mm for water}) \quad (1)$$

where γ is the surface tension of the liquid (Nm^{-1}), ρ is its density, and $g = 9.81$ ms^{-2}. Moreover, as it can be observed from dimensions, capillary forces linearly depend on length scale ($F \div \gamma \ell$), leading to the dominance of surface tension effects over other physical effects at small scale. Consequently, studying capillary forces is a good starting point to enter micro-engineering since it is a dominant

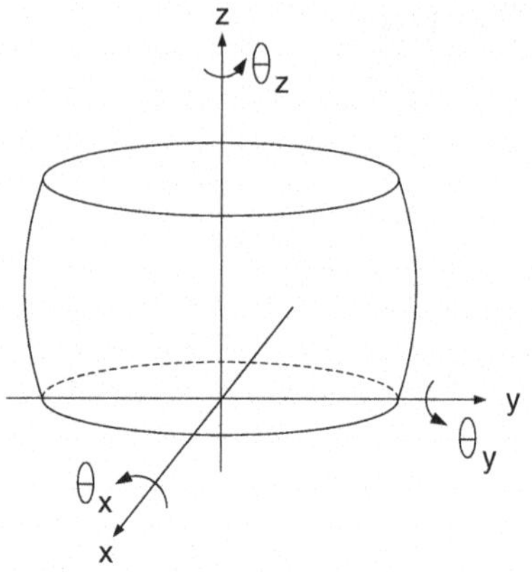

Fig. 1 Liquid joint with 6 DOF: 3 translations x, y, and z, 2 tilts θ_x and θ_y, and 1 rotation θ_z

effect. Additionally, they can be studied from the millimeter and below, i.e., at a scale sufficiently large to be addressed without dedicated experimental material.

The reader will be led from the underlying physics to various case studies, passing through the concept of fluidic joint, i.e., the mechanical description of a liquid bridge linking two solids as a mechanical joint with 6 DOF (Fig. 1). The static study outputs the stiffness and the dynamic study leads to the damping coefficients.

The models presented in this book find their far origin in the theory of capillary effects[1] which has existed for two centuries (Pierre-Simon de Laplace [1]): it used to focus essentially on the description of capillary filling of small gaps with liquids.

Nevertheless, Laplace went beyond phenomenological description and already wrote down the famous Laplace equation introducing the concept of surface tension from geometry (i.e. curvature) and mechanics (i.e. pressure). A few decades later, in 1873, Joseph Plateau (Professor at the University of Ghent, Belgium) described the equilibrium shape of droplets immersed in a bath with identical density [2]. These precursor works were restricted to equilibrium, such as the famous Jurin 's law [3] giving the rising height of a liquid in a thin capillary. Today, these effects can be simulated with powerful finite elements software such as Surface Evolver [4]. As early as 1921, Washburn [5] introduced an equation to model the dynamics of capillary rise. In 1936, Wenzel [6] described the effect of roughness of wettability and in 1944, Cassie and Baxter published their study on wettability of porous surfaces [7]. Nevertheless, dynamic contact angles were not studied before the end of the twentieth century by Voinov [8], Tanner [9], and Jiang [10]. Dynamic contact angles have been under study these last years [11–13].

With the introduction of numerical simulation and high speed cameras, dynamics became easier to study, yet it still focused on the shape of liquid bridges: Orr presented in 1975 a numerical simulation to compute the shape of liquid

[1] In French: 'Théorie des actions capillaires' [13].

bridges at equilibrium [14]; Edgerton published in 1937 the first results on the dynamics of drop formation [15]. Experimental results on liquid bridges dynamics was only published at the end of the twentieth century [16]. Besides the shape of the liquid bridges, another question is to study how these bridges acts on solids, i.e., provide forces which can be used in mechanical systems. The most famous model is probably Israelachvili's equation [17] describing the capillary force exerted by a meniscus linking a flat plane and a sphere.

Based on this background, this book clarifies some aspects of capillary forces calculation, extends the modeling from static to dynamic behaviors, and from axial to radial forces.

It introduces the concept of fluid mechanical joint and various case studies, which are briefly described in the following.

Problem Definition and Goals

This book does not try to address a particular industrial problem. On the contrary, the wide range of case studies proposed at the end of the book illustrates the generic interest of the models presented in the early chapters. Nevertheless, the reader will recognize many illustrations coming from micro-assembly domain and nanoscopic domains. For instance, capillary forces at the nanoscale are illustrated by the case of liquid delivery.

The justification for this proposal originates from the framework of micro-assembly and packaging.[2] Traditionally, the microrobotic community—which applies pick-and-place know-how to downscaled applications—makes use of surface science adhesion and mechanical contact models, because downscaling laws require an adequate understanding of the physical background. This improves design, simulation, haptic feedback, and automation of developed products. Since engineers are usually not familiar with scaling, good models are mandatory to ensure reliability, *sine qua non* condition for industrial perspectives of our research developments.

To this aim, we summarized, developed, and present here many capillary forces models that can be used in this field and beyond. For example, we recently published results on lateral dynamics of liquid bridges, which can serve as a basis toward design of flip-chip applications [18].

From an editorial point of view, the goal is to combine the knowledge of many specialists and present their most up-to-date research work in a way that this knowledge will not be out-of-date too quickly. We therefore tailor this book into different parts. The first two parts are expected to have a long-term validity, because they present capillary forces models in a generic way. The last part is

[2] Which includes a large variety of topics such as designing, manufacturing, feeding, positioning, joining, testing.

devoted to case studies, which obviously will be soon replaced by new research development. However, they show nowadays a broad and relevant picture of research trends toward designing micro-engineering applications with surface tension issues.

Methodology, Originalities, and Perspectives

Surface tension is a classical topic but is given many different insights: for example, the surface chemistry engineer is familiar with surface energies, the pharmaceutical industry researcher is familiar with surface tension and surfactant, tribologists know a lot about adhesion, physics teachers present laws as an appendix of their lectures, and so on. We hope to give a basis combining these aspects, actually the underlying concepts, which are required to understand how surface tension effects can be used to exert forces and torques on solids.

The key idea is therefore to write it 'as a textbook', even if the topic has not yet become a classical lecture. It will therefore include analytical developments, simulation recipes and results, description of experimental test beds, and, of course, experimental validation. In some case, the effort will be pushed until applicative demonstrators, such as for example in the case of the micro-robotic platform using three fluidic actuators combining surface tension and gas compressibility.

The results and data given in this book rely on literature reviews, mathematical and physical models, digital simulation, and experimental studies.

Its perspectives rely on the scope of the Belgian MicroMAST Network,[3] whose scientific objectives are driven by fundamental questions raised in microfluidics, interfacial science, and micromanipulation. The rational use of surface tension, surface stress, and capillary effects in micromanipulation is expected to be applied to highly relevant case studies, including capillary gripping, capillary filling, capillary alignment, capillary sealing, capillary self-assembly, and droplet manipulation (Fig. 2).

These fundamental questions can be grouped into three categories: the first category is partially addressed in this book while the two other can serve as future developments.

1. Fluid statics and dynamics: How much force is applied on solids by menisci and micro-flows in a given geometry? What happens if the solid bends when subject to these forces? Are the interfaces stable and what if not? What is the effect of an electric field? How can the microscopic description of wetting be translated into an adequate boundary condition at the macroscopic level?

---

[3] Microfluidics and Micromanipulation: Multi-Scale Applications of Surface Tension, www.micromast.be.

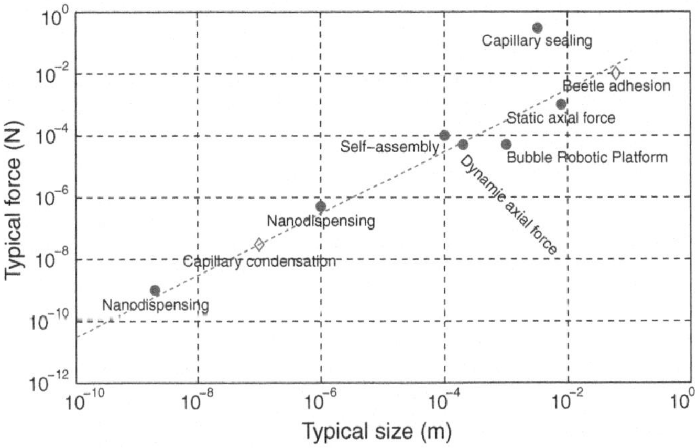

Fig. 2 Orders of magnitude in this book: from nano-dispensing at the nanoscale (Chap. 14) up to capillary-based sealing at the millimetric scale (Chap. 11). Self-assembly at 100 μm is described in Chap. 12, dynamic axial force models and measurements are led in Chap. 7, the bubble robotic platform devised in Chap. 5 and the static axial force in Chap. 2. Finally, capillary condensation and beetle adhesion have been reported elsewhere, respectively in [19] and [20]

2. Surface engineering: How does a contact line move on a rough surface? Can one pattern the surface microscopically to control this motion? How is the motion affected by evaporation, or by the presence of colloid particles in the liquid or at the interface? Do these particles interact with the micro-patterns on the surface? Can one create highly 3D patterns on the surface by using capillary forces?
3. Liquid engineering: How to measure the interfacial properties of complex liquids where apart from surface tension a surface viscoelastic response is present? How to infer macroscopic properties from the dynamics at the molecular scale? And how to engineer liquids and tailor them to the requirements arising from applications? Can one make a liquid that is biocompatible, and has a large surface tension and a low viscosity?

Overview

This book falls into a description of the related physical background (Part I), static and dynamic modeling of capillary forces and torques (Parts II and III), and an overview of relevant case studies (Part IV), such as a capillary micromotor (whose rotor follows a rotating surface energy gradient produced by electrowetting, see Chap. 10), a capillary-based sealing (whose limit is given by the capillary pressure, see Chap. 11), micro- and nano-assembly case studies using surface-tension-driven self-assembly (see Chap. 12), surface tension actuation and energy harvesting through transpiration actuation (Chap. 13) and finally, the illustration of surface tension effects in actuation and liquid nano-dispensing (Chap. 14).

All these case studies are positioned in a log–log diagram plotting the force order of magnitude as a function of the size order of magnitude (Fig. 2).

From Tokyo 2010 to Brussels 2013 Pierre Lambert

References

1. P.-S. de Laplace, Sur l'action capillaire. In *Mécanique Céleste*: *Supplément au Livre X*. (Courcier, Paris, 1805), pp. 349–498
2. J.A.F. Plateau, *Statique expérimentale et théorique des liquides soumis aux seules forces moléculaires*. (Gauthier-Villars, Paris, 1873)
3. J. Jurin, An account of some experiments shown before the royal society: with an enquiry into the cause of ascent and suspension of water in capillary tubes. Phil. Trans. Roy. Soc. **30**, 739–747 (1718)
4. K. Brakke, The surface evolver. Exp. Math. **1**(2), 141–65 (1992)
5. E.W. Washburn. The dynamics of capillary flow. Phys. Rev. **17**(3), 273–283 (1921)
6. R.N. Wenzel. Ind. Eng. Chem. **28**, 988 (1936)
7. A.B.D. Cassie, S. Baxter, Wettability of porous surfaces. Trans. Faraday Soc. (1944)
8. O.V. Voinov, Hydrodynamics of wetting. Fluid Dyn. **11**(5), 714–721 (1976)
9. L.H. Tanner, The spreading of silicone oil drops on horizontal surfaces. J. Phys. D **12**(9), 1473–1484 (1979)
10. T.-S. Jiang, O.H. Soo-Gun, J.C. Slattery, Correlation for dynamic contact angle. J. Colloid Interface Sci. **69**(1), 74–77 (1979)
11. M. Bracke, F. De Voeght, P. Joos, *The Kinetics of Wetting: The Dynamic Contact Angle*, vol. 79
12. R.G. Cox, Inertial and viscous effects on dynamic contact angles. J. Fluid Mech. **357**, 249–278 (1998)
13. Š. Šikalo, H.D. Wilhelm, I.V. Roisman, S. Jakirlić, C. Tropea, Dynamic contact angle of spreading droplets: experiments and simulations. Phys. Fluids **17**(6), 1–13 (2005)
14. F.M. Orr, L.E. Scriven, A.P. Rivas, Pendular rings between solids: meniscus properties and capillary force. J. Fluid Mech. **67**, 723–742 (1975)
15. A. de Lazzer, M. Dreyer, H.J. Rath, Particle–surface capillary forces. Langmuir **15**(13), 4551–4559 (1999)
16. J.F. Padday, G. Pétré, C.G. Rusu, J. Gamero, G. Wozniak, The shape, stability and breakage of pendant liquid bridges. J. Fluid Mech. **352**, 177–204 (1997)
17. J. N. Israelachvili, *Intermolecular and Surface Forces*, 2nd edn. (Academic Press, 1992)
18. P. Lambert, M. Mastrangeli, J.-B. Valsamis, G. Degrez, Spectral analysis and experimental study of lateral capillary dynamics (for flip-chip applications). *Microfluid Nanofluid*, (2010)
19. A. Chau, S. Régnier, A. Delchambre, P. Lambert, Theoretical and experimental study of the influence of AFM tip geometry and orientation on capillary force. J. Adhes. Sci. Technol. **24**, 2499–2510 (2010)
20. T. Eisner, D.J. Aneshansley, Defense by foot adhesion in a beetle (Hemisphaerota cyanea). PNAS. **97**(12), 6568–6573 (2000)
21. A.W. Adamson, A.P. Gast, *Physical Chemistry of Surfaces*, 6th edn. (John Wiley and Sons, 1997)

22. R.T. Borno, J.T. Steinmeyer, M.M. Maharbiz, Transpiration actuation: the design, fabrication and characterization of biomimetic mciroactuators driven by the surface tension of water. J. Micromech. Microeng. **16**, 2375–2383 (2006)
23. F. Bourgeois, *Vers la maîtrise de la qualité des assemblages de précision*. PhD thesis (Ecole Polytechnique Fédérale de Lausanne, 2007)
24. H. Bruus, *Theoretical Microfluidics*. (Oxford University Press, 2008)
25. D. Cheneler, M.C.L. Ward, M.J. Adams, Z. Zhang, Measurement of dynamic properties of small volumes of fluid using mems. Sens. Actuators, B. **130**(2), 701–706 (2008)
26. S.K. Cho, C.J. Kim, Particle separation and concentration control for digital microfluidics systems. (2003)
27. H.E. Edgerton, E.A. Hauser, W.B. Tucker, Studies in drop formation as revealed by the high-speed motion camera. J. Phys. Chem. **41**, 1017–1028 (1937)
28. J. Engmann, C. Servais, A.S. Burbidge, Squeeze flow theory and applications to rheometry: a review. J. Non-Newtonian Fluid Mech. **132**(1–3), 1–27 (2005)
29. J. Fang, K.F. Bohringer, Wafer-level packaging based on uniquely orienting self-assembly. J. Microelectromechanical Syst. **15**(3), 531 (2006)
30. A. De Greef, P. Lambert, A. Delchambre, Towards flexible medical instruments: review of flexible fluidic actuators. Precis. Eng. **33**, 311–321 (2009)
31. D.S. Haliyo, *Les forces d'adhésion et les effets dynamiques pour la micro-manipulation*. PhD thesis (Université Pierre et Marie Curie, 2002)
32. R.L. Hoffman, A study of the advancing interface i. interface shape in liquid–gas systems. J. Colloid Interface Sci. **50**(2), 228–241 (1975)
33. M. Kaneda, M. Yamamoto, K. Nakaso, T. Yamamoto, J. Fukai, Oscillation of a tilted circular pad on a droplet for the self-alignment process. Precis. Eng. **31**, 177–184 (2007)
34. J.M. Kim, Y.E. Shin, K. Fujimoto, Dynamic modeling for resin self-alignment mechanism. Microelectron. Reliab. **44**, 983–992 (2004)
35. P. Lambert, *A Contribution to Microassembly: A Study of Capillary Forces as a Gripping Principle*. PhD thesis (Université libre de Bruxelles, 2004)
36. P. Lambert, *Capillary Forces in Microassembly: Modeling, Simulation, Experiments, and Case Study*. Microtechnology and MEMS. (Springer, 2007)
37. P. Lambert, A. Chau, A. Delchambre, S. Régnier, Comparison between two capillary forces models. Langmuir **24**(7), 3157 (2008)
38. P. Lambert, A. Delchambre, Parameters ruling capillary forces at the submillimetric scale. Langmuir **21**, 9537–9543 (2005)
39. P. Lambert, P. Letier, A. Delchambre, Capillary and surface tension forces in the manipulation of small parts. In *Proceedings of International Symposium on Assembly and Tasks Planning (ISATP)*, pp. 54–59 (2003)
40. C. Lenders, *DEA: Study of valves used in microfluidics*. Master's thesis. (Université libre de Bruxelles, Brussels, 2006)
41. C. Lenders, *Study of Microbubbles Mechanical Behavior, Application to the Design of an Actuated Table for Micromanipulation in Liquid Media*. PhD thesis (Université libre de Bruxelles, 2010)
42. C. Lenders, M. Gauthier, P. Lambert, Meniscus-supported compliant table, 2009. patent request submitted: EP 09 172715
43. C. Lenders, M. Gauthier, P. Lambert, Parallel microrobot actuated by capillary effects. In *Accepted in ICRA Conference*, Shanghai, May 9–13 (2011)
44. C. Lin, F. Tsen, H. CHuan Kan, C.-C. CHieng, Numerical studies on micropart self-alignment using surface tension forces. *Microfluid Nanofluid* (2008)
45. W. Lin, S.K. Patra, Y.C. Lee, Design of solder joints for self-aligned optoelectronic assemblies. IEEE Trans. Adv. Packag. **18**(3), 543–551 (1995)
46. H. Lu, C. Bailey, Dynamic analysis of flip-chip self alignment. IEEE Trans. Adv. Packag. **28**(3), 475–480 (2005)

47. M.-H. Meurisse, M. Querry, Squeeze effects in a flat liquid bridge between parallel solid surfaces. J. Tribol. **128**(3), 575–584 (2006)
48. H. Oh, K. Kim, S. Kim, Characterization of deposition patterns produced by twin-nozzle electrospray. J. Aerosol Sci. **39**, 801–813 (2008)
49. D.E. Packham, Surface energy, surface topography and adhesion. Int. J. Adhes. Adhes. **23**(6), 437–48 (2003)
50. N.W. Pascarella, D.F. Baldwin, Compression flow modeling of underfill encapsulants for low cost flip chip assembly. In *Proceedings of 1998 International Symposium of Underfill Encapsulants for Low Cost Flip Chip Assembly*, (1998), p. 33
51. X. Pepin, D. Rossetti, S.M. Iveson, S.J.R. Simons, Modeling the evolution and rupture of pendular liquid bridges in the presence of large wetting hysteresis. J. Colloid Interface Sci. **232**, 289–297 (2000)
52. O. Pitois, P. Moucheront, X. Chateau, Liquid bridge between two moving spheres: an experimental study of viscosity effects. J. Colloid Interface Sci. **231**(1), 26–31 (2000)
53. Y.I. Rabinovich, J.J. Adler, M.S. Esayanur, A. Ata, R.K. Singh, B.M. Mougdil, Capillary forces between surfaces with nanoscale roughness. Adv. Colloid Interface Sci. **96**, 213–30 (2002)
54. Y.I. Rabinovich, M.S. Esayanur, B.M. Mougdil, Capillary forces between two spheres with a fixed volume liquid bridge: theory and experiment. Langmuir **21**, 10992–10997 (2005)
55. K.F. Bohringer, S.-H. Liang, X. Xiong, Towards optimal designs for self-alignment in surface tension driven self-assembly. In *Proceedings the 17th IEEE International Conference on MEMS*, pp. 9–12. IEEE, 2004
56. U. Srinivasan, D. Liepmann, R.T. Howe, Microstructure to substrate self-assembly using capillary forces. J. Microelectromech. Syst. **10**(1), 17–24 (2001)
57. T. Stifter, O. Marti, B. Bhushan, Theoretical investigation of the distance dependence of capillary and van der Waals forces in scanning force microscopy. Phys. Rev. B, **62**(20), 13667–13673 (2000)
58. H. Kikuchi, T. Fukushima, Y. Yamada, M. Koyanagi, New three-dimensional integration technology using self-assembly technique. *International Electron Devices Meeting*, (2005)
59. C.G. Tsai, C.M. Hsieh, and J.A. Yeh, Self-alignment of microchips using surface tension and solid edge. Sens. Actuat, A **139**, 343–349 (2007)
60. J.-B. Valsamis, *A Study of Liquid Bridges Dynamics: An Application to micro-Assembly*. PhD thesis (Université libre de Bruxelles)
61. N. van Veen, Analytical derivation of the self-alignment motion of flip chip soldered components. J. Electron. Packag. **121**, 116–121 (1999)

Acknowledgments

Writing and editing this book would not have been possible without all those who trusted me for so many years! I am clearly indebted to many colleagues, and particularly, I would like to thank very warmly colleagues who made this work possible, especially Prof. Alain Delchambre (Université libre de Bruxelles, Belgium), Prof. Jacques Jacot (Ecole Polytechnique Fédérale de Lausanne, Switzerland), Prof. Stéphane Régnier (Université Pierre et Marie Curie, France), Prof. Nicolas Chaillet (FEMTO-ST, CNRS, France), Prof. Hiroyuki Fujita (University of Tokyo, Japan), Dr. Michaël Gauthier (FEMTO-ST, CNRS, France). All the colleagues met respectively in Brussels, Lausanne, Paris, Besançon and Tokyo must obviously be associated to these words.

Naturally, I would like also to thank all my co-authors, who helped me turning a crazy idea into this book, in the alphabetic order Dr. Ruba T. Borno (University of California, USA), Dr. Michaël De Volder (KU Leuven, Belgium), Dr. Cyrille Lenders (Université libre de Bruxelles), Prof. Michel Maharbiz (University of California, USA), Dr. Massimo Mastrangeli (Ecole Polytechnique Fédérale de Lausanne, Switzerland), Dr. Thierry Ondarçuhu (CNRS-CEMES, Toulouse, France), Dr. Atsushi Takei (U. Tokyo, Japan) and Dr. Jean-Baptiste Valsamis (Université libre de Bruxelles).

Many Ph.D. students also took part to this adventure, among which Dr. Alexandre Chau, Dr. Aline De Greef, Dr. Marion Sausse-Lhernould, Dr. Vincent Vandaele. I am also grateful to the help provided by my master students and my colleagues from technical staffs. I cannot cite all the colleagues met here and abroad, who contributed to this work by their questions and interest.

The following institutions helped me to achieve this book: Université libre de Bruxelles, FNRS, FRIA, WBI, Emile Defay Funds (Belgium), CNRS (France), EPFL (Switzerland) and U. Tokyo (Japan). This book is also the key to the MicroMAST project, whose ongoing research has been funded by the Interuniversity Attraction Poles Programme (IAP 7/38 MicroMAST) initiated by the Belgian Science Policy Office.

Finally, I would like to thank friends and relatives who have supported me for all these years, both in my projects and doubts. Especially, how could I actually thank Virginie and our two droplets? Probably by stopping writing books...

Contents

Part I Physical Background

1 Physical Background 3
- 1.1 Forces Arising from Downscaling 3
- 1.2 Surface Tension 4
- 1.3 Important Non Dimensional Numbers and Similitude 5
- 1.4 Laplace's Law: from Surface Curvature to Surface Tension and Capillary Pressure 6
- 1.5 Interfacial Energy 6
- 1.6 Contact Angles 6
  - 1.6.1 Wettability: Contact Angle and Young (or Young-Dupré) Equation 7
  - 1.6.2 Contact Angle Hysteresis: Advancing and Receding Contact Angles 8
  - 1.6.3 Dynamic Contact Angle 8
- 1.7 Control Parameters: Temperature, Surfactants, Electric Fields, Evaporation 13
- 1.8 Conclusions 13
- References 14

Part II Static Modelling of Capillary Forces and Torques

2 Axial Capillary Forces 19
- 2.1 Introduction 19
- 2.2 Modeling Liquid Bridges Geometry: An Overview 20
  - 2.2.1 Energetic Method: Example of Two Parallel Plates 20
  - 2.2.2 Introduction to Surface Evolver 23
  - 2.2.3 Exact Resolution for Axially Symmetric Problems 26
  - 2.2.4 Geometrical Models 28

2.3 Pinned and Non Pinned Triple Lines ... 29
2.3.1 Non Pinned Menisci ... 29
2.3.2 Pinned Menisci ... 33
2.4 Conclusions ... 40
References ... 43

3 Lateral Capillary Forces ... 45
3.1 Introduction and Literature Overview ... 45
3.2 Experimental Measurement of Lateral Capillary Forces ... 48
3.2.1 Requirements and Constraints ... 48
3.2.2 The Sensing Principle ... 49
3.2.3 The Experimental Setup ... 50
3.2.4 Fluids ... 51
3.2.5 Data Processing ... 53
3.3 Modeling of Lateral Capillary Forces ... 54
3.3.1 Analytic Model ... 54
3.3.2 Finite-Element Model ... 56
3.4 Results and Discussion ... 58
3.4.1 Sources of Error ... 59
3.4.2 Hysteresis ... 60
3.5 Conclusions ... 62
References ... 67

4 Capillary Torque ... 71
4.1 Introduction ... 71
4.2 Experiment ... 72
4.2.1 Materials ... 73
4.2.2 Results ... 77
4.3 Mathematical Analysis ... 82
4.4 Conclusion ... 88
4.5 Appendix ... 89
References ... 90

5 Surface Tension Effects in Presence of Gas Compliance ... 93
5.1 Introduction ... 93
5.2 Modeling ... 94
5.2.1 Laplace Law ... 94
5.2.2 Surface Tension and Compressibility ... 94
5.2.3 Effect of Compressibility When Generating a Bubble from a Hole ... 96
5.2.4 Pressure Control ... 96
5.2.5 Volume Control ... 97

5.3 Dimensionless Analysis 104
5.3.1 Buckingham Theorem 104
5.3.2 Dimensionless Equation 104
5.3.3 Numerical Approximation 106
5.3.4 Analytical Approach 107
5.3.5 Stability Criterion 111
5.4 Experimental Validation of the Volume Controlled Bubble Generation Model 111
5.4.1 First Experimental Setup 111
5.4.2 Second Experimental Setup 118
5.4.3 Conclusion of Experimental Validation 121
5.5 Application to Bubble Robotics 121
5.6 Conclusions 124
References 125

Part III Dynamic Modelling of Capillary Forces

6 Introduction 129
6.1 Liquid Bridge as Mechanical Joint 129
6.2 Equations Governing Fluids 132
6.3 Fluid Dynamics Related Boundary Conditions 133
6.4 Overview of Next Chapters 135
References 136

7 Axial Capillary Forces (Dynamics) 137
7.1 Introduction 137
7.2 Problem Description and Methodology 138
7.3 Mathematical Background of the Kelvin-Voigt Model 140
7.4 Stiffness 141
7.4.1 Introduction 141
7.4.2 Numeric Approach 143
7.4.3 Negative Stiffness 144
7.4.4 Geometrical Assumptions 146
7.5 Damping Coefficient 148
7.6 Inertial Effect 152
7.7 Case Study and Numeric Validation of the Simplified Kelvin-Voigt Model 155
7.8 Gain Curves 158
7.9 Experimental Setup 162
7.9.1 Experimental Bench 163
7.9.2 Experimental Protocol 164
7.9.3 Image Processing Towards Geometrical Parameters Acquisition 165

7.10 Results . . . 165
7.10.1 Experiments List . . . 165
7.10.2 Gain and Phase Shift Curves . . . 166
7.11 Conclusions . . . 168
References . . . 177

8 Lateral Capillary Forces (Dynamics) . . . 179
8.1 State of the Art and Definition of the Problem . . . 179
8.2 Lateral Stiffness of Liquid Meniscus . . . 181
8.3 Coupled Problem Resolution with a Pseudo-Spectral Method Based on Chebyshev Polynomials . . . 185
8.4 Results and Discussion . . . 187
8.5 Conclusions . . . 192
References . . . 192

9 Conclusions . . . 195

Part IV Case Studies

10 Capillary Micro Motor . . . 199
10.1 Introduction . . . 199
10.2 Principle and Method . . . 201
10.2.1 Principal . . . 201
10.2.2 Fabrication . . . 202
10.3 Experiments . . . 204
10.3.1 Step Motion . . . 204
10.3.2 Continuous Rotation . . . 207
10.4 Conclusion . . . 208
References . . . 209

11 Capillary Based Sealing . . . 211
11.1 Introduction . . . 212
11.2 Operating Principle and Physics . . . 213
11.3 Fabrication Process and Assembly . . . 221
11.4 Experimental Section . . . 223
11.5 Conclusions and Outlook . . . 224
References . . . 225

12 Surface Tension-Driven Self-Assembly . . . 227
12.1 Introduction . . . 227
12.2 SA at Fluid-Fluid Interfaces . . . 229
12.3 Three-Dimensional SA . . . 232

12.4 Die-to-Substrate SA 234
12.4.1 Stochastic SA. 234
12.4.2 Hybrid Microhandling. 243
12.5 SA of Three-Dimensional MEMS. 246
12.5.1 Capillary Actuation of MEMS 246
12.5.2 Self-Assembled Polyhedra 246
12.6 Conclusions 248
References 249

13 Surface Tension Driven Actuation 255
13.1 Introduction 255
13.2 Transpiration Actuation 256
13.2.1 Fern Sporangium 257
13.2.2 Design and Modeling 258
13.2.3 Experimental Method 264
13.2.4 Experimental Results 265
13.3 Evaporation Driven Fluid Flow in Plants. 268
13.3.1 Transpiration in Plants and the Cohesion-Tension Theory 269
13.3.2 Energy and Rate of Evaporation. 270
13.3.3 Synthetic Leaf Design. 270
13.3.4 Electrical Parameter Model 271
13.3.5 Fabrication. 272
13.3.6 Testing Results: Evaporation Driven Flow. 273
13.3.7 Testing Results: Energy Scavenging Circuit. 273
13.3.8 Future Improvements 275
13.4 Conclusion. 275
References 276

14 Capillary Forces in Atomic Force Microscopy and Liquid Nanodispensing. 279
14.1 Introduction 279
14.2 Computation of Capillary Forces 281
14.3 Capillary Forces in Atomic Force Microscopy 283
14.3.1 Contact Mode 284
14.3.2 Oscillating Modes. 286
14.4 Capillary Forces in Liquid Nanodispensing 288
14.4.1 Experimental and Calculation Methods 290
14.4.2 Examples of Capillary Force Curves. 292
14.4.3 Interpretation of Experimental Force Curves 293
14.5 Conclusion. 301
References 302

Part V Conclusions and Perspectives

15 Conclusions and Perspectives 309
15.1 Conclusions 309
15.2 Current Trends 310
15.3 Perspectives 312
References 313

Appendix A: Navier-Stokes Equations 315

Appendix B: Vectorial Operators 319

Index 325

Acronyms

Greek Symbols

α	Atan(*u/h*), or reduced viscosity in lateral dynamics of chips
Δp	Pressure difference across the liquid–vapor interface, $\Delta p = p_{out} - p_{in}$
ε	Relative permittivity (no unit)
η	Dynamic viscosity (pa s)
γ	Surface tension (N/m) or surface free energy (Jm^{-2})
γ_{LV}	Liquid–vapor interfacial energy, i.e. surface tension γ (Jm^{-2})
γ_{SL}	Solid–liquid interfacial energy (Jm^{-2})
γ_{SV}	Solid–vapor interfacial energy (Jm^{-2})
λ	Dimensionless number representing the molecular length over the tube radius
μ	Dynamic viscosity (pa s)
μ_0	Magnetic permeability (Chap. 4)
v	Kinematic viscosity
Π	Disjoining pressure
ρ	Density (kg m^{-3}), curvature radius (m)
σ	Non dimensional number predicting bubble generation (in) stability (Chap. 5)
Σ	Liquid–vapor interface area (m^2)
τ	Characteristic time
τ_c	Dimensional damping time
τ_d	Non dimensional damping time
θ	Contact angle (°) or alternatively angular position in polar system (rad)
θ_a	Advancing contact angle (°)
θ_d	Dynamic contact angle, by comparison with θ_s (°)
θ_m	In Cox's theory, denotes the macroscopic contact angle, by comparison with θ_w (°)
θ_r	Receding contact angle (°)
θ_s	Static contact angle, by comparison with θ_d (°)
θ_w	In Cox's theory, denotes the microscopic contact angle, by comparison with θ_m (°)

ϕ	Rotational displacement of a floating plate (Chap. 4)
ω	Pulsation (rad s^{-1})

Roman Symbols

A	Surface area of a droplet (also noted $\Sigma(m^2)$) or Hamaker's constant
A_{SL}	Solid–liquid interface area (m^2)
A_{SV}	Solid–vapor interface area (m^2)
b	Beam width (m), damping coefficient
c	Linear damping coefficient (kg s^{-1})
D	Pad diameter (m)
E	Young modulus
f	Oscillation frequency
F	Force (N)
F_L	Capillary force or Laplace force, i.e. part of capillary forces due to the pressure gradient Δp
F_T	Tension or surface tension force, i.e. part of capillary forces due to the tensile action of surface tension on the contact line
g	Gravitational acceleration constant (9.81 ms^{-2})
h	Height, meniscus height, immersion height, meniscus gap, bubble height (m)
H	Mean curvature, i.e. $H = \frac{1}{2}\left(\frac{1}{\rho_1} + \frac{1}{\rho_2}\right)$ (m^{-1})
H_{ext}	External Magnetic field (Chap. 4)
I	Second moment of inertia
k, K	Stiffness (of springs, cantilevers and menisci as indicated by respective subscripts (N/m))
ℓ	Typical system size (m)
l_a	Amplitude of the perturbation of the floating plate (Chap. 4)
l_b	Amplitude of the perturbation of the base plate (Chap. 4)
L	Cantilever length
L_c	Capillary length (2.7 mm for water)
m	Mass, with additional index when needed, equivalent mass or inertia
$\tilde{m}$	Reduced mass: $\tilde{m} = (2m)/(\rho Sh)$
M_{eff}	Effective mass
M	Magnetic moment (Chap. 4)
n	Moles number (in gas law), periods number (related to capillary torque)
p	Parasitic (vertical) displacement
p	Pressure
p_{atm}	Atmospheric pressure
p_0	Reference pressure, for instance p_{atm}
$p_{in,\ out}$	Pressure inside/outside the liquid meniscus
$q(z)$	Element of displacement
r	Radial coordinate

r_b	Pad radius (m)
R	Pad radius, bubble radius, sphere radius (m)
R_a	Diameter of the floating plate (Chap. 4)
R_b	Diameter of the base plate (Chap. 4)
R_g	Gas constant (8.314 J $mol^{-1}K^{-1}$)
s	Channel diameter, related to bubble generation (m)
S	Piston section, related to bubble generation (m)
T	Torque in general expression (Chap. 11), temperature (K)
T_c	Critical temperature (K)
T_s	Torque generated by surface tension (Chap. 4)
T_m	Torque generated by magnetic field (Chap. 4)
u	Relative displacement between pads ($=\|x^t - x^b\|$, the lateral shift) or alternatively velocity profile
v	Velocity
V	Volume
W	Energy, interfacial energy
W_{LV}	Liquid–vapor interface energy
W_{SL}	Solid–liquid interface energy
W_{SV}	Solid–vapor interface energy
x^t	x coordinate of top pad, with additional pedex when needed
x^b	x coordinate of bottom pad, with additional pedex when needed
y	y coordinate
z	z coordinate, or axial coordinate in a cylindrical coordinates system, separation distance

Usual or Non-Usual Non Dimensional Numbers

Bo	Bond number
Ca	Capillary number
Re	Reynolds number
σ	Unamed non-dimensional number involved in bubble generation

Acronyms

AFM	Atomic Force Microscope
CA	Contact Angle (°)
CMOS	Complementary Metal-Oxide-Semiconductor
DOF	Degree Of Freedom
DPN	Dip Pen Nanolithography
DRIE	Deep Reactive Ion Etching
FIB	Focused Ion Beam
FSRM	Fondation Suisse pour la Recherche en Microtechnique
ITO	Indium Tin Oxide
LED	Light-Emitting Diode

MEMS	Micro Electro-Mechanical Systems
NADIS	NAnoDISpensing Technique
PAMMA	Poly-Methyl-Methacrylate
RF	Radio Frequency
SA	Self-Assembly
SAM	Self-Assembled Monolayer
SEM	Scanning Electron Microscope
STM	Scanning Tunneling Microscope
SOI	Silicon On Insulator
SQUID	Superconducting Quantum Interference Device

Contributors

Ruba T. Borno Electrical Engineering and Computer Science, University of California, 646 Sutardja Dai Hall, Berkeley, CA 94720, USA, e-mail: rborno@gmail.com

Pierre Lambert BEAMS Department CP, Université libre de Bruxelles, 165/56, Avenue F.D. Roosevelt, 50, Brussels 1050, Belgium, e-mail: pierre.lambert@ulb.ac.be

Cyrille Lenders BEAMS Department CP, Université libre de Bruxelles, 165/56, Avenue F.D. Roosevelt, 50, Brussels 1050, Belgium, e-mail: cyrille.lenders@gmail.com

Michel Malharbiz Electrical Engineering and Computer Science, University of California, 646 Sutardja Dai Hall, Berkeley, CA 94720, USA, e-mail: maharbiz@eecs.berkeley.edu

Massimo Mastrangeli Distributed Intelligent Systems and Algorithms Laboratory–Microsystems Laboratory, Ecole Polytechnique Fédérale de Lausanne, GR A2 475–BM 3.202, Station 2–Station 17, Lausanne 1015, Switzerland, e-mail: mastrangelim@gmail.com

Thierry Ondarçuhu CEMES-CNRS, Rue Jeanne Marvig 29, Toulouse 31055, Cedex 4, France, e-mail: ondar@cemes.fr

Atsushi Takei Department of Mechano-Informatics, Graduate School of Information Science and Technology, University of Tokyo, 7-3-1 Hongo, Bunkyo-ku, Tokyo 113-8656, Japan, e-mail: atakei@iis.utokyo.ac.jp

Jean-Baptiste Valsamis BEAMS Department CP, Université libre de Bruxelles, 165/56, Avenue F.D. Roosevelt, 50, Brussels, 1050, Belgium, e-mail: jvalsami@ulb.ac.be

Michaël De Volder Department of Mechanical Engineering–IMEC, Katholieke Universiteit Leuven, Kapeldreef 75, Leuven 3001, Belgium, e-mail: Michael.DeVolder@mech.kuleuven.be

Part I
Physical Background

Chapter 1
Physical Background

J.-B. Valsamis, M. De Volder and P. Lambert

Abstract This chapter introduces the key underlying parameters which are requested in this book, such as for example scaling laws, surface tension and its control through surfactants and temperature, surface energies and contact angles.

1.1 Forces Arising from Downscaling

As the lengthscales of a system are decreased to the micrometer domain, the relative importance of physical laws drastically change. Consider for instance a cube with a 1 m side. The total length of the ribs is 12 m (there are 12 edges), the total area of the sides is 6 m^2, and the volume is 1 m^3. If the sides are scaled down to 1 mm, the length becomes 12×10^{-3} m, the surface 6×10^{-6} m^2, and the volume 1×10^{-9} m^3. Importantly, forces that are proportional to a length, such as surface tension (see further), are 1000 times smaller in a 1 mm cube, while forces related to surfaces such as pressure became 10^6 smaller, and volume forces such as gravity 10^9 smaller. An illustration of this principle is that surface tension is able to support water insects, while this is impossible for humans. This phenomenon, known as the scaling laws, gives rise to particular engineering challenges, but also great opportunities at the microscale. Taking advantage of the importance of surface tension has led to new

J.-B. Valsamis · P. Lambert (✉)
Université libre de Bruxelles, BEAMS Department CP 165/56,
Avenue F.D. Roosevelt, 50, 1050 Brussels, Belgium
e-mail: jvalsami@ulb.ac.be

P. Lambert
e-mail: pierre.lambert@ulb.ac.be

M. De Volder
Katholieke Universiteit Leuven, Department of Mechanical Engineering—IMEC,
Kapeldreef 75, 3001 Leuven, Belgium
e-mail: Michael.DeVolder@mech.kuleuven.be

P. Lambert (ed.), *Surface Tension in Microsystems*, Microtechnology and MEMS,
DOI: 10.1007/978-3-642-37552-1_1, © Springer-Verlag Berlin Heidelberg 2013

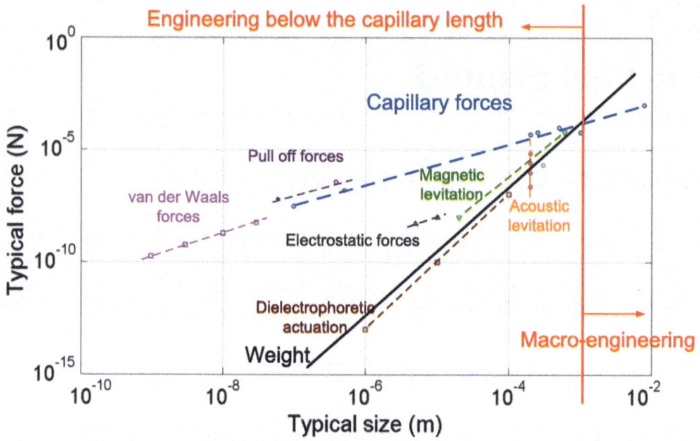

Fig. 1.1 Forces arising from downscaling: all the points (○) represent different experimental data, while the dashed straight lines (– –) represent scaling laws of various phenomena. Surface tension effects, i.e. capillary forces, can be seen as dominant at the microscale. The boundary between gravity and surface tension driven scales is known as the capillary length

microassembly methods and actuators which will be discussed throughout this book. A more exhaustive overview of how other physical laws scale with size is summarized in Fig. 1.1. Circles (○) indicate experimental results which have been published previously for various physical effects (weight, magnetic and electrostatic forces, acoustic and magnetic levitation, capillary forces, dielectrophoretic effects...), and dashed lines (– –) figure out the scaling law in a log-log diagram plotting the force as a function of system size, ℓ.

1.2 Surface Tension

Figure 1.1 again nicely demonstrates how surface tension is of great importance for micro- and nanoscale systems, since it clearly shows how capillary forces dominate down to $100\,\text{nm}-1\,\mu\text{m}$. Let us now present the concepts involved in capillary phenomena.

A first representation of the three states of the matter is to consider that a solid is characterized by a volume and a shape and conversely, a gas does not have any own volume or own shape. A liquid is somewhere in between, having its own volume but no own shape. Usually, if a liquid is not contained, it spreads out. However, when we look at soap bubbles or small water droplets, we observe that they behave as if their surface was an elastic membrane, characterized by a 'surface tension' that acts against their deformations.

This surface tension is presented in a didactic way in [19]. Intuitively, surface tension arises from an unbalance of the forces acting on molecules at a liquid interface.

For instance, a water molecule at the air interface is experiencing a strong cohesion force from other water molecules in the bulk liquid, and a much weaker interaction with air. This creates an inward pressure on the liquid, and is responsible for shaping water in droplets, or other configurations with a minimal surface area.

This introduces the concept of surface energy (or surface tension), which has the dimensions of an energy by surface unit ($\mathrm{J\,m^{-2}}$). The mechanical point of view considers the surface tension a tensile force by length unit ($\mathrm{N\,m^{-1}}$). The surface tension is denoted by γ and its numerical value depends on the molecular interactions: in most oils, the molecular interaction is van der Waals interaction, leading to quite low surface tensions ($\gamma \approx 20\ \mathrm{mN\,m^{-1}}$). As far as water is concerned, due to the hydrogen bonding, the molecular attraction is larger ($\gamma \approx 72\ \mathrm{mN\,m^{-1}}$). Typical values for conventional liquid range from $20\ \mathrm{mN\,m^{-1}}$ (silicone oil) to $72\ \mathrm{mN\,m^{-1}}$ (water at $20\,^{\circ}\mathrm{C}$). For example, [19] gives the following values for ethanol ($23\ \mathrm{mN\,m^{-1}}$), acetone ($24\ \mathrm{mN\,m^{-1}}$), glycerol ($63\ \mathrm{mN\,m^{-1}}$).

Surface tension is an important parameter in the perspective of a downscaling of the assembly equipment because the force it generates linearly decreases with the size while the weight decreases more quickly. Let us now emphasize the role of non-dimensional numbers in the next section.

1.3 Important Non Dimensional Numbers and Similitude

Dimensional analysis include dimensions homogeneity, non dimensional numbers and similitude laws [71]. Non dimensional numbers are of the utmost importance to compare the relative importance different physical phenomena, for instance:

1. The capillary length L_C relates surface tension effects to gravity: $L_C = \sqrt{\frac{\gamma}{\rho g}}$, where γ is the surface tension, ρ is the density and $g = 9.81\ \mathrm{m\,s^{-1}}$ ($L_C \approx 2\ \mathrm{mm}$ for water).
2. The Bond number compared the relative importance of gravity and surface tension: $\mathrm{Bo} = \frac{\rho g \ell^2}{\gamma}$. When the typical size $\ell = L_C$, $\mathrm{Bo} = 1$.
3. The Weber number describes the importance of dynamic effects from flow inertia relatively to surface tension: $\mathrm{We} = \frac{\rho v^2 L}{\gamma}$, where v is a typical velocity.
4. The capillary number Ca compares the influence of viscous drag to surface tension: $\mathrm{Ca} = \frac{\mu v}{\gamma}$
5. The Péclet number[1] relates the convective force to the diffusion rate force: $\dfrac{\rho v L}{D}$

[1] The Péclet number is used in transport phenomena. It is a kind of ratio between the force entering in a mesh element and how this force is diffused inside this element (by the means of the viscous effect). It presents also an interest in numeric computational: it can measure the numerical stability. In this case, it takes a particular form, as a local Reynolds number where the length is the mesh element size [14].

6. A number σ presented by Lenders (Chap. 5): $\sigma = P_0 s^4/\gamma V_0$, where P_0 and V_0 are the pressure and volume of a gas container filled under atmospheric conditions, and s is the nozzle diameter through which a single gas bubble is generated. The threshold between stable and unstable bubbles growth is given by $\sigma = 128/(27\pi)$

1.4 Laplace's Law: from Surface Curvature to Surface Tension and Capillary Pressure

When an interface separating two fluid phases present a mean curvature H, there exist a relationship (the so-called Laplace law [21]) between the interface curvature H, the interface surface tension γ and a pressure gap accross the interface Δp, the latter being defined as $\Delta p = p_{\text{in}} - p_{\text{out}}$:

$$\Delta p = 2H\gamma \tag{1.1}$$

The curvature in the case of an axially symmetric interface will be given by (2.24) in Sect. 2.2.3. When the curvature is positive (convex shape), the pressure difference is positive too, i.e. $p_{\text{in}} > p_{\text{out}}$ (and conversely). Δp is sometimes called the capillary pressure. The capillary pressure of a system can also represents its volumetric energy density (see Sect. 13.3.4 for a example applied to biomimetic evapotranspiration).

1.5 Interfacial Energy

Surface tension introduced in previous section can also be described as the energy by unit area of the liquid-vapor interface ($\text{Nm}^{-1} = \text{Jm}^{-2}$). From this perspective, surface tension is the energy required to stretch the liquid-vapor interface by ine unit area. Two other interfacial energies can also be defined at the interfaces between a liquid and a solid (γ_{SL}) and between a solid and a vapor (γ_{SV}). Typical values of γ_{SV} are summarized in [38]: Nylon (Polyamid) 6.6 (41.4 mN m^{-1}), PE High density (30.3–35.1 mN m^{-1}), PE Low density (32.1–33.2 mN m^{-1}), PET (40.9–42.4 mN m^{-1}), PMMA (44.9–45.8 mN m^{-1}), PP (29.7), PTFE (20.0–21.8 mN m^{-1}).

We refer to [1] for further information on this topic.

1.6 Contact Angles

The situation is getting more complex at a liquid-solid-gas interface. For instance, at the interface with a container water typically rises up to wet the container or depreciates to avoid contact. This behavior arises from the inbalance between the liquid and solid energies, and is quantified by the contact angle. Energetically favourable wetting leads to the well-known fluid rise in capillary tubes. Importantly, the contact angle is not characteristic of the liquid as such, but depends the specific material combination. And to make things worse, it is also influenced by the surface topography, wetting direction, and dynamics.

In the following, we give some extensions to the concepts of wettability, contact angle, receding and advancing contact angles, contact angles hysteresis, dynamic contact angle.

1.6.1 Wettability: Contact Angle and Young (or Young-Dupré) Equation

When liquids are in contact with solids, the total surface energy must also include the energy of the solid/vapour interface, and of the solid/liquid interface. These two additional surface energies are also represented by a surface tension γ_{SL} and γ_{SV} respectively.

The result of this minimal energy state is indicated by the slope of the air/liquid interface with respect to the liquid/solid interface. This is characterised by the contact angle θ, as shown in Fig. 1.2.

- If $\theta < \frac{\pi}{2}$, the liquid wets the surface. This is the case of silicon oils on most of surfaces because of their relatively low surface tension [65].
- If $\theta > \frac{\pi}{2}$, the liquid is non-wettable. An example is the water on the teflon, or the mercury on most of the surfaces.

The contact angle on the triple line at equilibrium is given by the Young (or Young-Dupré) equation [1, 29]. This equation describes the tangential balance of s along the solid surface (see Fig. 1.3):

$$\gamma_{LV} \cos\theta + \gamma_{SL} = \gamma_{SV} \tag{1.2}$$

At equilibrium, this angle is called the static contact angle θ_s. The value of this contact angle is generally not exactly fixed, and presents hysteresis. This hysteresis is variable and depends on several parameters as roughness, coating, etc. [6, 10, 39].

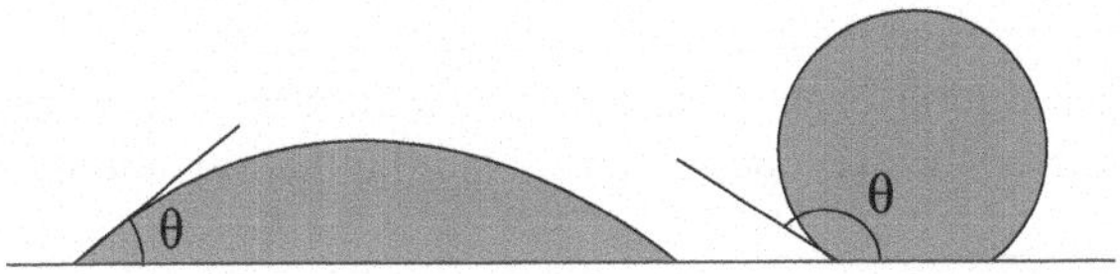

Fig. 1.2 Illustration of wettability: on *left side*, a good wettability, on *right side*, a poor wettability

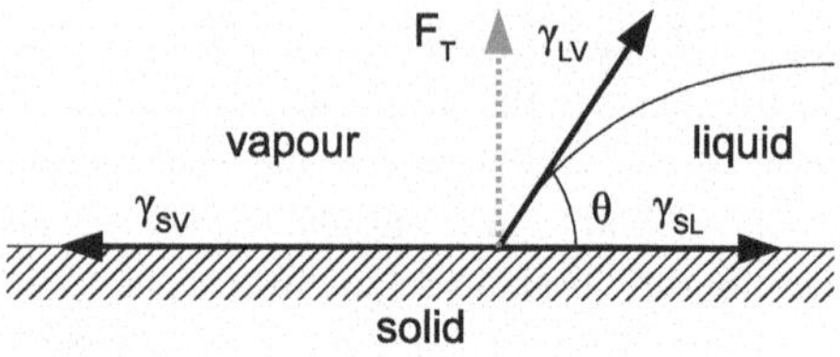

Fig. 1.3 Illustration of Young-Dupre equation

1.6.2 Contact Angle Hysteresis: Advancing and Receding Contact Angles

Due to hysteresis, contact angles are not uniquely defined:

- when the volume of liquid increases, the contact angle increases too, up to a limit called the advancing contact angle θ_a above which the triple line moves forward (i.e. to the solid-vapor interface);
- when the volume of liquid decreases, the contact angle decreases, down to a limit called the receding contact angle θ_r below which the triple line moves backward (i.e. to the solid-liquid interface).

A model has been proposed by Zisman (see Adamson, [1]) who observed that $\cos\theta_A$ (advancing angle) is usually a monotonic function of γ. Henceforth he proposed the following equation :

$$\cos\theta_A = a - b\gamma \tag{1.3}$$

The difference between the advancing and the receding contact angles is called the contact angle hysteresis, which is influenced by surface heterogeneities, surface roughness and surface texturation. A recent review paper on the topic has been written by Bhushan [3], from the precursors of the domain (Cassie, Wenzel, Baxter) to the most recent developments on natural and biomimetic artificial surfaces for superhydrophobicity, self-cleaning, low adhesion and drag reduction.

1.6.3 Dynamic Contact Angle

The study of the contact angle started in the early 1800's. It is only in the second half of the 19th century that researches paid attention to the dynamic of the contact angles. If the dependency between the dynamic contact angle and the velocity was observed, no appropriate description existed. The hydrodynamic effects[2] were supposed to play a role [27]. Moreover, the existence of a precursor liquid film of uniform thickness is

[2] Hoffman [28] considers two categories of dynamic systems: (1) the systems in which the advancing interface is driven at constant velocity over the solid substrate; (2) the systems in which the interface shape is in nonequilibrium state and changes as a result of its recent displacement to a new position. Therefore, the studies of driven interface fall into *dynamic* studies and *relaxation* studies [28, 33].

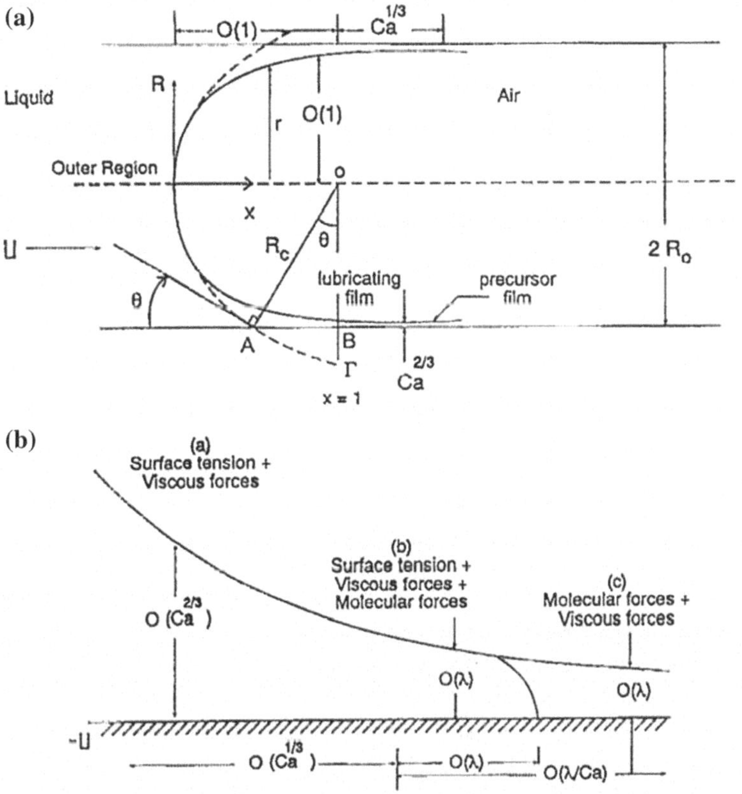

Fig. 1.4 Real and apparent contact angle. Reprinted with permission from [33]. Copyright 1994, American Institute of Physics. **a** Difference between the apparent and real contact angles. Apparent contact angles are the extrapolation of the interface at macroscopic scale. **b** Forces in the thin film. λ is a dimensionless number representing the molecular length on the tube radius R_0, Ca is the capillary number (see below). The $O(x^y)$ means the order of the dependency of the parameter x in power y

pointed out by Kalliadasis and Chang in [33]. They present the difference between apparent and real contact angles in the case of a moving air/liquid interface for liquid with high wettability with a solid, highlitghing the existence of an invisible lubrication film in front of meniscus, as shown in 1.4a. Usually, the experimental contact angle is the measure of an extrapolation of the *outer region* to the wall. In capillary rise for example, this extrapolation consists of approximating the interface by a spherical cap. Hence, apparent and real contact angles can be related to macroscopic and microscopic contact angles developed by Cox [16]. In the region of the film, the molecular forces become important and are added in the model. There are three regions (1.4b): the first one is driven only by viscosity and forces, the second is subject to viscous, tension and molecular forces and in the third one the forces disappear. The Laplace pressure is modified as follows:

$$\Delta p = 2\gamma H + \Pi \tag{1.4}$$

where $\Pi = \frac{A}{6\pi z^3}$ is the disjoining pressure (z is the film thickness and A is Hamaker's constant). The disjoining pressure is a pressure exerted on the fluid by the wall because of the van der Waals forces.

The existence of a precursor film makes the definition of the dynamic contact angle slightly ambiguous. An old definition is given by Voinov, defining the dynamic contact angle as "(…) the angle of the slope of the free surface α if this angle weakly changes with height h above the solid surface." [76] (see 1.5). This means that the dynamic contact angle verifies:

$$h \left| \frac{\mathrm{d}\alpha}{\mathrm{d}h} \right| \ll \alpha \tag{1.5}$$

He adds that "If this condition is not satisfied then the surface is significantly curvilinear and it is impossible to observe the contact angle." [76]. The contact angle should be taken at the limit h_m several times the size of a molecule (corresponding to the value of 1000 Å according to Hansen [27])[3]. The work of Voinov focused on the shape of the interface rather than on defining α_m. The flow near the triple line has been modeled by Moffat who was interested to describe the flow at sharp corner [54]. The problem has been studied in a similar way by Hansen and Toong in [27]. As Voinov, they did not provide any law for the dynamic contact angle.

Later on, Cox developed a theory, dividing the fluid into three regions in which he apply particular assumptions on the flow (viscous, inviscid, …), as illustrated in Fig. 1.6. He made also some assumptions on the slip condition in these regions, mainly on the behaviour stress inside the fluid (see [15, 16]).

Literature also provides many experimental laws which are based on macroscopic (or at least mesoscopic) quantities.

1. For small capillary numbers Ca, let us cite the Tanner's law [72]:

$$\theta_d \sim \mathrm{Ca}^{\frac{1}{3}} \tag{1.6}$$

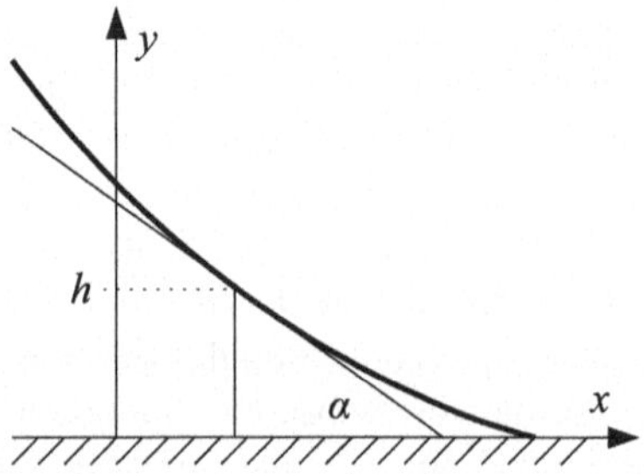

Fig. 1.5 Definition of the dynamic contact angle by Voinov

[3] Below h_m, the hydrodynamic description is not valid anymore.

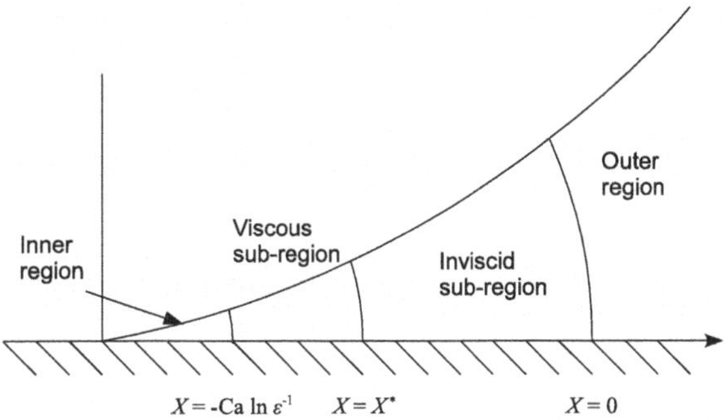

Fig. 1.6 The region subdivision in the case of middle Reynold number [16]. Near the triple line, the *inner region* at low Reynold number. Far from the triple line, the *outer region* in which no assumption is made on the fluid. Between the inner region and the outer region, the *sub-regions* with two particular states of the fluid. Adapted from [16], with permission of Cambridge University Press (Journal of Fluid Mechanics)

2. A second relationship providing the dynamic contact angle at a glance is the experimental Hoffman-Voinov-Tanner (HVT) law defined by:

$$\theta_d^3 = \theta_s^3 + 72\text{Ca} \tag{1.7}$$

3. Jiang et al wrote down an experimental-empirical law [30]:

$$\frac{\cos\theta_s - \cos\theta_d}{\cos\theta_s + 1} = \tanh\left(4.96\text{Ca}^{0.702}\right) \tag{1.8}$$

that can be reads as

$$\theta_d = \arccos\left[\cos\theta_s - (\cos\theta_s + 1)\tanh\left(4.96\text{Ca}^{0.702}\right)\right] \tag{1.9}$$

4. Another experimental law has been found by Bracke et al in [8] is

$$\theta_d = \arccos\left(\cos\theta_s - 2\,(1 + \cos\theta_s)\sqrt{\text{Ca}}\right) \tag{1.10}$$

5. Kistler developed another law [37]:

$$\theta_d = f_H(\text{Ca} + f_H^{-1}(\theta_s)) \tag{1.11}$$

with

$$f_H(x) = \arccos\left[1 - 2\tanh\left(5.16\left(\frac{x}{1 + 1.31x^{0.99}}\right)^{0.706}\right)\right] \tag{1.12}$$

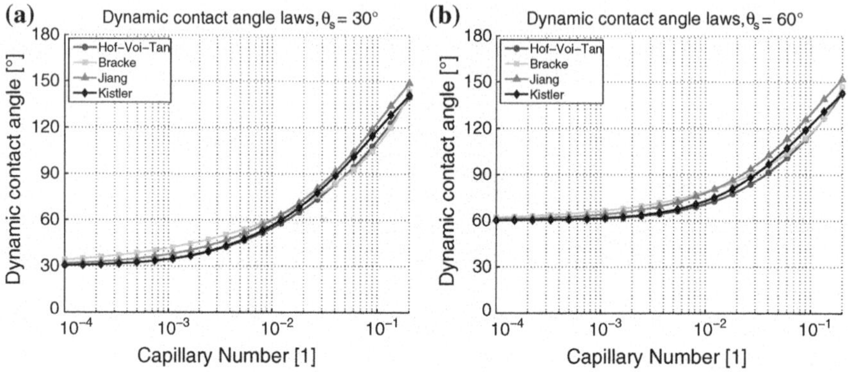

Fig. 1.7 Dynamic contact angles as a function of the capillary number Ca, showing the good agreement among the literature models. **a** Dynamic laws with $\theta_s = 30°$. **b** Dynamic laws with $\theta_s = 60°$

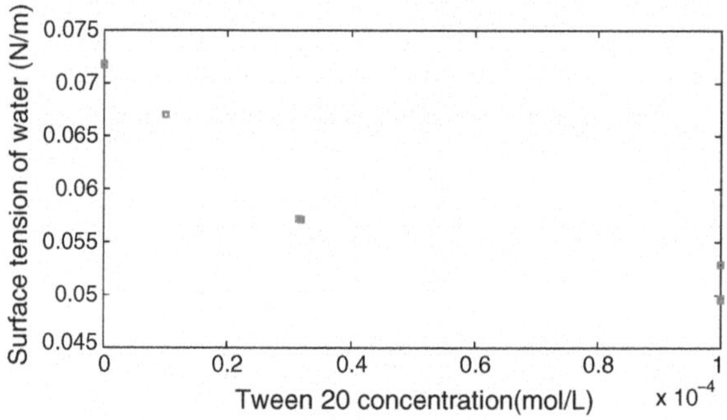

Fig. 1.8 Surface tension of water as a function of anionic surfactant Tween 20 concentration (own experimental results, [18])

Theoretically, these laws are valid over a wide range of capillary number: all of them converge to the static contact angle when Ca $\rightarrow$ 0. The validity at high capillary number is limited. Typical limit are in the order of Ca $\sim$ 10 [68]. All these models are plotted in Figs. 1.7a and 1.7b for two values of the static contact angle. One can see that these models give very similar values of dynamic contact angle, as function of the capillary number.

Beside, there also exists a molecular-kinetic theory impulsed by Blake in 1969, but this falls out of the scope of this book.

1.7 Control Parameters: Temperature, Surfactants, Electric Fields, Evaporation

Surface tension and contact angles depend on (and can somehow be controlled by) several paramters:

1. The dependence of surface tension on temperature is given by Guyon [25]:

$$\gamma(T) = \gamma(T_0)\,(1 - b\,(T - T_0)) \tag{1.13}$$

   where b ranges from 10^{-2} to $10^{-1}\mathrm{K}^{-1}$.
2. Surface tension is decreased when adding surfactant, such as for instance the anionic surfactant Tween 20 (see Fig. 1.8)
3. Contact angle can be apparently[4] modified by applying an electric field: this is known as the electrowetting effect [2, 44]:

$$\cos\theta(V) = \cos\theta(V = 0) + \frac{1}{2}\frac{\varepsilon\varepsilon_0}{\gamma h}V^2 \tag{1.14}$$

   where V is the voltage applied between a conductive liquid sessile drop and the conductive substrate, h and ε are the thickness and relative permittivity of the hydrophobic and insulative layer coated on the substrate to prevent hydrolysis.
4. Evaporation plays also a role by modifying the volume of liquid and by inducing temperature changes. Due to volume decrease, the contact angle slightly decreases during evaporation, down to the receding contact angle. The triple line displacement then triggers a geometry change. As reported by Parneix, non uniform evaporation can structure films of nanometric thickness [60].

1.8 Conclusions

This first chapter stipulates the dominance of surface tension effects when downscaling, which make them very promising in engineering below the capillary length. We then introduced the key physical parameters of this book: surface tension, Laplace's law, interfacial energy, contact angles and control parameters. The section on contact angles reported static, advancing, receding and dynamic contact angles, comparing several models giving the dynamic contact angle as a function of the capillary number Ca. Finally, from an engineering point of view, it is essential to provide control: to this aim, the last section reported different possible effects such as temperature, surfactant, evaporation or electric fields.

---

[4] This is subjected to discussion, as reported by Jones who gives an electromechanical interpretation of electrowetting [31] or Mugele who reported a study on equilibrium drop surface profiles in electric fields [55].

References

1. A.W. Adamson, A.P. Gast, *Physical Chemistry of Surfaces*, 6th edn. (John Wiley and Sons, New York, 1997)
2. B. Berge, J. Peseux, Variable focal lens controlled by an external voltage: an application of electrowetting. Eur. Phys. J. E. **3**, 159–163 (2000)
3. B. Bhushan, Y.C. Jung, Natural and biomimetic artificial surfaces for superhydrophobicity, self-cleaning, low adhesion and drag reduction. Prog. Mater Sci. **56**, 1–108 (2011)
4. K.F. Bohringer, S-H. Liang, X. Xiong, Towards optimal designs for self-alignment in surface tension driven self-assembly. In *Proceedings the 17th IEEE International Conference on MEMS*, (IEEE, 2004), pp. 9–12
5. R.T. Borno, J.T. Steinmeyer, M.M. Maharbiz, Transpiration actuation: the design, fabrication and characterization of biomimetic mciroactuators driven by the surface tension of water. J. Micromech. Microeng. **16**, 2375–2383 (2006)
6. N. Boufercha, J. Sägebarth, M. Burgard, N. Othman, D. Schlenker, W. Schäfer, H. Samndamier, Fluidassem—a new method of fluidic-based assembly with surface tension. In *Micro-Assembly Technologies and Applications*. International Precision Assembly Seminar (IPAS), (Springer, 2008)
7. F. Bourgeois, Vers la maîtrise de la qualité des assemblages de précision. Ph.D. thesis, Ecole Polytechnique Fédérale de Lausanne, 2007
8. M. Bracke, F. De Voeght, P. Joos, The kinetics of wetting: the dynamic contact angle. Progr. Colloid Pol. Sci. **79**, 142–149 (1989)
9. K. Brakke, The surface evolver. Exp. Math. **1**(2), 141–165 (1992)
10. H. Bruus, *Theoretical Microfluidics* (Oxford University Press, New York, 2008)
11. R. Burley, B.S. Kennedy, An experimental study of air entrainment at a solid/liquid/gas interface. Chem. Eng. Sci. **31**(10), 901–911 (1976)
12. D. Cheneler, M.C.L. Ward, M.J. Adams, Z. Zhang, Measurement of dynamic properties of small volumes of fluid using mems. Sens. Actuators B, **130**(2), 701–706 (2008)
13. S.K. Cho, C.J. Kim, Particle separation and concentration control for digital microfluidics systems, In *Proceedings of the Sixteenth Annual Conference on MEMS*, 2003
14. COMSOL AB, *COMSOL Multiphysics V3.5a: User's Guide*, 2008
15. R.G. Cox, The dynamics of the spreading of liquids on a solid surface. part 1. viscous flow. J. Fluid Mech. **168**, 169–194 (1986)
16. R.G. Cox, Inertial and viscous effects on dynamic contact angles. J. Fluid Mech. **357**, 249–278 (1998)
17. A. de Lazzer, M. Dreyer, H.J. Rath, Particle-surface capillary forces. Langmuir **15**(13), 4551–9 (1999)
18. B. Daunay, P. Lambert, L. Jalabert, M. Kumemura, R. Renaudot, V. Agache, H. Fujita, Effect of substrate wettability in liquid dielectrophoresis (LDEP) based droplet generation: theoretical analysis and experimental confirmation. Lab Chip **12**, 361–368 (2012)
19. P.G. de Gennes, F. Brochard-Wyart, D. Quéré, *Capillarity and Wetting Phenomena: Drops, Bubbles, Pearls, Waves*, 1st edn. (Springer, New York, 2003)
20. A. De Greef, P. Lambert, A. Delchambre, Towards flexible medical instruments: review of flexible fluidic actuators. Precis. Eng. **33**, 311–321 (October 2009)
21. P.S. de Laplace, Sur l'action capillaire. In *Mécanique Céleste— Supplément au Livre X*, (Courcier, Paris, 1805), pp. 349–498
22. H.E. Edgerton, E.A. Hauser, W.B. Tucker, Studies in drop formation as revealed by the high-speed motion camera. J. Phys. Chem. **41**, 1017–1028 (1937)
23. J. Engmann, C. Servais, A.S. Burbidge, Squeeze flow theory and applications to rheometry: a review. J. Non-Newtonian Fluid Mech. **132**(1–3), 1–27 (2005)
24. J. Fang, K.F. Bohringer, Wafer-level packaging based on uniquely orienting self-assembly. J. Microelectromechanical Syst. **15**(3), 531 (2006)
25. E. Guyon, J.-P. Hulin, L. Petit, *Hydrodynamique Physique*, 3rd edn. (CNRS Editions, Paris, 2012)

26. D.S. Haliyo, Les forces d'adhésion et les effets dynamiques pour la micro-manipulation. Ph.D. thesis, Université Pierre et Marie Curie, 2002
27. R.J. Hansen, T.Y. Toong, Dynamic contact angle and its relationship to forces of hydrodynamic origin. J. Colloid Interface Sci. **37**(1), 196–207 (1971)
28. R.L. Hoffman, A study of the advancing interface i. interface shape in liquid-gas systems. J. Colloid Interface Sci. **50**(2), 228–241 (1975)
29. J.N. Israelachvili, *Intermolecular and Surface Forces*, 2nd edn. (Academic Press, London, 1992)
30. T.S. Jiang, O.H. Soo-Gun, J.C. Slattery, Correlation for dynamic contact angle. J. Colloid Interface Sci. **69**(1), 74–77 (1979)
31. T.B. Jones, An electromechanical interpretation of electrowetting. J. Micromech. Microeng. **15**, 1184–1187 (2005)
32. J. Jurin, An account of some experiments shown before the royal society: With an enquiry into the cause of ascent and suspension of water in capillary tubes. Phil. Trans. Roy. Soc. **30**, 739–747 (1718)
33. S. Kalliadasis, H-C. Chang, Apparent dynamic contact angle of an advancing gas-liquid meniscus. Phys. Fluids **6**(1), 12–23 (1994)
34. M. Kaneda, M. Yamamoto, K. Nakaso, T. Yamamoto, J. Fukai, Oscillation of a tilted circular pad on a droplet for the self-alignment process. Precis. Eng. **31**, 177–184 (2007)
35. H. Kikuchi, T. Fukushima, Y. Yamada, M. Koyanagi, New three-dimensional integration technology using self-assembly technique. Int. Electron Devices Meeting (Baltimore, MD, USA, 2005)
36. J.M. Kim, Y.E. Shin, K. Fujimoto, Dynamic modeling for resin self-alignment mechanism. Microelectron. Reliab. **44**, 983–992 (2004)
37. S.F. Kistler, *Hydrodynamics of Wetting* (Marcel Dekker, New York, 1993), pp. 311–420
38. P. Lambert, A Contribution to Microassembly: a Study of Capillary Forces as a gripping Principle. Ph.D. thesis, Université libre de Bruxelles, Belgium, 2004
39. P. Lambert, *Capillary Forces in Microassembly: Modeling, Simulation, Experiments, and Case Study*. Microtechnology and MEMS. (Springer, Boston, 2007)
40. P. Lambert, A. Chau, A. Delchambre, S. Régnier, Comparison between two capillary forces models. Langmuir **24**(7), 3157 (2008)
41. P. Lambert, A. Delchambre, Parameters ruling capillary forces at the submillimetric scale. Langmuir **21**, 9537–9543 (2005)
42. P. Lambert, P. Letier, A. Delchambre, Capillary and surface tension forces in the manipulation of small parts. In *Proceedings of the International Symposium on Assembly and Tasks Planning (ISATP)*, (2003), p. 54–59
43. P. Lambert, M. Mastrangeli, J.B. Valsamis, G. Degrez, Spectral analysis and experimental study of lateral capillary dynamics (for flip-chip applications). Microfluid Nanofluid, Published online, 2010
44. J. Lee, C.-J. Kim, Surface-tension-driven microactuation based on continuous electrowetting. J. Microelectromechanical Syst. **9**(2), 171–180 (2000)
45. S.H. Lee, K.C. Lee, S.S. Lee, Fabrication of an electrothermally actuated electrostatic microgripper. Int. J. Nonlinear Sci. Numer. Sim. **3**, 789–793 (2002)
46. C. Lenders, DEA: Study of valves used in microfluidics. Master's thesis, Université libre de Bruxelles, Brussels, Belgium, 2006
47. C. Lenders, Study of Microbubbles Mechanical Behavior, Application to the Design of an Actuated Table for Micromanipulation in Liquid Media. Ph.D. thesis, Université libre de Bruxelles, 2010
48. C. Lenders, M. Gauthier, P. Lambert, Meniscus-supported compliant table, 2009. patent request submitted: EP 09 172715
49. C. Lenders, M. Gauthier, P. Lambert, Parallel microrobot actuated by capillary effects. In *Accepted in ICRA Conference*, Shanghai, May 2011, pp. 9–13
50. W. Lin, S.K. Patra, YC. Lee, Design of solder joints for self-aligned optoelectronic assemblies. IEEE Trans. Adv. Packag. **18**(3), 543–551 (1995)

51. C. Lin, F. Tsen, H.C. Kan, C-C. CHieng, Numerical studies on micropart self-alignment using surface tension forces. Microfluid. Nanofluid. **6**, 63–75 (2008)
52. H. Lu, C. Bailey, Dynamic analysis of flip-chip self alignment. IEEE Trans. Adv. Packag. **28**(3), 475–480 (2005)
53. M.H. Meurisse, M. Querry, Squeeze effects in a flat liquid bridge between parallel solid surfaces. J. Tribology **128**(3), 575–584 (2006)
54. H.K. Moffat, Viscous and resistive eddies near a sharp corner. J. Fluid Mech. **18**(1), 1–18 (1964)
55. F. Mugele, J. Buehrle, Equilibrium drop surface profiles in electric fields. J. Phys. Condens. Matter **19**, 375112 (2007)
56. H. Oh, K. Kim, S. Kim, Characterization of deposition patterns produced by twin-nozzle electrospray. J. Aerosol Sci. **39**, 801–813 (2008)
57. F.M. Orr, L.E. Scriven, A.P. Rivas, Pendular rings between solids: meniscus properties and capillary force. J. Fluid Mech. **67**, 723–42 (1975)
58. D.E. Packham, Surface energy, surface topography and adhesion. Int. J. Adhes. Adhes. **23**(6), 437–48 (2003)
59. J.F. Padday, G. Pétré, C.G. Rusu, J. Gamero, G. Wozniak, The shape, stability and breakage of pendant liquid bridges. J. Fluid Mech. **352**, 177–204 (1997)
60. C. Parneix, P. Vandoolaeghe, V.S. Nikolayev, D. Quéré, J. Li, B. Cabane, Dips and rims in dried colloidal films. Phys. Rev. Lett. **105**, 266103 (2010)
61. N.W. Pascarella, D.F. Baldwin, Compression flow modeling of underfill encapsulants for low cost flip chip assembly. In *Proceedings of 1998 International Symposium of Underfill Encapsulants for Low Cost Flip Chip Assembly*, (1998), p. 33
62. X. Pepin, D. Rossetti, S.M. Iveson, S.J.R. Simons, Modeling the evolution and rupture of pendular liquid bridges in the presence of large wetting hysteresis. J. Colloid Interface Sci. **232**, 289–97 (2000)
63. O. Pitois, P. Moucheront, X. Chateau, Liquid bridge between two moving spheres: an experimental study of viscosity effects. J. Colloid Interface Sci. **231**(1), 26–31 (2000)
64. J.A.F. Plateau, *Statique Expérimentale et Théorique des Liquides Soumis aux Seules Forces Moléculaires* (Gauthier-Villars, Paris, 1873)
65. D. Quéré, Rough idea on wetting. Physica A **313**, 32–46 (2002)
66. Y.I. Rabinovich, M.S. Esayanur, B.M. Mougdil, Capillary forces between two spheres with a fixed volume liquid bridge: theory and experiment. Langmuir **21**, 10992–10997 (2005)
67. Y.I. Rabinovich, J.J. Adler, M.S. Esayanur, A. Ata, R.K. Singh, B.M. Mougdil, Capillary forces between surfaces with nanoscale roughness. Adv. Colloid Interface Sci. **96**, 213–30 (2002)
68. Š. Šikalo, H.D. Wilhelm, I.V. Roisman, S. Jakirlić, C. Tropea, Dynamic contact angle of spreading droplets: Experiments and simulations. Phys. Fluids **17**(6), 1–13 (2005)
69. U. Srinivasan, D. Liepmann, R.T. Howe, Microstructure to substrate self-assembly using capillary forces. J. Microelectromechanical Syst. **10**(1), 17–24 (2001)
70. T. Stifter, O. Marti, B. Bhushan, Theoretical investigation of the distance dependence of capillary and van der Waals forces in scanning force microscopy. Phys. Rev. B **62**(20), 13667–73 (2000)
71. T. Szirtes, *Applied Dimensional Analysis and Modeling* (MacGraw-Hill, New-York, 1997)
72. L.H. Tanner, The spreading of silicone oil drops on horizontal surfaces. J. Phys. D **12**(9), 1473–1484 (1979)
73. C.G. Tsai, C.M. Hsieh, J.A. Yeh, Self-alignment of microchips using surface tension and solid edge. Sens. Actuators A **139**, 343–349 (2007)
74. J.B. Valsamis, A Study of Liquid Bridges Dynamics: an Application to micro-Assembly. Ph.D. thesis, Université libre de Bruxelles, 2010
75. N. van Veen, Analytical derivation of the self-alignment motion of flip chip soldered components. J. Electron. Packag. **121**, 116–121 (1999)
76. O.V. Voinov, Hydrodynamics of wetting. Fluid Dyn. **11**(5), 714–721 (1976)
77. M. De Volder, J. Peirs, D. Reynaerts, J. Coosemans, R. Puers, O. Smal, B. Raucent, A novel hydraulic microactuator sealed by surface tension. Sens. Actuators A **123–124**, 547–554 (2005)
78. E.W. Washburn, The dynamics of capillary flow. Phys. Rev. **17**(3), 273–283 (1921)
79. R.N. Wenzel, cited by [56]. Ind. Eng. Chem. **28**, 988 (1936)

Part II
Static Modelling of Capillary Forces and Torques

Chapter 2
Axial Capillary Forces

Pierre Lambert and Jean-Baptiste Valsamis

Abstract This chapter presents classical techniques to compute the shape of a meniscus using the energetic method and its numerical implementation using Surface Evolver. It provides the numerical solution of the Laplace equation for axially symmetric configuration and some useful analytical approximations (circular or toroidal approximation, parabolic approximation). Formal equivalence between these approaches is given and results are provided as rules of thumbs for the designer.

2.1 Introduction

The contribution of this chapter is to model the force exerted on solids by capillary bridges. These liquid bridges can be seen as mechanical joints with 6 degrees-of-freedom. This chapter focuses on the axial degree of freedom, i.e. on the forces developed along the symmetry axis of the liquid bridges.

As experimentally confirmed in the previous chapter (as shown in Fig. 1.1), the very first model of a phenomenon usually relies on its scaling law, acknowledging in this case the linear dependence of capillary forces on the size of the set up, and more generally, the importance of dimensional analysis and scaling laws.

Nevertheless, targeting the design of devices using surface tension effects, it is useful to get a more detailed insight on the parameters ruling capillary forces. This is the starting point of models development. The axial capillary forces are of interest in all micromanipulation case studies, to know for example the amount of picking force in a microassembly application. At the nanoscale, the capillary force is an

P. Lambert (✉) · J.-B. Valsamis
BEAMS Department, Université libre de Bruxelles, CP 165/56,
Avenue F.D. Roosevelt, 50, 1050 Brussels, Belgium
e-mail: pierre.lambert@ulb.ac.be

J.-B. Valsamis
e-mail: jvalsami@ulb.ac.be

P. Lambert (ed.), *Surface Tension in Microsystems*, Microtechnology and MEMS,
DOI: 10.1007/978-3-642-37552-1_2,

important contribution to adhesion, such as for example adhesion in atomic force microscopy or stiction in RF MEMS. The lateral capillary forces models are more dedicated to self-assembly problems or to the dynamics of components floating on solder paste menisci, such as in flip-chip assembly. The dynamics of a chip in contact with a fluid meniscus has been studied as a classical second order system, including inertial, viscous and stiffness effects. Only the stiffness term depends on capillary forces, while the viscous term depends on the shear stress on the component, i.e. on the liquid viscosity and the liquid flow inside the meniscus.

In order to make our results useful to readership we try to propose analytical models or to present numerical results in the form of maps and graphs.

2.2 Modeling Liquid Bridges Geometry: An Overview

Literature highlights two different ways to compute capillary forces. The first one consists in computing first the surface energy of the liquid bridge and deriving it with respect to the degree-of-freedom of interest[1] (see Sect. 2.2.1). This approach is particularly well suited for systems smaller than the capillary length, because in this case, gravity energy and inertial effects are typically smaller than surface energy. The second approach directly gives the force from the liquid bridge geometry (Sect. 2.2.3). Note that in both cases finding the right liquid bridge geometry is the key point, leading to some useful approximations (Sect. 2.2.4). Computing the force from surface energy is quite straightforward and will be done in the following. On the contrary, beside this approach based on energy, a second approach based on forces directly was highlighted in literature: basically, it consists in computing separately the effect of surface tension and the effect of pressure gap induced by the curvature of the liquid bridge. We contributed to that point by giving formal and experimental evidences of the equivalence of both approaches [1]. This will be detailed later on in Sect. 2.4. Consequently, in the current section, we don't focus on force computation but rather on liquid bridge geometry. The link between geometry and force will be introduced later on.

Additional details can be found in our related publications [1–3].

2.2.1 *Energetic Method: Example of Two Parallel Plates*

As an introduction, we propose the case of a meniscus between two parallel plates, with a contact angle $\theta = \pi/2$. This method consists in:

- writing the interfacial energy W of the system as a function of the parameters defining the geometry of the system;

---

[1] E.g. the force along the z-direction is the derivative of the energy with respect to z.

- deriving this energy with respect to one of the parameters (the separation distance z is often used) in order to calculate the capillary force as a function of this parameter;
- estimating the derivative of the other parameters with respect to the chosen parameter by assuming a mathematical relationship (for example the conservation of the liquid volume).

This approach can be illustrated by the case of two parallel plates linked by a meniscus, such as represented in Fig. 2.1:

The system has three phases (S: solid, L: liquid, V: vapor) and three interfaces (LV;liquid–vapor, SL: solid–liquid, SV: solid–vapor) leading to a total energy equal to:

$$W = W_{SL} + W_{SV} + W_{LV} = \gamma_{SL} S_{SL} + \gamma_{SV} S_{SV} + \gamma \Sigma \tag{2.1}$$

where:

$$W_{SL} = 2\gamma_{SL}\pi r^2 \tag{2.2}$$
$$W_{SV} = 2\gamma_{SV}(\pi r_0^2 - \pi r^2) \tag{2.3}$$
$$W_{LV} = \gamma\Sigma = \gamma 2\pi r z \tag{2.4}$$

In these equations, r_0 is an arbitrary constant radius, larger than r and $\gamma_{SL}(\gamma_{SV})$ represents the interfacial energy between solid and the liquid (vapor). Σ represents the area of the liquid-vapor interface (the lateral area of the meniscus). This leads to:

$$\begin{aligned} W &= 2\gamma_{SL}\pi r^2 + 2\gamma_{SV}(\pi r_0^2 - \pi r^2) + \gamma 2\pi r z \\ &= \underbrace{2\gamma_{SV}\pi r_0^2}_{\text{constant}} - 2\pi r^2 \underbrace{(\gamma_{SV} - \gamma_{SL})}_{\gamma \cos\theta = 0} + \gamma 2\pi r z \end{aligned}$$

As we try to get the expression of the force F acting on one of the plates along the vertical axis z as a function of the separation distance z, the latter equation must be derived with respect to z:

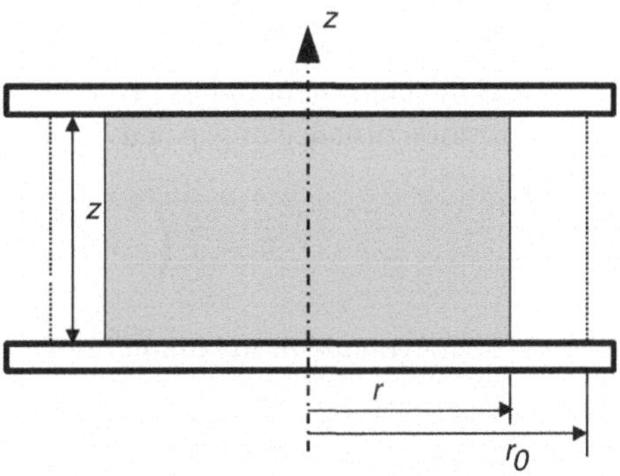

Fig. 2.1 Example of the energetic method: case of two parallel plates. z is the gap between plates, r the wetting radius i.e. the radius of the wetting circle, and r_0 an arbitrary radius for area computation (see the related explanation in the text)

$$F = -\frac{\mathrm{d}W}{\mathrm{d}z} = -2\pi r\gamma - \underbrace{2\pi z\gamma\frac{\mathrm{d}r}{\mathrm{d}z}}_{\text{requires an additional assumption}} \tag{2.5}$$

In order to calculate all the derivatives involved in this expression, an additional assumption must be stated. The assumption is that the volume $V = \pi r^2 z$ of the meniscus remains constant (we consequently do not consider the evaporation of the liquid), and its conservation leads to:

$$\frac{\mathrm{d}V}{\mathrm{d}z} = 2\pi rz\frac{\mathrm{d}r}{\mathrm{d}z} + \pi r^2 = 0 \tag{2.6}$$

Finally, combining both latter equations, the force can be written as:

$$F = -\pi r\gamma \tag{2.7}$$

Of course, this method only gives an exact analytical result in the very restrictive case of two parallel plates and a contact angle equal to $\pi/2$. When the liquid-vapor interface cannot be estimated analytically, it is necessary to turn oneself toward a software such as Surface Evolver (see next section).

Israelachvili [4] applied this method to calculate the capillary force between a sphere and a flat surface[2]:

$$F = -4\pi R\gamma\cos\theta \tag{2.8}$$

where R is the sphere radius, γ is the surface tension, and $\cos\theta$ the mean cosine of contact angles θ_1 and θ_2 on sphere and plate respectively.

Let us add a recently published model [6] giving an analytical expression for the capillary force between two spheres with radii R_1 and R_2, as a function of the separation distance z:

$$F_{\text{sphere/sphere}} = -\frac{2R\cos\theta}{1 + z/(2h)} \tag{2.9}$$

where R is the equivalent radius given by $R = \frac{2R_1R_2}{R_1+R_2}$, $2\cos\theta = \cos\theta_1 + \cos\theta_2$, z is the separation distance or gap and h is the immersion height, approximately given by [6]:

$$h = \frac{z}{2}\left(-1 + \sqrt{1 + 2V/(Rz^2)}\right) \tag{2.10}$$

where V is the volume of the liquid bridge.

[2] As it can be seen this expression does not depend on the volume of liquid. This approximation is only valid for small volumes. More rigorous expressions, valid for large volumes are given by [5].

2.2.2 Introduction to Surface Evolver

Surface Evolver is a simulation software which computes minimal energy surfaces.[3] Therefore, in this software, constraints on contact angles, pinning lines, volume of liquid must be defined in a text file which is processed to evolve the vapor-liquid interface toward an energy minimum (Fig. 2.2).

In general, the energy can be written as (see previous section):

$$W = \text{constant} - \underbrace{A_{\mathrm{SL}}\underbrace{(\gamma_{\mathrm{SV}} - \gamma_{\mathrm{SL}})}_{\gamma\cos\theta}}_{I} + \gamma\Sigma \tag{2.11}$$

The surface energy $\gamma\Sigma$ is easy to compute as soon as the liquid-vapor is meshed. But usually in Surface Evolver, what is meshed is **only** this surface and **not** the solid-liquid interface (Fig. 2.3).

Therefore, the energetic content of this interface $-A_{\mathrm{SL}}\gamma\cos\theta$ has to be computed from the single elements of this interface to be meshed, which in this case is its boundary, i.e. the triple line. This is achieved using the Stokes theorem:

$$\iint \mathrm{rot}\bar{w} \cdot \mathrm{d}\bar{S} = \oint \bar{w} \cdot \mathrm{d}\bar{l} \tag{2.12}$$

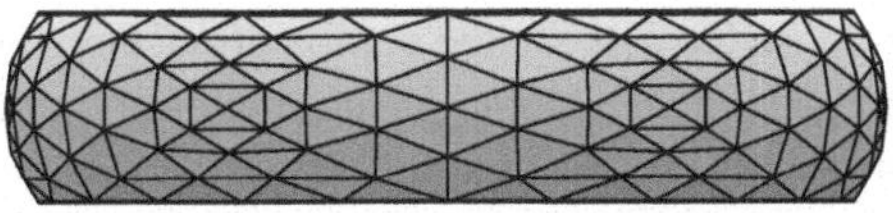

Fig. 2.2 Example of a meniscus meshed in surface evolver

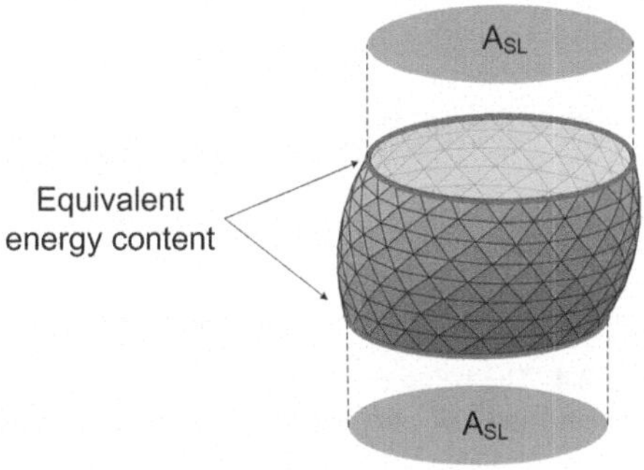

Fig. 2.3 The surface energy of the meniscus must be computed from the meshed elements, i.e. from the triple line

[3] Available at http://www.susqu.edu/brakke/evolver/evolver.html.

where rot$\bar{w}$ is equal to $\bar{\nabla} \times \bar{w}$:

$$\bar{\nabla} \times \bar{w} = (\partial_y w_z - \partial_z w_y)\bar{1}_x + (\partial_x w_z - \partial_z w_x)\bar{1}_y \ldots \\ \ldots + (\partial_x w_y - \partial_y w_y)\bar{1}_z \tag{2.13}$$

$$= \left(\frac{1}{r}\partial_\theta w_z - \partial_z w_\theta\right)\bar{1}_r + (\partial_z w_r - \partial_r w_z)\bar{1}_\theta \ldots \\ \ldots + \frac{1}{r}(\partial_r(r w_\theta) - \partial_\theta w_r)\bar{1}_z \tag{2.14}$$

The integral I defined in (2.11) can be developed using $\bar{\nabla} \times \bar{w} = \bar{1}_z$:

$$I = -2\gamma\cos\theta \underbrace{\iint \bar{1}_z \cdot \mathrm{d}\bar{S}}_{A_{SL}} = -2\gamma\cos\theta \oint \bar{w} \cdot \mathrm{d}\bar{l} \tag{2.15}$$

with:

$$\bar{w} = x\bar{1}_y = r\sin\theta\cos\theta\bar{1}_r + r\cos^2\theta\bar{1}_\theta \tag{2.16}$$

and:

$$\mathrm{d}\bar{l} = r\mathrm{d}\theta\bar{1}_\theta \tag{2.17}$$

The vector $\bar{w}$ must be defined in the input file of Surface Evolver, so that the software can compute the energetical content of the non-meshed interface.

Similarly, we saw in the previous section that the condition of constant volume was imposed to achieve the computation. Here again, Surface Evolver needs information to compute a volume V of which only a part of its boundary is meshed. We refer now to the divergence theorem (Fig. 2.4):

$$\iiint_V \mathrm{div}\bar{u}\mathrm{d}V = \oint\!\!\!\oint_{\partial V} \bar{u} \cdot \mathrm{d}\bar{S} \tag{2.18}$$

Choosing a vector $\bar{u}$ whose divergence is equal to 1, the volume can be written as:

$$V = \iiint_V \mathrm{d}V$$

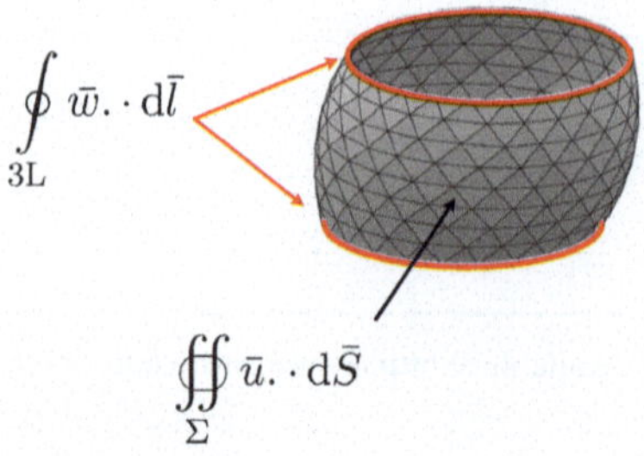

Fig. 2.4 The volume of the meniscus must be computed from the meshed elements, i.e. by summing a contribution from the lateral area Σ and a contribution from the triple line (3L)

$$
\begin{aligned}
&= \oiint_{\Sigma} \bar{u}. \cdot \mathrm{d}\bar{S} + \oiint_{SL} \underbrace{\bar{u}}_{\mathrm{rot}\bar{w}}. \cdot \mathrm{d}\bar{S} \\
&= \oiint_{\Sigma} \bar{u}. \cdot \mathrm{d}\bar{S} + \oint_{3L} \bar{w}. \cdot \mathrm{d}\bar{l} \qquad (2.19)
\end{aligned}
$$

where Σ denotes the meshed liquid-vapor interface, SL the solid-vapor interface (not meshed) and 3L the triple line contours (on top and bottom of the meniscus). The vector $\bar{w}$ is defined by $\mathrm{rot}\bar{w} = z(x, y)\bar{1}_z$ where $z(x, y)$ is the equation of the solid-liquid interface.

Example: Truncated Cone

Let us take the case of a truncated cone limited by the planes $z = 0$ *and* $z = z_H$ *(such as depicted in Fig. 2.5). The cone equation is:*

$$
r(z) = r_H + (z_H - z)\tan a \qquad (2.20)
$$

and the normal vector is given by:

$$
\bar{n} = \cos a\,\bar{1}_r + \sin a\,\bar{1}_z \qquad (2.21)
$$

With the surface element $\mathrm{d}S = \frac{r}{\cos a} dz d\theta$, *the volume is computed as follows:*

$$
\begin{aligned}
V &= \oiint_{\Sigma} z\bar{1}_z. \cdot \mathrm{d}\bar{S} + \oint_{3L} \bar{w}. \cdot \mathrm{d}\bar{l} \\
&= \oiint_{\Sigma} z\bar{1}_z. \cdot \mathrm{d}\bar{S} + \oint_{3L} z_H r_H \cos\theta(\sin\theta\bar{1}_r + \cos\theta\bar{1}_\theta) \cdot r_H \mathrm{d}\theta\bar{1}_\theta
\end{aligned}
$$

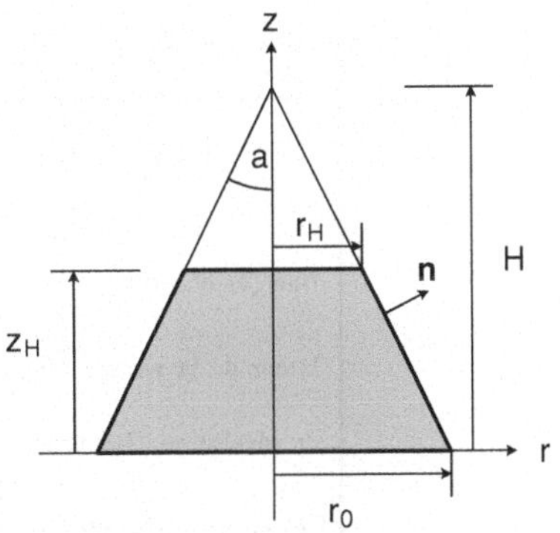

Fig. 2.5 Truncated cone illustrating the computation of volume is surface evolver

$$= \int_0^{2\pi} \mathrm{d}\theta \int_0^{z_H} \tan a r(z) z \mathrm{d}z + z_H r_H^2 \int_0^{2\pi} \cos^2 \theta \mathrm{d}\theta$$

$$= \pi r_H z_H^2 \tan a + \frac{\pi}{3} z_H^3 \tan^2 a + \pi r_H^2 z_H \tag{2.22}$$

which is obviously equal to the difference of two cone volumes $(r_0 = r(z = 0))$*:*

$$V = \frac{\pi}{3} r_0^2 H - \frac{\pi}{3} r_H^2 (H - z_H) \tag{2.23}$$

Surface Evolver has been used to deduce lateral capillary forces and capillary forces due to capillary condensation [7].

2.2.3 Exact Resolution for Axially Symmetric Problems

For axially symmetric menisci of equation $r = r(z)$, the curvature $2H$ can be written as follows (see [3] for details):

$$2H = -\frac{r''}{(1 + r'^2)^{\frac{3}{2}}} + \frac{1}{(1 + r'^2)^{\frac{1}{2}}} = \frac{p_{\text{in}} - p_{\text{out}}}{\gamma} \equiv \frac{\Delta p}{\gamma} \tag{2.24}$$

where p_{in} and p_{out} are the pressures inside and outside the liquid bridge, and γ is the surface tension of the considered liquid. Δp is the pressure gap across the liquid-gas interface: as it will be seen later on, it is one of the drivers of capillary forces. This equation was first derived by Pierre-Simon Laplace in 1805 [8] (Fig. 2.6).

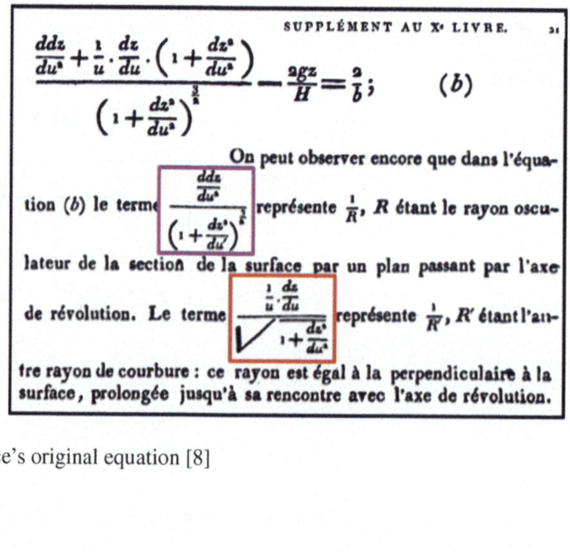
SUPPLÉMENT AU Xe LIVRE. 21

$$\frac{\frac{ddz}{du^2} + \frac{1}{u} \cdot \frac{dz}{du} \cdot \left(1 + \frac{dz^2}{du^2}\right)}{\left(1 + \frac{dz^2}{du^2}\right)^{\frac{3}{2}}} - \frac{2gz}{H} = \frac{2}{b}; \qquad (b)$$

On peut observer encore que dans l'équation (*b*) le terme $\frac{\frac{ddz}{du^2}}{\left(1+\frac{dz^2}{du^2}\right)^{\frac{3}{2}}}$ représente $\frac{1}{R}$, *R* étant le rayon osculateur de la section de la surface par un plan passant par l'axe de révolution. Le terme $\frac{\frac{1}{u} \cdot \frac{dz}{du}}{\sqrt{1+\frac{dz^2}{du^2}}}$ représente $\frac{1}{R'}$, *R'* étant l'autre rayon de courbure : ce rayon est égal à la perpendiculaire à la surface, prolongée jusqu'à sa rencontre avec l'axe de révolution.

Fig. 2.6 Laplace's original equation [8]

As illustrated in Fig. 2.7, it can be seen that the first term of (2.24) is the curvature of the planar curve defined by the meniscus profile, known to be $\frac{d\theta}{ds}$ where θ is the angle between the meniscus tangent and a fixed direction, for example the horizontal direction. ds is the infinitesimal intrinsic length along the meniscus. Consequently, the first curvature radius is $\frac{ds}{d\theta}$. The second curvature radius is the distance between the surface and the symmetry axis along the normal vector, given by $r/\sin\theta$.

Let us now put (2.24) as a system of two first-order differential equations:

$$\begin{cases} u = \frac{dr}{dz} \\ \frac{du}{dz} = \frac{1+u^2}{r} - \frac{\Delta p}{\gamma}(1+u^2)^{\frac{3}{2}} \end{cases} \tag{2.25}$$

To solve these equations, Δp must be known and boundary conditions have to be set (Fig. 2.8): let us assume that we know the point P of the meniscus in contact with the gripper. Therefore z_P and $r_P = r(z_P)$ are given by the initial coordinates of P and the slope of the meniscus in P is given by:

$$u_P = \frac{dr}{dz}|_{z_P} = \begin{cases} \frac{1}{\tan(\phi_P+\theta_2)}, & \text{if}(\theta_2+\phi_P) \neq \frac{\pi}{2} \\ 0, & \text{if}(\theta_2+\phi_P) = \frac{\pi}{2} \end{cases} \tag{2.26}$$

where the contact angle θ_2 and ϕ_P depends on the gripper geometry.

In our problem (how to determine the meniscus for given contact angles θ_1 and θ_2 and liquid volume V), only θ_2 is known. Indeed, Δp and the position of P are a priori

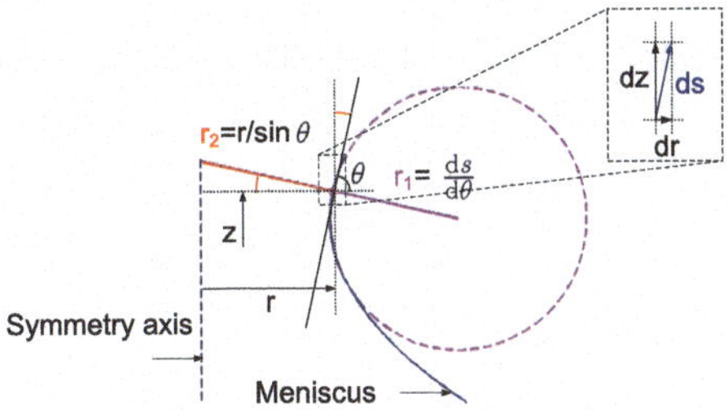

Fig. 2.7 Curvature radii of an axially symmetric surface: r_1 is the azimuthal curvature radius (in a plane containing the symmetry axis) and r_2 is the radial curvature radius (distance between the curve and the symmetry axis, measured along the normal)

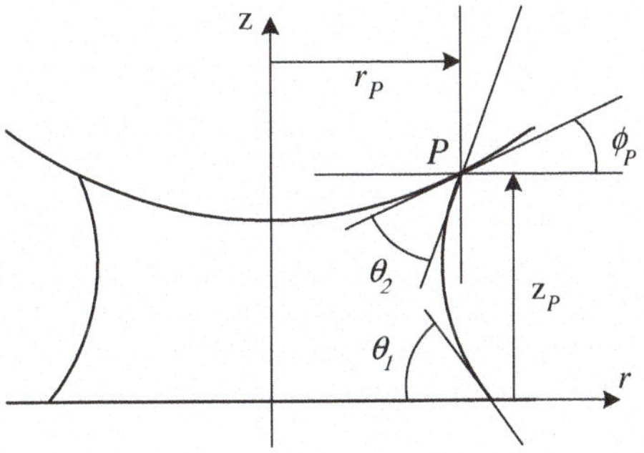

Fig. 2.8 Boundary conditions depend on the geometry and materials: the *upper line* states for the gripper, while the *lower straight line* states for the component

unknown. De Lazzer et al. [9] already suggested to iterate on Δp in order to adjust θ_1 to the prescribed value. Indeed, increasing pressure difference (i.e. more negative Δp) leads to a more curved meniscus, and, consequently, to smaller θ_1. P is still unknown and the condition on V has not yet been used. Therefore, a second iteration loop is used [10] to determine P: an initial position of P is guessed in order to solve the first iteration loop (i.e. determine a meniscus that would be correct as far as contact angles are concerned), leading to a candidate whose volume is computed. If this volume is smaller (larger) than the prescribed one, P is moved away (closer) from the symmetry axis (this is achieved by dichotomous search). This double iterative scheme is actually an application of the so-called shooting method. It is graphically illustrated in Fig. 2.9.

2.2.4 Geometrical Models

It has been seen in the previous section that solving (2.24) requires the knowledge of two initial values (position and slope of the meniscus) and the pressure difference parameter. Therefore, three conditions of the problem were used: the volume of liquid, the initial slope and the final slopes. These three conditions can alternatively be used to find the three parameters of a circle (two center coordinates and the radius) or the three coefficients of the parabola $r(z) = a_0 + a_1 z + a_2 z^2$ (Fig. 2.10).

These models are physically incorrect, but they provide a good estimate of the meniscus geometry from which the force is computed. Moreover, the parabolic model provides a full analytical resolution, as shown by Valsamis in his doctoral dissertation [11]. He also provided two maps (see Sect. 2.3.2.4) giving the relative error of these models with respect with the exact numerical resolution of (2.24).

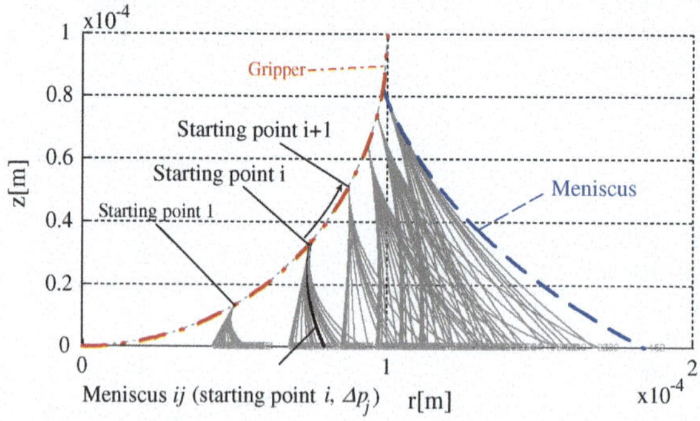

Fig. 2.9 Graphical illustration of the double iterative scheme for a sphere of 100 μm radius, contact angles equal to 3°, a volume of liquid equal to 45 nL and a gap equal to zero. Meniscus ij is obtained with the ith starting point and the jth pressure difference (Reprinted with permission from [1]. Copyright 2005 American Chemical Society)

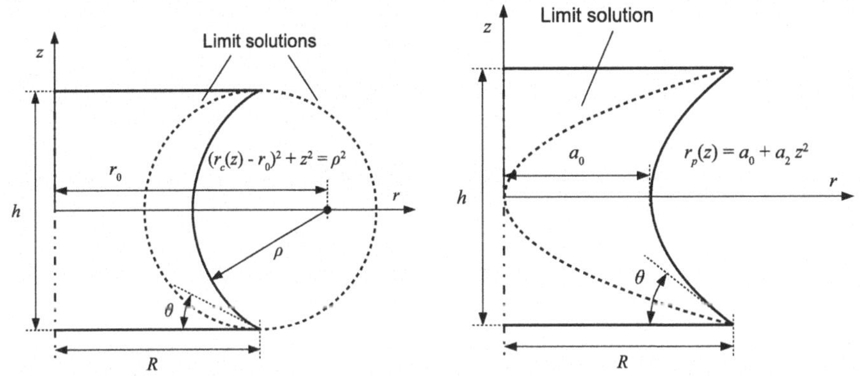

Fig. 2.10 Circular and parabolic models (Courtesy of J.-B. Valsamis [11])

As a conclusion, the meniscus shape may be assumed circular, which is valid when the radius r of the triple line is much larger than the meniscus height, or it may me assumed parabolic, which is valid for contact angles near $\pi/2$. The parabolic model has been used to get analytical estimates of the axial stiffness and in combination with gas law.

Remarks We have seen different methods to compute the geometry of a liquid bridge at equilibrium. When using surface energy minimization, the surface energy is known and the link to capillary force is obvious (the force is found by derivation). For both other methods, the link between geometry on the one hand and capillary force on the other hand has to be explained: this is the focus of next section.

2.3 Pinned and Non Pinned Triple Lines

The pinning of the triple line is made possible at the intersection between two surfaces with different surface energies or on sharp edges.

2.3.1 Non Pinned Menisci

The double iterative scheme briefly presented in Fig. 2.9 allowed us to study the influence of the pertinent parameters on the capillary forces between two solids: surface tension γ, contact angle θ, volume of liquid V, separation distance between both solids z. All these numerical experiments were performed with axial symmetry, solving (2.24).

Preliminarily, in order to validate the developed simulation code, we have studied the case of two parallel plates separated by a distance b and for a difference of pressure

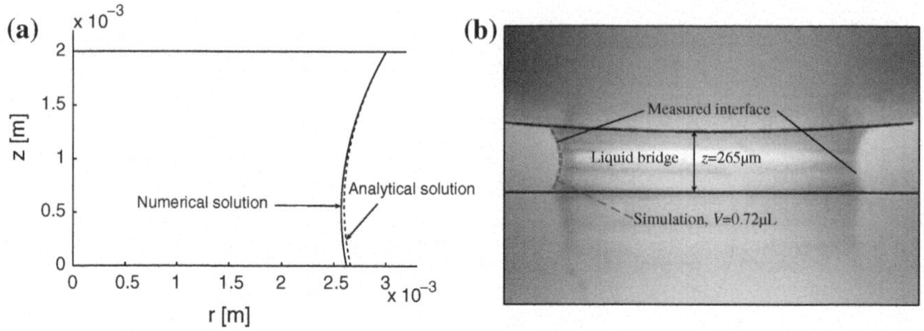

Fig. 2.11 Study of the meniscus shape **a** Comparison between the numerical and analytical meniscus shape, with $\Delta p = 0$, $a = 3\,\text{mm}$, $b = 2\,\text{mm}$ and $\theta_2 = 60°$; **b** Comparison between the simulation and the experimental meniscus shape (water, steel flat component, steel sphere ($R = 6.35\,\text{mm}$))(Reprinted with permission from [1]. Copyright 2005 American Chemical Society)

equal to zero, leading to the analytical equation of a catenary curve [3] (Fig. 2.11a). For $z = 0$, the relative error between the numerical radius and the analytical one is about 1.5 %.

A second case has been tested, namely the case of a meniscus between two parallel plates, with contact angles equal to 90°, leading to a cylindrical meniscus (with radius R) whose principal curvature radii are $R_1 = \infty$ and $R_2 = R$.

The last verification operated on the meniscus shape is the comparison between the output profile and the picture of the meniscus, as presented in Fig. 2.11b. This picture corresponds to a 12.7 mm diameter sphere linked to a steel component with 0.72 μL water. The gap in this case is 265 μm. The dashed line states for the meniscus output by the simulation tool.

The numerical computation of the force has also been compared with the analytical approximation (2.8) of the capillary force between a plane and a sphere (radius R), for a gap equal to zero and an equivalent contact angle $\theta \left(\text{if } \theta_1 \neq \theta_2, \cos\theta \equiv \frac{\cos\theta_1 + \cos\theta_2}{2} \right)$.

$$F = 4\pi R \gamma \cos\theta \tag{2.27}$$

Figure 2.12 plots the force as a function of the equivalent contact angle θ for a 26 mm diameter sphere and for two liquids: the upper curve shows the force for water and the lower one that for silicone oil (R47V50). It can be seen on this picture that the results of the simulations tend to the analytical approximations (solid and dashed lines).

After these preliminary validations, the influence of the following parameters was studied:

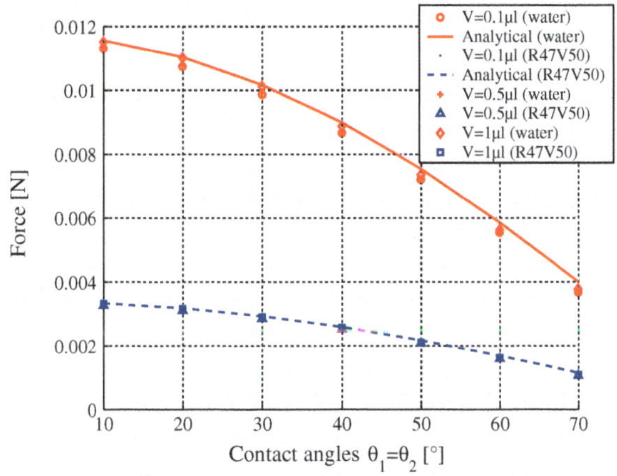

Fig. 2.12 Comparison between the simulation results and the analytical approximation $F = 4\pi\gamma R\cos\theta$ for a 13.0 mm diameter sphere, $\gamma = 72 \times 10^{-3}$ N m^{-1} (water) and $\gamma = 20.8 \times 10^{-3}$ N m^{-1} (silicone oil R47V50). The results are presented for different volumes (0.1, 0.5 and 1 μL) and different contact angles simulating different materials; the simulation points tend to the analytical approximations for water (*solid line*) and silicone oil (*dashed line*) (Reprinted with permission from [1]. Copyright 2005 American Chemical Society)

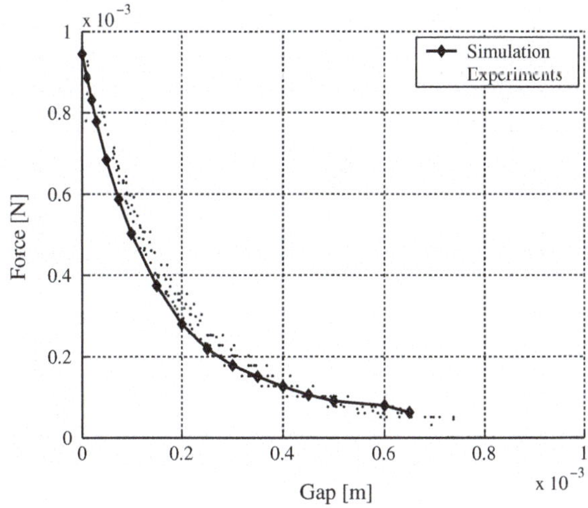

Fig. 2.13 Force-Distance curve for $V = 0.5\,\mu$L R47V50, Si-component and GS-St-7.9 (Both reprinted with permission from [1]. Copyright 2005 American Chemical Society)

1. **Separation distance**: The curve of Fig. 2.13 plots the capillary force exerted by a 7.9 mm diameter sphere (see [3] for details) on a silicon component. The force is exerted by a 0.5 μL silicone oil droplet (R47V50). This curve has been measured with an almost zero velocity (equilibrium curve). As far as the simulation is concerned, receding contact angles have been input since the meniscus is stretched

by moving the sphere upwards. The correspondence between simulation and experiment can be seen on this picture although the rupture distance predicted by the simulation is a little smaller than the measured one.
This result indicates that the simulation tool can predict the capillary force with separation distances different from zero: this comes as a complement to the previous validations made by comparing the simulation with the analytical approximations, that was valid only at contact.

2. **Pressure term versus tension term**: The simulation results of Fig. 2.13 also allow to calculate and compare the importance of the interfacial tension force and the Laplace term of the capillary force (see Fig. 2.14). The results presented in this figure justify some approximations found in the literature, neglecting the 'tension' term for small gaps. Nevertheless, this assumption is no longer valid for larger gaps where the tension term even becomes dominant.
3. **Surface tension**: Additionally, the force has been shown to be proportional to surface tension γ which ranges from 20 to 72 mNm^{-1} for usual liquids such as alcohols, silicone oils or water. Solder pastes can exhibit surface tensions as high as a few hundreds of mNm^{-1}, leading to an important surface tension effect in solder paste assembly processes. An increase of surface tension nevertheless increases the contact angle (i.e. reduces the wettability), leading to a decrease of the force: the cut-off between both effects cannot be decided without numerical quantification.
4. **Volume of liquid**: The influence of the volume of liquid depends on the geometry of the objects. It can be surprising indeed in (2.8) to observe that the force does not depend on the volume of liquid. Hence this formula is widely used in capillary adhesion studies, since the exact volume of liquid got by capillary condensation

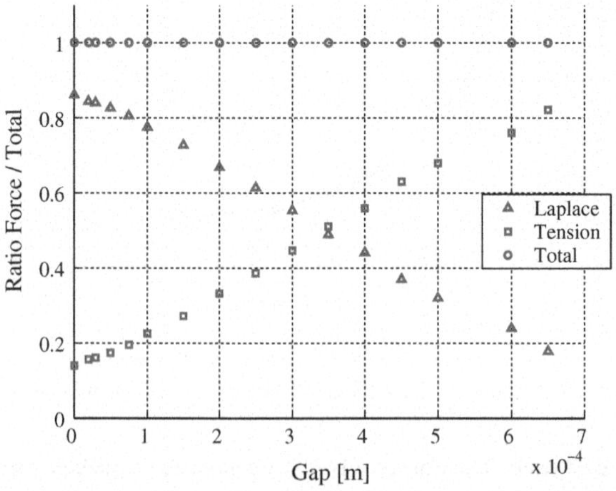

Fig. 2.14 Respective contribution of the 'tension' and 'Laplace' terms in the total amount of the force—Ratios without dimensions (Reprinted with permission from [1]. Copyright 2005 American Chemical Society)

is usually unknown. This formula is nevertheless only valid for small volumes of liquid (the so-called *immersion height* must be small). Other geometries lead to different dependences on the volume of liquid: for conical tips, the force increases with increasing volume of liquid.

As a conclusion of this part of the work, we set up a numerical tool ready to compute the force in whatever axially symmetric geometry. Three main limitations arise: (1) finding an optimum with a numerical tool is very tedious; (2) non axially symmetric geometries cannot be studied with this tool: the case of lateral capillary forces will be considered in Chap. 3; (3) finally, the dynamical aspects will be studied in Part III.

2.3.2 *Pinned Menisci*

2.3.2.1 Parabolic Model

As briefly exposed previously, we must distinguish between non-pinned menisci with imposed contact angle and pinned menisci with imposed wetting radius. In the latter case, such as depicted in Fig. 2.15, analytical approximations can be set, which is of the utmost interest to serve as benchmarks or to be exploited in design. According to Sect. 2.2.4 on geometrical approximations, it must be chosen between circular and parabolic models.

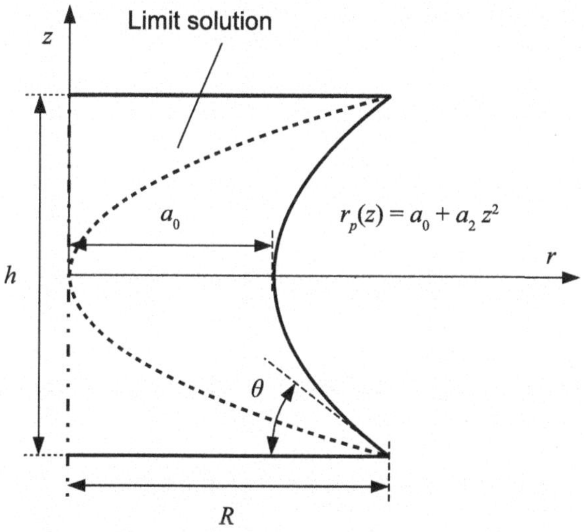

Fig. 2.15 Parabola model: as usual, three information (pinned contact angle, pad radius and gap) are used to determine the three coefficients of a parabola

This assumption of parabolic profile does not rely on any physical consideration but corresponds to the meniscus shape observed experimentally. The equation of this parabola is given by:

$$r_p(z) = a_0 + a_1 z + a_2 z^2 \tag{2.28}$$

where r and z are the axis shown in Fig. 2.15. The geometric conditions on the parabola are:

1. the symmetry with the r-axis imposes the vertex to be at $z = 0$,
2. the slope at $z = \frac{h}{2}$ is given by the contact angle,
3. the radius at $z = \frac{h}{2}$ is R.

Therefore

$$\begin{aligned} a_1 &= 0 \\ a_2 h &= \cot\theta \\ a_0 + a_2 \frac{h^2}{4} &= R \end{aligned}$$

The parabola is:

$$r_p(z) = R - \frac{h\cot\theta}{4} + \frac{\cot\theta}{h} z^2 \tag{2.29}$$

Since the parabola represents a meniscus, $r_p(z)$ must be positive. The condition for its existence is then:

$$R > \frac{h\cot\theta}{4} \tag{2.30}$$

The volume is:

$$\begin{aligned} V_p &= \pi \int_{-\frac{h}{2}}^{\frac{h}{2}} r_p^2(z) dz \\ &= \pi R^2 h - \frac{\pi R h^2 \cot\theta}{3} + \frac{\pi h^3 \cot^2\theta}{30} \end{aligned} \tag{2.31}$$

The relation between the volume and the contact angle can be inverted. The volume can be considered as an input parameter instead of the contact angle. The local curvature computed from (2.24) is:

$$2H(z) = 2h^2 \frac{2h^2 - 4Rh\cot\theta + h^2\cot^2\theta + 4\cot^2\theta\, z^2}{(4Rh - h^2\cot\theta + 4\cot\theta\, z^2)(h^2 + 4\cot^2\theta\, z^2)^{\frac{3}{2}}} \tag{2.32}$$

The curvatures at the apex (subscript $_0$) and at the triple line (subscript $_3$) are:

$$2H_0 = -\frac{2\cot\theta}{h} + \frac{4}{4R - h\cot\theta} \tag{2.33}$$

$$2H_3 = \sin\theta \frac{h - 2R\sin\theta\cos\theta}{hR} \tag{2.34}$$

From this volume equation, we can determine a_2 and its derivative with respect to h (V is assumed to be constant), consequently, the force F_z can be determined from its both components :

$$\begin{aligned} F_z &= F_L + F_T \\ &= -\pi\gamma R^2 2H_0 + 2\pi\gamma R\sin\theta \end{aligned} \tag{2.35}$$

In the particular case of $\theta = \frac{\pi}{2}$ (corresponding to a cylindrical meniscus), the force is equal to the well-known value $-\pi R\gamma$ (2.7).

The model validation has been performed by comparing the parabolic model output with the force benchmark $F = -\pi R\gamma$ for a cylindrical meniscus and by comparing it with numerical results obtained with Surface Evolver. This has been done for a pad of radius $R = 100\,\mu\text{m}$, a surface tension $\gamma = 72\,\text{mNm}^{-1}$ and a volume of liquid $V = \pi 10^{-12}\,\text{m}^{-3} = \pi\,\text{nL}$, which corresponds to a separation distance between the pad and the component equal to $h_0 = 100\,\mu\text{m}$ for a cylindrical cylinder.

We see in Fig. 2.16 the good correspondence between the parabolic model and the software results, for a large range of separation distances, between 80 and 150 % of the separation distance h_0.

As a conclusion, this analytical model can be used to compute the axial force and the axial stiffness of a liquid meniscus between two circular pads. The validity

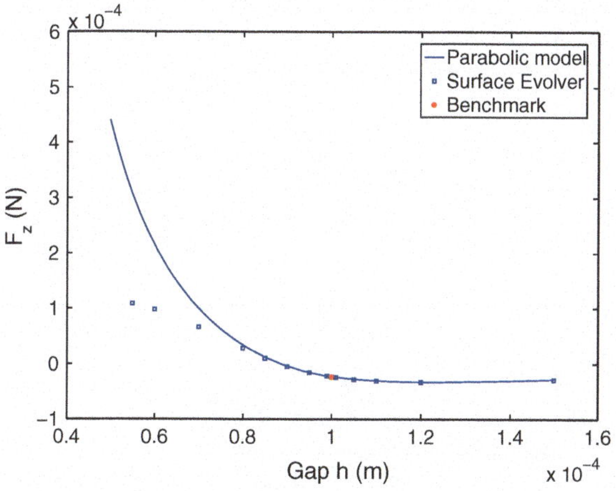

Fig. 2.16 Benchmarking of the force with our parabola model (2.35) with (2.7) and numerical simulation (surface evolver)

domain of this model—deduced from the comparison with numerical results output by Surface Evolver—ranges from 80 to 150 % of the separation distance h_0 for which the meniscus is a cylindre. Let us well note that the lower limit of the physics seems to be about 55 % of h_0. Below this level, the Surface Evolver does not converge anymore, which is related to the non physical existence of such a meniscus.

2.3.2.2 Circular Model

This model assumes a circular profile of the interface shape in the plane rz, of centre (r_c, z_c) and radius ρ (see Fig. 2.17):

$$(r_c(z) - r_0)^2 + (z - z_0)^2 = \rho^2 \tag{2.36}$$

The conditions on the arc of circle are:

1. the symmetry with the r-axis imposes the centre of circle to be on the r-axis,
2. the slope at $z = \frac{h}{2}$ is given by the contact angle,
3. the radius at $z = \frac{h}{2}$ is R.

The symmetry imposes $z_0 = 0$. By differentiating the expression with respect to z, we have:

$$2(r - r_0)\frac{\mathrm{d}r_c}{\mathrm{d}z} + 2z = 0$$

With the above relation, the contact angle can be imposed through the derivative of r_c. With the conditions at $z = \frac{h}{2}$, the position of the centre is:

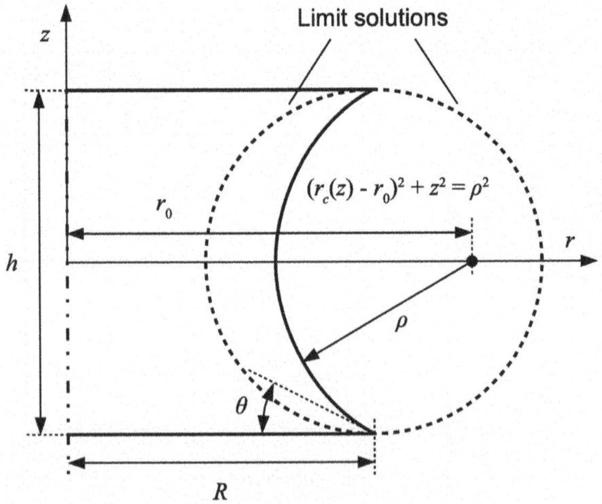

Fig. 2.17 Circular model: as usual, three information (pinned contact angle, pad radius and gap) are used to determine the three coefficients of a parabola

$$r_0 = R + \frac{h}{2}\tan\theta \tag{2.37}$$

Using the third condition:

$$\rho = \frac{h}{2|\cos\theta|} \tag{2.38}$$

With the above conditions, 2.36 may be rewritten under an explicit form $r(z)$. However, the sign of the square root differs when the contact angle is lower or higher than $\frac{\pi}{2}$:

$$r_c(z) = -\sqrt{\frac{h^2}{4\cos^2\theta} - z^2} + R + \frac{h}{2}\tan\theta \qquad \text{if } \theta < \frac{\pi}{2}$$

$$r_c(z) = +\sqrt{\frac{h^2}{4\cos^2\theta} - z^2} + R + \frac{h}{2}\tan\theta \qquad \text{if } \theta > \frac{\pi}{2}$$

More conveniently, by extracting the $\cos\theta$ from the square root, the two relations become:

$$r_c(z) = \frac{-1}{2\cos\theta}\sqrt{h^2 - 4\cos^2\theta\, z^2} + R + \frac{h}{2}\tan\theta \tag{2.39}$$

The reader will notice the singularity at $\theta = \frac{\pi}{2}$: when $\theta = 0$, the centre is at $r = R$. When θ increases, the centre moves towards $+\infty$ (occurring for $\theta = \frac{\pi}{2}$). When $\theta = \pi$, the centre is at $r = R$. The centre moves towards $-\infty$ when θ decreases. The jump of θ from $\frac{\pi}{2} - \varepsilon$ to $\frac{\pi}{2} + \varepsilon$ is accompanied of an abrupt change of the centre of circle from $+\infty$ to $-\infty$. The relation still converges to R when $\theta \to \frac{\pi}{2}$ since:

$$\lim_{\theta \to \frac{\pi}{2}} \left(\tan\theta - \frac{1}{\cos\theta}\right) = 0$$

The volume is:

$$\begin{aligned} V_c &= \pi \int_{-\frac{h}{2}}^{\frac{h}{2}} r_c^2(z)dz \\ &= \pi R^2 h + \frac{\pi R h^2 \tan\theta}{2} + \frac{\pi h^3}{4\cos^2\theta} - \frac{\pi h^3}{12} - \frac{\pi h^2}{4\cos^2\theta}\left(\frac{\pi}{2} - \theta\right)(2R + h\tan\theta) \end{aligned} \tag{2.40}$$

The reader will remark that the contact angle cannot be expressed as a function of volume as for the parabolic model. The local curvature is:

$$2H(z) = -\frac{2\cos\theta}{h} + \frac{2\cos\theta\sqrt{h^2 - 4\cos^2\theta\, z^2}}{-h\sqrt{h^2 - 4\cos^2\theta\, z^2} + 2Rh\cos\theta + h^2\sin\theta} \tag{2.41}$$

The curvature at the apex and at the triple line are:

$$2H_0 = -\frac{2\cos\theta}{h} + \frac{2\cos\theta}{2R\cos\theta + h\sin\theta - h} \tag{2.42}$$

$$2H_3 = -\frac{2\cos\theta}{h} + \frac{\sin\theta}{R} \tag{2.43}$$

2.3.2.3 Comparison of the Analytic Models for Axial Forces

The circular model is established following the same development presented above. For a symmetric case (the plane containing the neck is a plane of symmetry so that top and bottom contact angles as well as top and bottom radii are equal), the analytical expressions of the shape, the volume and the curvature are given in Table. 2.1. Input parameters are the radius of the solid liquid interface R, the gap h and the contact angle θ. The volume can also be considered as an input parameter (instead of the contact angle for example).

Table 2.1 Analytical expressions from the parabolic model and the circular model

Parabolic model	$\theta \to \frac{\pi}{2}$
$r_p(z) = R - \frac{h\cot\theta}{4} + \frac{\cot\theta}{h} z^2$	$\to R$
$V = \pi R^2 h - \frac{\pi R h^2 \cot\theta}{3} + \frac{\pi h^3 \cot^2\theta}{30}$	$\to \pi R^2 h$
$\theta(V) = \frac{5\pi R h^2 - \sqrt{30\pi h^3 V_p - 5\pi^2 R^2 h^4}}{\pi h^3}$	
$2H(z) = 2h^2 \frac{2h^2 - 4Rh\cot\theta + h^2\cot^2\theta + 4\cot^2\theta\, z^2}{(4Rh - h^2\cot\theta + 4\cot\theta\, z^2)(h^2 + 4\cot^2\theta\, z^2)^{\frac{3}{2}}}$	$\to \frac{1}{R}$
$2H_0 = -\frac{2\cot\theta}{h} + \frac{4}{4R - h\cot\theta}$	$\to \frac{1}{R}$
$2H_3 = \sin\theta \frac{h - 2R\sin\theta\cos\theta}{hR}$	$\to \frac{1}{R}$
Circular model	
$r_c(z) = \frac{-1}{2\cos\theta}\sqrt{h^2 - 4\cos^2\theta\, z^2} + R + \frac{h}{2}\tan\theta$	$\to R$
$V = \pi R^2 h + \frac{\pi R h^2 \tan\theta}{2} + \frac{\pi h^3}{4\cos^2\theta} - \frac{\pi h^3}{12} - \frac{\pi h^2}{4\cos^2\theta}(\frac{\pi}{2} - \theta)(2R + h\tan\theta)$	$\to \pi R^2 h$
$\theta(V) =$ No analytic solution	
$2H(z) = -\frac{2\cos\theta}{h} + \frac{2\cos\theta\sqrt{h^2 - 4\cos^2\theta\, z^2}}{-h\sqrt{h^2 - 4\cos^2\theta\, z^2} + 2Rh\cos\theta + h^2\sin\theta}$	$\to \frac{1}{R}$
$2H_0 = -\frac{2\cos\theta}{h} + \frac{2\cos\theta}{2R\cos\theta + h\sin\theta - h}$	$\to \frac{1}{R}$
$2H_3 = -\frac{2\cos\theta}{h} + \frac{\sin\theta}{R}$	$\to \frac{1}{R}$

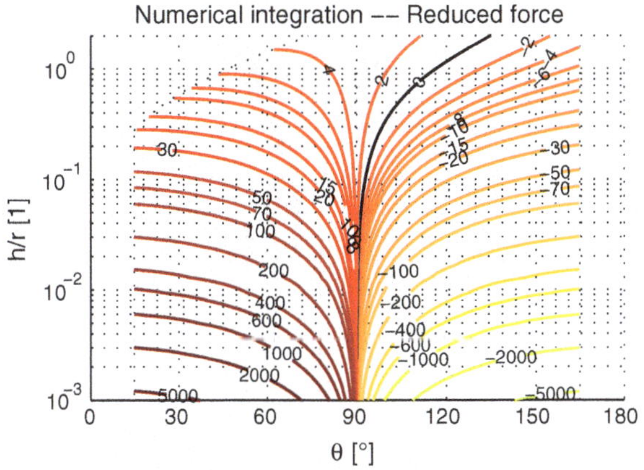

Fig. 2.18 Map of the reduced force $\hat{F} = F/r\gamma$ according to the contact angle θ and the reduced height $\hat{h} = h/r$ computed by numeric integration

The advantages of the parabola model are twofold: switching from negative to positive curvatures is smoothly done when $a_2 = 0$, while the circular approximation is not continuous: the center of the circle abruptly switches from $-\infty$ to $+\infty$ (the mathematical expressions present alocal singularity at $\theta = \frac{\pi}{2}$). Secondly, the volume of liquid can be analytically expressed as indicated by equation, having the advantage to skip the knowledge of the contact angle (contact angles are sometimes difficult to measure).

2.3.2.4 Error Maps for Circular and Parabolic Models

The reduced force $\hat{F} = F/r\gamma$ is represented on a map according to the contact angle θ and the reduced height $\hat{h} = h/r$. On Fig. 2.18, the shape $r(z)$ from 2.24 has been integrated numerically, providing an exact estimation of the curvature and contact angle. The zero force curve does not superpose on the curve $\theta = \frac{\pi}{2}$: the Laplace F_L term must cancel the surface tension term F_T. At a fixed contact angle θ, the curvature decreases with the reduced gap $\hat{h}$. Therefore when the gap increases, the contact angle must also increase to balance the surface tension force.

To evaluate the accuracy of the analytical approximation of the interface, we present in Fig. 2.19a and b the relative error of the force for both models:

$$\varepsilon_p = \frac{F_p - F}{F} \tag{2.44}$$

$$\varepsilon_c = \frac{F_c - F}{F} \tag{2.45}$$

The parabolic model presents a relative error below 20 % for an angle between 60 and120° (Fig. 2.19a). On the left side of the graph, the error increases near the zero-

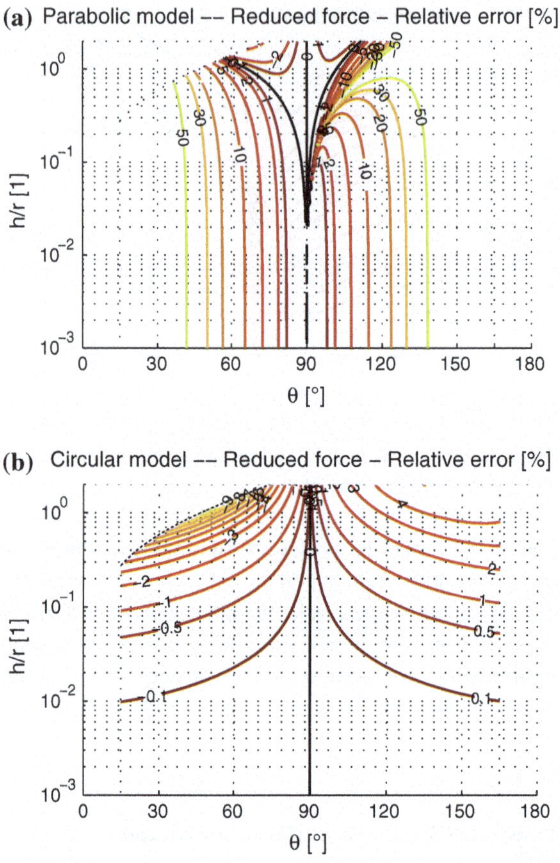

Fig. 2.19 Map of relative error of the reduced force $\hat{F} = F/r\gamma$ according to the contact angle θ and the reduced height $\hat{h} = h/r$

force curve of the exact solution (Fig. 2.18) because there is no perfect match between the parabolic model and the exact solution. Fig. 2.19b shows that the circular model is very accurate, especially at small gap.

2.4 Conclusions

This chapter recalled the fundamentals to calculate capillary forces for simple geometries, either by using the energetic method (Sect. 2.2.1), or by solving the equations exactly (Sect. 2.2.3). Geometrical approximations were also provided, assuming a circular or a parabolic meniscus shape. This preliminary description was limited to equilibrium configurations: axial dynamics will be presented in Chaps. 6, 7.

Acknowledgments This work has been funded by the F.R.I.A.—*Fonds pour la Formation et la Recherche dans l'Industrie et l'Agriculture.*, and developments were achieved among the framework of the European project

Appendix

Capillary Force Developed by a Meniscus at Equilibrium

Let us consider the meniscus depicted in Fig. 2.20. At equilibrium, the sum of forces acting on any slice of the meniscus must be equal to zero. On each face, capillary forces can be split into two contributions: the so-called tension force F_T, due to the action of surface tension along the tangent to liquid-gas interface, and the Laplace or (also called) capillary force F_L originating from the pressure acting on the face.

The force exerted on the bottom face of the slice is equal to:

$$\bar{F}(z) = \left(\underbrace{\pi r^2 \Delta p}_{F_L} - \underbrace{2\pi r \gamma \cos u}_{F_T} \right) \bar{1}_z \tag{2.46}$$

while the force acting on the top face of the slice is given by:

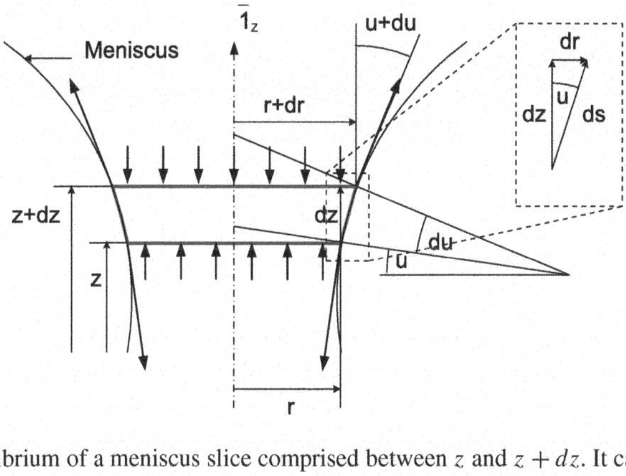

Fig. 2.20 Equilibrium of a meniscus slice comprised between z and $z + dz$. It can be shown that the capillary force computed at height z exactly balance the the capillary force computed at height $z + dz$

$$\bar{F}(z+\mathrm{d}z) = \left(-\pi(r+\mathrm{d}r)^2\Delta p + 2\pi(r+\mathrm{d}r)\gamma\cos(u+\mathrm{d}u)\right)\bar{1}_z$$
$$= \left[-\pi r^2\Delta p + 2\pi r\gamma\cos u + 2\pi\underbrace{(-r\mathrm{d}r\Delta p + r\gamma\sin u\mathrm{d}u + \gamma\cos u\mathrm{d}r))}_{I}\right]\bar{1}_z \tag{2.47}$$

The underbraced expression I can be shown to be equal to zero by expressing the Laplace law (Eq. (2.24)):

$$\Delta p = 2H\gamma$$
$$= \left(-\frac{r''}{(1+r'^2)^{3/2}} + \frac{1}{r(1+r'^2)^{1/2}}\right)\gamma \tag{2.48}$$
$$= \left(\frac{\mathrm{d}u}{\mathrm{d}s} + \frac{\cos u}{r}\right)\gamma \tag{2.49}$$
$$= \left(\frac{\mathrm{d}u}{\mathrm{d}r}\sin u + \frac{\cos u}{r}\right)\gamma \tag{2.50}$$

leading to:

$$I = -r\mathrm{d}u\sin u\gamma - \mathrm{d}r\cos u\gamma + r\gamma\sin u\mathrm{d}u + \gamma\cos u\mathrm{d}r = 0 \tag{2.51}$$

Consequently, the forces $\bar{F}(z) + \bar{F}(z+\mathrm{d}z)$ balance, and the capillary force given by $F(z)$ can be computed at any value of z. This means that in the case of two solids linked by a liquid meniscus, the force can be computed on the top component or on the bottom component. For the sake of convenience, it can also be computed at the neck in case the latter exists (it may not exist if the extremum radius of the meniscus corresponds to one of both wetting radii).

Equivalence of Formulations

A lot of work has been reported on capillary forces modeling (see for example [1, 4–6, 9, 12–14]), based on the energetic method (i.e. derivation of the total interface energy) or a direct force computation from the meniscus geometry, the latter being either determined exactly through the numerical solving of the so-called Laplace equation or approximated by a predefined geometrical profile such as a circle (i.e. toroidal approximation) or a parabola. The energetic approach is usually quite clear on its approximations: the liquid-vapor interface energy is sometimes neglected in order not to compute the exact shape of the meniscus, but an exact solution can be found if the lateral is computed for example by mean of a finite element solver such as Surface Evolver. At the contrary, literature results are not so clear as far as the other method is concerned. For example, some authors neglect the so-called tension term with respect to the Laplace term . This sometimes pertinent assumption has led many author authors to add the tension term to the result obtained by deriving the interface energy, i.e. to mix both methods. A recent by one of the authors [15] contributed to clarify this situation by showing that the capillary force obtained by deriving the interfacial energy is exactly equal to the sum of the Laplace and tension terms. The equivalence is considered with three qualitative arguments, and an analytical argument is developed in the case of the interaction between a prism and a place. Experimental results also contributed to show this equivalence.

Mathematically, the equivalence between the energetic approach and the direct formulation based on the Laplace and the tension terms can be shown:

$$F = F_{\rm L} + F_{\rm T} = -\frac{{\rm d}W}{{\rm d}z} \tag{2.52}$$

where $F_{\rm L}$ and $F_{\rm T}$ are given by (2.46), W by (2.1). z is the separation distance between both solids.

As it is shown that both approaches are equivalent, it means that the energetic approach already involves the tension term and the Laplace term on an implicit way. Consequently, the energetic approach as proposed by Israelachvili (see (2.8)) includes both terms, even if, for zero separation distance, the pressure term usually dominates the tension one. For axially symmetric configurations, the method based on the Laplace equation will be preferred because it can be easily numerically solved.

References

1. P. Lambert, A. Delchambre, Parameters ruling capillary forces at the submillimetric scale. Langmuir **21**, 9537–9543 (2005)
2. P. Lambert, M. Mastrangeli, J.-B. Valsamis, G. Degrez, Spectral analysis and experimental study of lateral capillary dynamics (for flip-chip applications). Microfluid. Nanofluid. (Published online) (2010)
3. P. Lambert, *Capillary Forces in Microassembly: Modeling, Simulation, Experiments, and Case Study. Microtechnology and MEMS* (Springer, Boston, 2007)
4. J.N. Israelachvili, *Intermolecular and Surface Forces*, 2nd edn. (Academic Press, San Diego, 1992)
5. F.M. Orr, L.E. Scriven, A.P. Rivas, Pendular rings between solids: meniscus properties and capillary force. J. Fluid Mech. **67**, 723–742 (1975)
6. Yakov I. Rabinovich, Madhavan S. Esayanur, Brij M. Mougdil, Capillary forces between two spheres with a fixed volume liquid bridge: theory and experiment. Langmuir **21**, 10992–10997 (2005)
7. A. Chau, S. Régnier, A. Delchambre, P. Lambert, Theoretical and experimental study of the influence of AFM tip geometry and orientation on capillary force. J. Adhes. Sci. Technl. **24**, 2499–2510 (2010)
8. P.-S. de Laplace, Sur l'action capillaire, in *Mécanique Céleste—Supplément au Livre X*, pp. 349–498, Courcier, Paris (1805)
9. A. de Lazzer, M. Dreyer, H.J. Rath, Particle-surface capillary forces. Langmuir **15**(13), 4551–4559 (1999)

10. P. Lambert, P. Letier, A. Delchambre, Capillary and surface tension forces in the manipulation of small parts, in *Proceedings of International Symposium on Assembly and Tasks Planning (ISATP)*, pp. 54–59 (2003)
11. J.-B. Valsamis, A study of liquid bridges dynamics: an application to micro-assembly. Ph.D. thesis, Université libre de Bruxelles
12. X. Pepin, D. Rossetti, S.M. Iveson, S.J.R. Simons, Modeling the evolution and rupture of pendular liquid bridges in the presence of large wetting hysteresis. J. Colloid Interface Sci. **232**, 289–297 (2000)
13. Y.I. Rabinovich, J.J. Adler, M.S. Esayanur, A. Ata, R.K. Singh, B.M. Mougdil, Capillary forces between surfaces with nanoscale roughness. Adv. Colloid Interface Sci. **96**, 213–230 (2002)
14. T. Stifter, O. Marti, B. Bhushan, Theoretical investigation of the distance dependence of capillary and van der Waals forces in scanning force microscopy. Phys. Rev. B **62**(20), 13667–13673 (2000)
15. P. Lambert, A. Chau, A. Delchambre, S. Régnier, Comparison between two capillary forces models. Langmuir **24**(7), 3157 (2008)

Chapter 3
Lateral Capillary Forces

Massimo Mastrangeli and Pierre Lambert

Abstract Lateral capillary forces ensuing from perturbed fluid menisci are pivotal to many important technologies, including capillary self-alignment and self-assembly of heterogeneous microsystems. This chapter presents a comprehensive study of the quasi-statics of lateral capillary forces arising from a constrained cylindrical fluid meniscus subjected to small lateral perturbations. After a contextual literature review, we describe a novel experimental apparatus designed to accurately characterize such a fundamental system. We then reproduce our experimental data on lateral meniscus forces and stiffnesses by means of both a novel analytical model and a finite element model. The agreement between our measurements and our models validate earlier reports and provides a solid foundation for the applications of lateral capillary forces to microsystems handling and assembly.

3.1 Introduction and Literature Overview

Capillary and surface-tension-related phaenomena are ubiquitous in everyday life. As explained in earlier chapters, their thermodynamic origin lies in the minimization of the free energy of physical systems containing non-rigid interfaces [1, 12]. Fundamental in bio-chemistry [20], quasi-static capillary phaenomena, as the one studied

M. Mastrangeli (✉)
MTM Department KULeuven and IMEC, Leuven, Belgium
e-mail: mastrangelim@gmail.com

P. Lambert
BEAMS Department, Université Libre de Bruxelles, 1050 Bruxelles, Belgium
e-mail: pierre.lambert@ulb.ac.be

P. Lambert (ed.), *Surface Tension in Microsystems*, Microtechnology and MEMS, DOI: 10.1007/978-3-642-37552-1_3, © Springer-Verlag Berlin Heidelberg 2013

in this chapter, are presently at the core of many important technologies, as well, such as e.g. switchable lenses [2], tightness systems [11], precision manufacturing [22], metal bonding and soldering [51], microfabrication and microfluidics [16], and integration and packaging of heterogeneous [6].

With specific focus on the latter, the molten solder-driven self-aligning motion of IC dies on top of pre-processed substrates was first exploited in so-called *flip-chip* assembly to achieve highly-accurate passive die registration [15, 19, 21, 46]. Capillarity is one powerful enabler of the three-dimensional deployment and actuation of articulated microelectromechanical structures of unprecedented complexity [17, 44] and it is one of the physical mechanisms driving the growing class of packaging techniques based on self-assembly [30]. Surface tension-driven aggregation was first adopted for the self-assembly of heterogeneous functional systems by the Whitesides' Group [5, 47, 48]. It was later adapted to part-to-substrate assembly tasks by the teams of Howe [41, 42], Böhringer [53], Koyanagi [13, 14], Parviz [33, 43] and Jacobs [8, 57]— to mention but a few examples.[1]

Most aforementioned assembly techniques share, at least partly, the same underlying mechanism. Upon contact between the functional (i.e. pre-conditioned to enable further processing) side of the *part* to be assembled (hereby representing e.g. an IC die, a microdevice, a MEMS component) and the highly-energetic mating surface—composed by a fluid, such as e.g. hydrocarbons [42], water-based solutions [14] or molten solders [37]—of the corresponding *binding site* on the substrate, capillary torques[2] and forces—both perpendicular (i.e. vertical) and parallel (i.e. lateral) to the substrate—act on the part and drive it to self-align with the underlying binding site [31]. Floating on the liquid meniscus and aligning to the pattern of the binding site, the part achieves its rest (i.e. minimal energy) configuration. The same capillary forces intervene to oppose any sufficiently-small displacement of the part from its rest position—therefore, they are also referred to as *restoring* forces in this context. The accuracy of registration between part and binding site critically depends on lateral capillary forces—the object of the present contribution. Capillary torque and vertical capillary meniscus forces were investigated, among others, by Takei et al. [45] and by Lambert [22], respectively.

Early theoretical investigations on lateral capillary forces addressed the shapes and corresponding alignment performance of fluid drops (representing e.g. molten-solder bumps) vertically-constrained by parallel plates and laterally-confined by the geometrical patterns, of contrasting wetting properties, defined on the same plates (i.e. the binding sites) [35, 36, 55]. Finite-element quasi-static numerical simulations of a similar physical system—where the top plate was represented by a part of finite dimensions and unconstrained in all translational and rotational degrees of freedom (DOFs)—further illustrated the dependency of lateral capillary forces on the

[1] See Chap. 12 for a review of capillarity-driven assembly.

[2] Capillary torque is addressed in Chap. 4.

physical properties of the fluid meniscus and on the geometry of the confining geometrical patterns [25, 27]. Briefly, the forces are proportional to the surface tension of the fluid and inversely proportional to the height of the meniscus (also referred to as the *gap* in the following). Moreover, they follow the (a)symmetry of the patterns of the confining sites: for equal relative displacements, those along shorter sides of the sites originate stronger forces. Böhringer proposed a geometrical model of confined capillary menisci that greatly simplified energy and force calculations [4]. Built around the two-dimensional convolution of the patterns of the confining surfaces, the model easied the theoretical investigations on geometry-dependent self-alignment performance of binding sites [24, 54]. Nonetheless, being essentially two-dimensional, Böhringer's model accurately reproduces known results only for relative displacements from the rest position larger than the meniscus height. Thus it is not suitable to model the alignment performance in close proximity to the rest position—which is conversely of utmost importance for accurate positioning such as in flip-chip-like assembly of dies with very-small interconnection pitches (e.g. few microns). Particularly, the model neglects the curvature of the surface of the fluid meniscus; the curvature affects the capillary restoring forces especially for small meniscus perturbations [27, 50]. Given fixed geometrical boundary conditions, the meniscus curvature is directly influenced by the volume of the fluid: hence the need for accurate control of the fluid volume to achieve high process reliability and reproducibility [29].

As compared to the extensive modeling literature, experimental quasi-static investigations of lateral capillary forces were up to now overall unsatisfactory. A few works reported on the determination of the ultimate registration accuracy achievable between part and binding site by means of capillary self-alignment: measurements were performed either optically [39, 40, 42] or by analyzing assembly cross-sections by scanning electron microscopy [13]. Best claimed figures were of the order of 1 μm or lower—amenable to advanced packaging applications. Other researchers [3, 26] performed blowing tests [10] to estimate the maximum adhesion force binding flat parts onto confined fluid drops. However, in such conditions—where the direction of displacement of the floating part was not constrained to be parallel to the substrate—the estimated force values may hardly be attributed *only* to lateral force components. To our knowledge, only Zhang reported experimental measurements and numerical simulations of lateral capillary forces for the same physical system [56]. The system, immersed in water, was composed by a 450 μm $\times$ 250 μm flat thin silicon piece floating on a matching rectangular site confining a drop of hydrophobic fluid. Lateral capillary forces were measured in situ by means of a micromachined optical encoder featuring laser-illuminated, calibrated optical gratings and a horizontal probe that displaced the part from its rest position along a direction parallel to the substrate. Anyway, though the proposed numerical model showed reasonable match with experimental data, no mention was thereby to be found of e.g. surface tension and volume of the fluid and of the meniscus height—pivotal parameters to reproduce and appreciate the numerical results. Moreover, the way the physical system was set up, though consistent with their proposed application, was not easily controllable; and only one experimental measurement was shown.

In this chapter we present a comprehensive quasi-static study of lateral capillary meniscus forces, in which experimental, numerical and analytical analyses performed on the same system are compared. We describe the conceptually-simple experimental setup that we designed to quasi-statically measure the lateral restoring forces arising from small perturbation of fluid menisci confined between two parallel plates (Sect. 3.2). Easy to use and macroscopic, our setup still allowed to control the position of the movable plates with 1 μm accuracy, and to resolve forces as low as about 1 μN. By fixing the dimensions of the confining pads and the meniscus fluid and height beforehand, the setup allowed for repeatable experiments. Our setup may be roughly likened to an upscaled version of the flexure-based microgap rheometer developed by Clasen et al. to study microgap-dependent flow behavior of complex fluids [9]. The experimental results obtained for circular pads were then supported by both an analytical (Sect. 3.3.1) and a finite-element model (Sect. 3.3.2) predicting restoring forces for small perturbation of cylindrical fluid menisci with good accuracy (Sect. 3.4). In view of our results, our setup may constitute a reference testbed for further investigations on confined fluid menisci (Sect. 3.5).

Finally, this chapter focuses on *statics* only. The *dynamics* of capillary self-alignment [23] is investigated in Part 3 of this book.

3.2 Experimental Measurement of Lateral Capillary Forces

3.2.1 Requirements and Constraints

In designing our sensing apparatus, we tried to satisfy several requirements at once:

1. simple and precise spatial manipulations;
2. repeatability of experiments: all relevant meniscus parameters of each experiment had to be reliably known (before or after the experiment);
3. measuring lateral capillary forces of the same order of magnitude of those arising in flip-chip-like assembly, i.e. forces reportedly of the order of tens of μN [25];
4. measuring *only* lateral capillary forces;
5. sampling the capillary force-versus-lateral displacement curve close to its origin, i.e. for lateral displacements smaller that the meniscus height. Finally
6. using fluids of low volatility, so to allow for up to tens of minutes-long quasi-static measurements.

3.2.2 The Sensing Principle

Our experimental setup was built around a fluid meniscus confined between two horizontal, parallel circular *pads* and monitored by a mechanical *spring* (as detailed in Sect. 3.2.3). The setup embodied a sensing principle based on the balance between *capillary* and *elastic* lateral forces (Fig. 3.1).

As initial conditions, the vertical superposition and distance of the centers of bottom and top pads (with coordinates x^b and x^t, respectively) were imposed. This enforced a cylindrical meniscus profile (Fig. 3.2a). In this state, both the fluid meniscus and the spring were in their rest configurations $x_0^b = x_0^t = 0$ (Fig. 3.1a).

The measurements were performed by imposing a known lateral displacement x_d^b to the bottom pad (Figs. 3.1b and 3.2b). As a consequence of this perturbation, two opposing lateral forces acted on the top pad: the capillary force, tending to restore the rest position of the meniscus, and the elastic spring force, tending to restore that of the spring. The equilibrium of the forces determined the actual displacement of the top pad: this was x_d^t with respect to the rest position of the spring, and $u = x_d^b - x_d^t$ relative to the bottom pad. Hence, the restoring lateral meniscus force corresponding to the net displacement u of the top pad relative to the bottom pad could be calculated by multiplying the absolute displacement x_d^t of the top pad times the bending stiffness of the spring K—whose estimation is described in Sect. 3.5.

By imposing series of temporally-isolated displacements to the bottom pad, we could measure the lateral restoring force as function of the relative lateral displacement of our system. The lateral stiffness of each fluid meniscus K_m^L was obtained by first-order polynomial fitting of this curve. In all experiments the same set $\{x_{d_i}^b\}$ of bottom pad displacements was imposed in both directions (i.e. forward and backward scans) around the initial position, in order to ascertain eventual asymmetries and/or hysteretic phaenomena.

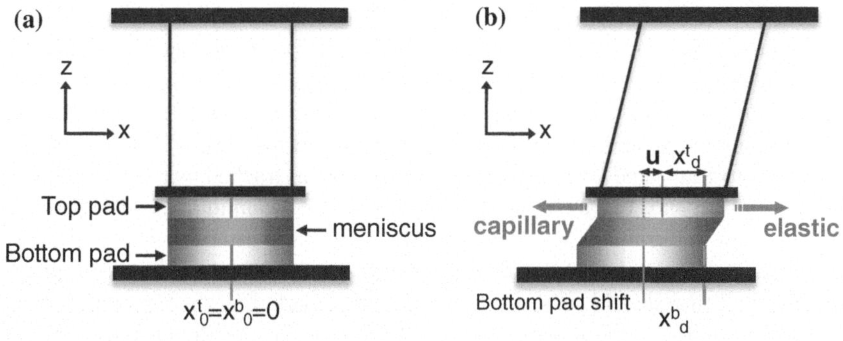

Fig. 3.1 The sensing principle, based on the balance between capillary and elastic lateral forces. **a** Rest configuration. **b** Perturbed configuration . Relative dimensions are out of scale for illustration purposes. (Actual examples of rest and perturbed configurations are shown in Fig. 3.2).

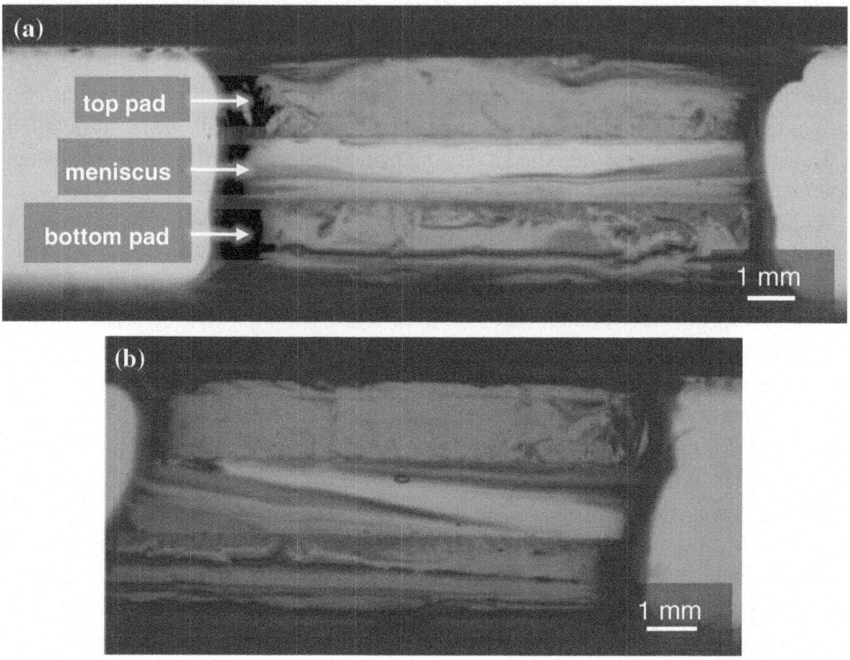

Fig. 3.2 Measurement of lateral capillary forces of a cylindrical meniscus. **a** Rest configuration. *Top and bottom* pads are aligned, the height of the meniscus (1.205 μm) is trimmed to get a cylindrical profile. **b** Perturbed configuration. After imposing a displacement ($x_d^b = 812\,\mu\text{m}$) to the *bottom* pad, the lateral force balance between capillary and elastic spring forces determines the equilibrium displacement ($u = 660\,\mu\text{m}$) of the *top* relative to the *bottom* pad

3.2.3 The Experimental Setup

The schematic diagram of and the actual experimental setup are shown in Figs. 3.3 and 3.4, respectively. Two equal glass pads (diameter $D = 2 \cdot R = 9.4\,\text{mm}$, thickness $= 1.6\,\text{mm}$) confined and shaped the liquid menisci under test. The bottom pad was solidal to an anchored 6-DOFs stage (Newport M-562-XYZ) that could be manually roto-translated with micrometric resolution, as tracked by a laser displacement sensor (Keyence LC-2440) working within its linear output range. The sensor's controller unit (Keyence LC-2400W) displayed real-time-averaged displacements. The top pad was solidal to the bottom surface of a finely-machined aluminum parallelepiped (the *shuttle*, of mass $m_{sh} = 12.748\,g$ including the top pad), in turn held by two equal steel cantilevers (Precision Brand) of certified dimensions and total mass $m_b = 2 \cdot 2.85\,g$ hanging from an overarching solid bridge (not shown). This double-cantilever-supported shuttle constituted the actual sensing apparatus, i.e. the *spring*. The sliding motion of the shuttle (i.e. the bending motion of the cantilevers) was thus constrained to take place exclusively along the chosen lateral measurement direction (see also Footnote 2 of this chapter), thereby tracked by an independent

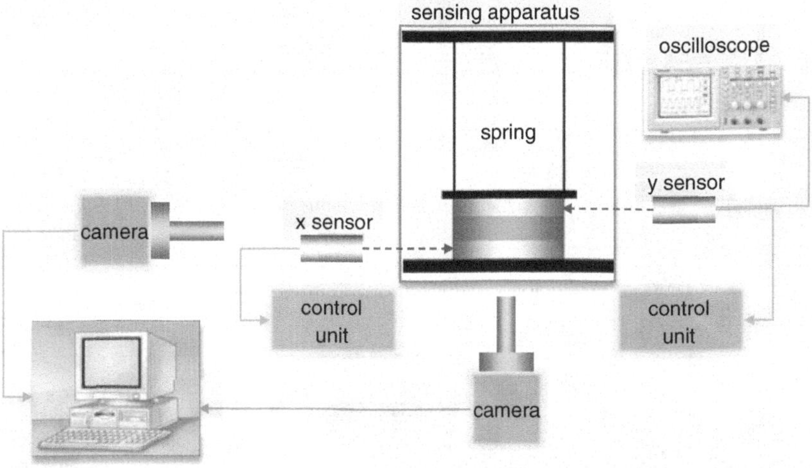

Fig. 3.3 Schematic representation of the experimental setup. Relative dimensions are out of scale for representation purposes

laser displacement sensor (and control unit). The hanging point and the length of the spring cantilevers (and so the vertical position of the top pad) were never modified after initial calibration (described below): we used exactly the same apparatus in all experiments. Accordingly: (1) the height h of the fluid menisci was set using only the controlled vertical motion of the translation stage; (2) the vertical components of the capillary force were excluded from measurements (requirement (iv)), being balanced by the vertical vincular reactions of the rigid supports.

To check pad alignment, plane-parallelism of pad surfaces and fluid meniscus profile, top and bottom pads were tracked by PC-connected cameras: a USB camera pointed along the measurement direction, and a high-speed camera (Photron Fastcam SA3 120K) pointed along a perpendicular direction intersecting the former at the initial position of the bottom pad. The pads were illuminated by a flat backlight (LDL-TP-83x75, with CCS PD-3012 power supply unit) with uniform red light emission profile and unnoticeable heating power. For every experiment, pictures of the menisci were taken with the high-speed camera right after the initial meniscus trimming and at the end of measurements. The 1024 pixels-wide pictures spanned a field of view of about 20 mm, providing a positioning accuracy of less than 20 μm.

3.2.4 Fluids

Homologous silicone oils were used as reference fluids (Table 3.1). All the experiments were performed in a laboratory environment: room temperature varied between 25 and 28 °C, the relative air humidity was about 38 %.

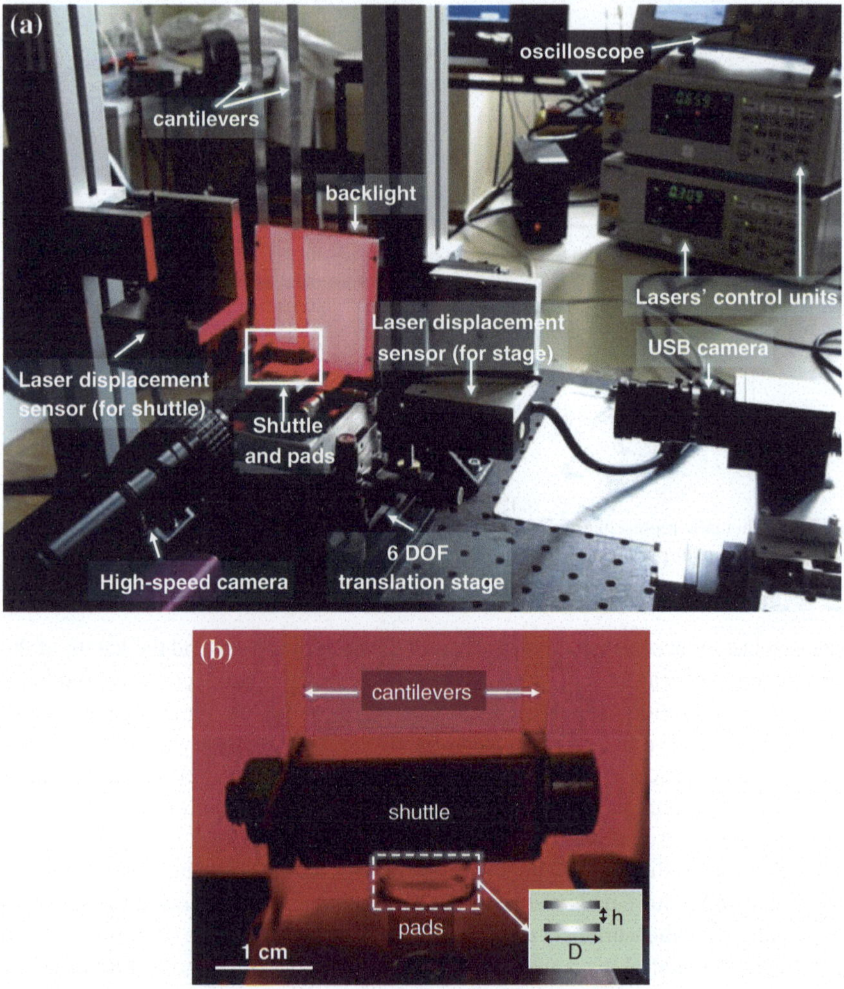

Fig. 3.4 The experimental setup for lateral capillary force measurements

The $L_c = \sqrt{\frac{\gamma}{\rho g}}$ of the oils was about 1.49 mm. The imposed heights of the menisci were smaller than L_c in all experiments, so that gravity could be neglected as compared to capillary forces (requirement (iv)). For each fluid, we performed a series of experiments with progressively-smaller meniscus heights. In the first experiment of each series, the meniscus height was calibrated to exactly 1 mm by means of a ceramic slip gage of certified thickness (Mitutoyo). The contact angle (CA or θ in the following) of silicone oils on clean hydrophilic surfaces is reportedly rather small [1]. This, combined with the geometric constraint captured by the *Gibbs' inequality* [34] (discussed in Sect. 3.4.2) set an upper bound on the maximum perturbation of the fluid meniscus to avoid fluid overflow beyond the edges of the

Table 3.1 Fluids used in the experiments

Name	Density [Kg/m^3]	Surface tension [N/m]	Supplier
Oil 1	970	0.0211	Rhodorsil R47V500
Oil 2	973	0.0211	Rhodorsil R47V5000
Oil 5	934	0.0201	Dow Corning DC200FLUID10
Oil 6	960	0.0209	Dow Corning DC200FLUID100
Oil 7	971	0.0212	Dow Corning DC200FLUID1000
Water	1000	0.072	Tap water

bottom pad (and consequent change of the volume of the meniscus). The range of relative pad displacements still allowed was fully-coherent with our focus on small lateral meniscus perturbations (i.e. requirement (v)). Effects of oil evaporation were negligible throughout all our experiments, consistently with their low volatility. On the contrary, in some explorative experiments we also used water as meniscus fluid ($\gamma = 72\,\text{mJ}\cdot\text{m}^{-2}$, $L_c^{H_2O} = 2.72\,\text{mm}$); as expected, its high volatility made our quasi-static measurements problematic and difficult to reproduce (see Sect. 3.4.2).

3.2.5 Data Processing

The fluids were manually dispensed between the pads from calibrated pipettes. Hardly-controllable tip pinch-off effects made dispensing not precise enough. Hence we calculated the actual volumes V and height h of the fluid menisci by digital post-processing of the pictures taken at the beginning of the experiments. Images calibrated through the 1 mm-thick ceramic gage led to a spatial resolution of $14.2\,\mu\text{m}\cdot\text{pixel}^{-1}$. Our semi-automatic numerical algorithm computed V, h and D for each experiment assuming axial symmetry of the menisci. We experimentally enforced the meniscus axisymmetry, as assumed in the analytical model—see Sect. 3.3.1. That is, for each dispensed fluid drop we set an initial cylindrical meniscus profile, as judged visually (see e.g. Fig. 3.2a). We made 3 numerical estimates on each calibration picture, and we *a posteriori* discarded all experiments for which the ratio between the average values of V and h and their standard deviation was smaller than 10. The uncertainty on the heights amounted typically to 2 pixels, i.e. about $28\,\mu\text{m}$. Thus, we kept the heights of the menisci always higher than $280\,\mu\text{m}$. Moreover, by comparing the reconstructed value of D to its caliber-measured value, we could assess the accuracy of the numerical estimates of D, V, and h (see Table 3.2).

We remark that, although we used only circular pads in our experiments, fluid menisci shaped by pads of *arbitrary* geometries can in principle be investigated with our setup, as well. The exact value of the dispensed fluid volume may be harder to calculate for non-axisymmetric menisci.

Table 3.2 Summary of results (Exp = experimental, SE = Surface Evolver, An = analytical)

Experiment	Geometry	Fluid	K $[N/m]$	Modeling error (%)
	height: 1.138 mm		Exp: 0.2466	
1	volume: 85.6 nL	Oil 2	SE: 0.2415	2.07
	diameter: 9.186 mm (RE: 2.28 %)		An: 0.2797	13.42
	height: 1 mm		Exp: 0.2827	
2	volume: 74.2 nL	Oil 6	SE: 0.2940	3.99
	diameter: 9.63 mm (RE: 2.45 %)		An: 0.3152	11.48
	height: 0.852 mm		Exp: 0.3392	
3	volume: 57.1 nL	Oil 6	SE: 0.3456	1.89
	diameter: 9.41 mm (RE: 0.11 %)		An: 0.3698	9.01
	h: 1 mm		Exp: 0.2407	
4	V: 65.5 nL	Oil 2	SE: 0.2648	10.01
	D: 9.29 mm (RE: 1.21 %)		An: 0.3182	32.19
	h: 0.906 mm		Exp: 0.2962	
5	V: 54.2 nL	Oil 2	SE: 0.3285	10.90
	D: 9.15 mm (RE: 2.64 %)		An: 0.3510	18.50
	h: 0.720 mm		Exp: 0.3485	
6	V: 47.3 nL	Oil 2	SE: 0.3989	14.46
	D: 9.44 mm (RE: 0.42 %)		An: 0.4417	26.75

The relative errors of the SE and An models compared to the experimental estimates are shown in the last column. The error (relative to the measured value; indicated by RE in the second column) in the reconstruction of the pad diameter by the image post-processing algorithm was used to assess the accuracy of the estimates on meniscus heights and volumes

3.3 Modeling of Lateral Capillary Forces

3.3.1 Analytic Model

We developed an analytic model to estimate the lateral stiffness of a cylindrical fluid meniscus confined between two circular pads (Fig. 3.5). Perfectly-cylindrical meniscus profiles were assumed for analytical closed-form tractability—though such assumption is admittedly ideal, and not always satisfied experimentally.

First, we computed the lateral area of a tilted cylinder of height h, radius R and relative lateral shift between pads $u < h$. In this configuration, the cylinder axis is not perpendicular to both pads but inclined with an angle α given by $\tan\alpha = u/h$.

The equation of this cylinder is given by:

$$S \equiv \bar{OP} = z\bar{1}_z + u(z)\bar{1}_x + R\bar{1}_r \quad (3.1)$$

$$= \left(\frac{z}{h}u + R\cos\theta\right)\bar{1}_x + (R\sin\theta)\bar{1}_y + z\bar{1}_z \quad (3.2)$$

Computing the area element $\mathrm{d}\Sigma$ as:

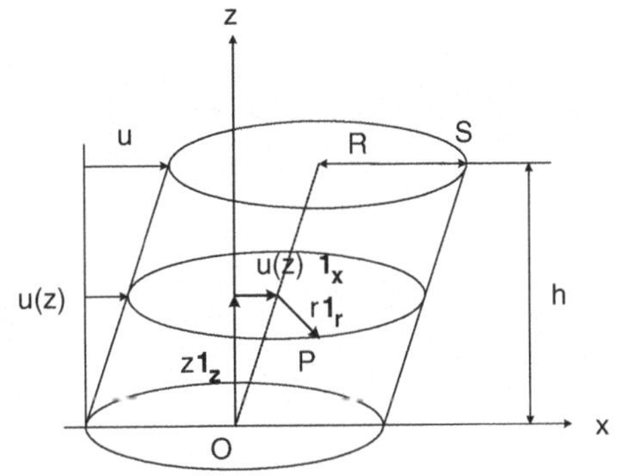

Fig. 3.5 Geometrical model of a fluid meniscus of height h confined between circular parallel pads of radius R laterally offset by u

$$d\Sigma = ||\overline{N}||d\theta dz \tag{3.3}$$

with $\overline{N} = \frac{\partial \bar{S}}{\partial \theta} \times \frac{\partial \bar{S}}{\partial z}$, we finally find:

$$d\Sigma = \mathrm{R}\sqrt{1 + \cos^2\theta \frac{u^2}{h^2}} d\theta dz \tag{3.4}$$

The lateral area is consequently equal to:

$$\Sigma = R \int_0^h dz \int_0^{2\pi} \sqrt{1 + \cos^2\theta \frac{u^2}{h^2}}\, d\theta \tag{3.5}$$

Equation (3.5) can be solved in closed-form invoking elliptic integrals [23]—a solution that still needs numerical evaluations, though. On the other hand, consistently with our stated focus on small relative meniscus perturbations $\frac{u}{h} < 1$, we make use of the well-known linear approximation $(1 + x)^n \approx 1 + nx$, holding for $x \ll 1$, to replace the square root by $1 + \frac{1}{2}\frac{u^2}{h^2}\cos^2\theta$, finally obtaining:

$$\Sigma \approx 2\pi Rh \left(1 + \frac{u^2}{4h^2}\right) \tag{3.6}$$

Since the total (variable) energy of the system is here equal to:

$$W = \gamma\Sigma \tag{3.7}$$

(where γ is the surface tension), the lateral restoring force is equal to (see Sect. 2.2.2):

$$F = -\frac{\partial W}{\partial u} = -\pi R\gamma \frac{u}{h} \tag{3.8}$$

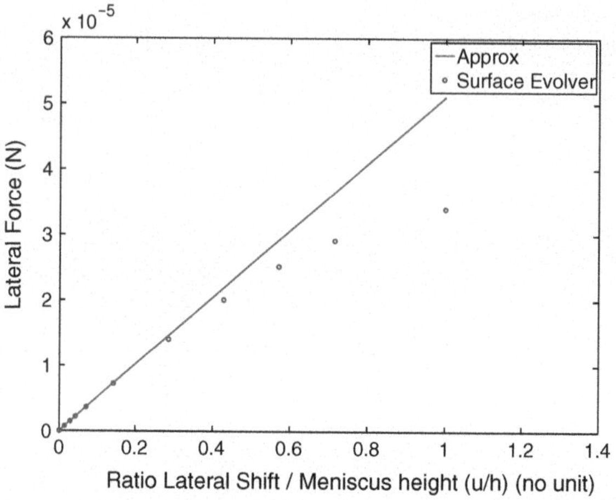

Fig. 3.6 Comparison of analytical and numerical (SE) results for the benchmark meniscus configuration, with $R = 50\,\mu\text{m}$, $h = 70\,\mu\text{m}$ and $\gamma = 0.325\,\text{Nm}^{-1}$. (Reproduced with permission from [23], ©(2010) Springer Science + Business Media)

which correspond to a *constant* lateral meniscus stiffness K_{m}^{L} given by:

$$K_{\text{m}}^{\text{L}} = -\frac{\partial F}{\partial u} = \frac{\pi R \gamma}{h} \tag{3.9}$$

This formulation was benchmarked using our Surface Evolver model, in the case of $r = 50\,\mu\text{m}$, $h = 70\,\mu\text{m}$, $\gamma = 0.325\,\text{Nm}^{-1}$ and a volume of liquid given by $V = \pi r^2 h$. The comparison is plotted in Fig. 3.6. Good agreement between the models was achieved for small relative shifts (i.e. small u/h values)—coherently with our working hypotheses and with those induced in our experiments.

3.3.2 Finite-Element Model

We modeled our experiments on cylindrical fluid menisci with Surface Evolver (SE) [7] (Fig. 3.7). As introduced in Sect. 2.2.2, Surface Evolver is a powerful finite-element software dedicated to the computation of energy-minimizing surfaces defined by geometrical and physical constraints [28]. Its mathematical solver works essentially in 2D, i.e. the energies to be minimized are quantities that may be defined as integrals over surfaces. This is straightforward for the surface tension of the triangular facets of the body. To account for the gravitational potential energy E_g of a body of volume Ω, surface $\partial\Omega$ and uniform density ρ, the required surface integral (computed for each facet bounding the body) is derived from the divergence theorem as follows [52]:

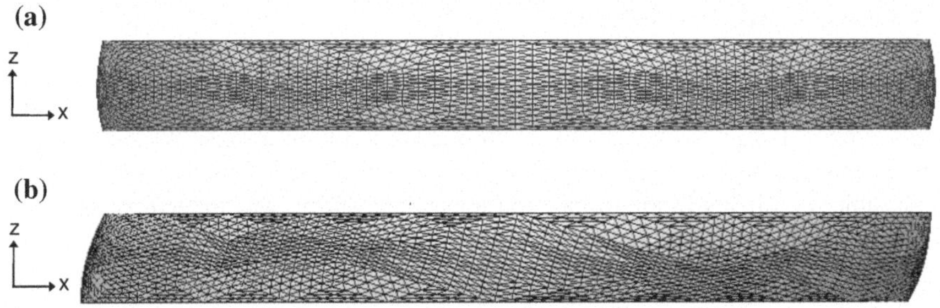

Fig. 3.7 Finite-element model for the fluid meniscus of Experiment 2 of Table 3.2. **a** Rest configuration. **b** Maximally-perturbed configuration ($u = 273\,\mu\text{m}$)

$$E_g = g\rho \int\int\int_{\Omega} z\,dV \underset{\text{divergence}}{=} g\rho \int\int_{\partial\Omega} \frac{z^2}{2}\overrightarrow{k}\cdot\overrightarrow{dS} \tag{3.10}$$

where $\overrightarrow{k}$ and $\overrightarrow{dS}$ are the unit vectors parallel to the vertical direction of the local frame of reference and normal to the facet surfaces, respectively.

Our code was adapted from the *column* example wrote by SE's developer K. Brakke. All geometrical (radius of the pads R, meniscus height and volume as obtained by image post-processing for each experiment) and physical (fluid density and surface tension) parameters reproduced those of the experiments chosen to be modeled. We imposed the pinning of the fluid triple contact-line along all the circular edge of both pads. This condition was enforced at the beginning of each experiment, as already said, and was satisfied for all measurements of small meniscus perturbations, as judged by visual inspection. The angle between the lateral meniscus surface and the pad surfaces at the triple contact-line was therefore not imposed, rather a consequence of the physical (i.e. fluid surface tension) and geometrical (fluid volume and pad areas) boundary conditions of the problem.

The capillary forces were calculated by the method of virtual works (see e.g. [49], p. 25) using central differences. Our script for mesh refinement and geometry evolution made repeated use of the Hessian of the energy function, achieving energy convergence to its 5th significant digit and a satisfactory match (<1 %) against the benchmark given by the axial force produced by a perfectly-cylindrical fluid meniscus—analytically given by $F = -\pi\gamma R$ (see (2.7) or [22]). In simulating the force-versus-displacement curve of few of our experiments, we input the same set $\{u_i\}$ of relative displacements that were observed in the experiments. We finally extrapolated the simulated lateral stiffness of the fluid menisci by first-order polynomial fitting of the curves.

3.4 Results and Discussion

When performing the experiments, the magnitude of the successive bottom pad displacements $\{x^b_{d_i}\}$ was empirically determined as a compromise between two competing needs: (i) to accurately sample the curves of interest, and (ii) to avoid as much as possible the effects of environmental noise. The spring was indeed as sensitive as to be clearly perturbed by the movement of the surrounding air. This was the main source of noise, together with the impulsive vibration of the translation stage (induced by floor vibrations, partly-attenuated by an absorbing plastic multilayer set underneath the apparatus). As a consequence, the spring could eventually undergo stochastic swinging movements (as large as few tens of micrometers, at maximum) before settling to the equilibrium position imposed by the boundary conditions. To cope with this: (1) we spaced the successive positions $\{x^b_{d_i}\}$ of the bottom pad up to 50 μm apart; (2) we moved the bottom pad slowly enough in between the prefixed positions to avoid inducing excessive air and fluid flows; (3) we waited up to several tens of seconds after each prefixed position was reached to let transient air and fluid flows, and fluid velocity gradients and shear stresses inside the meniscus, go extinct: at each step, the spring settled in its new equilibrium position after undergoing few decaying oscillations and within a time dependent on the imposed shuttle velocity and the fluid viscosity (see [23] for the detailed analysis of the system's transient dynamics); and (4) we averaged (131072 times) the interferometric laser signals to cancel out fluctuations and thus achieve the best force resolution possible. Note that the relatively-wide spacing between sampling points, as compared to the full range of imposed displacements, made no harm to faithfully reconstruct the desired curves: close to the origin, the behavior of the fluid menisci was indeed expected to be essentially linear [25, 27]—as we experimentally confirmed it to be *a posteriori* (see e.g. Fig. 3.8).

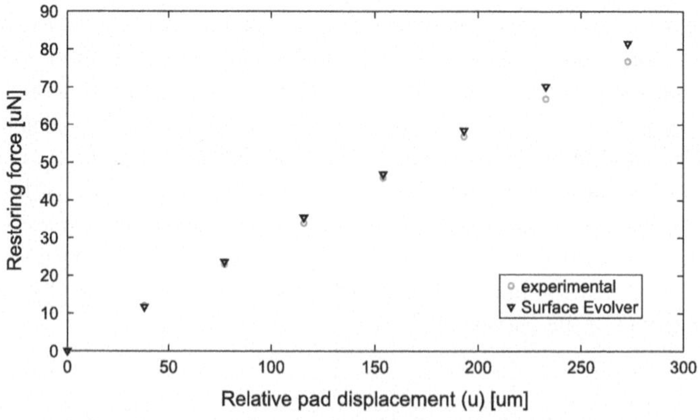

Fig. 3.8 Experimental and numerical (SE) results for experiment 2 in Table 3.2

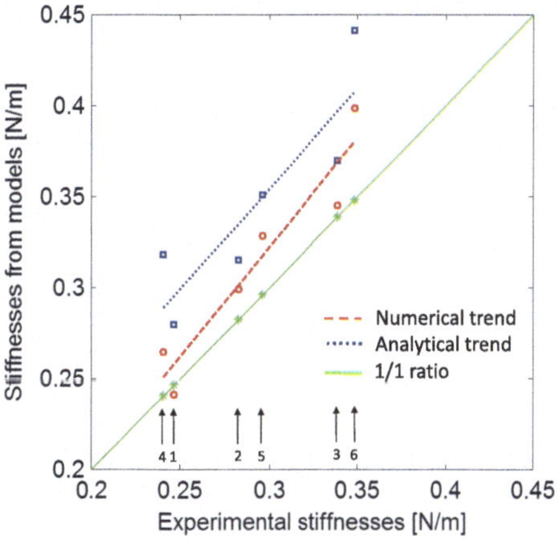

Fig. 3.9 Comparison of lateral stiffness of 6 cylindrical menisci (labels refer to Table 3.2; values ordered along abscisses), as estimated from experimental measurements and from numerical and analytical models. The *trend lines* represent first-order least-mean-squares fits of model data versus experimental data. It is apparent that both models tend to overestimate the meniscus stiffness

Table 3.2 summarizes the lateral stiffnesses of 6 cylindrical fluid menisci, as resulting from experiments and both numerical and analytical models, together with the physical and geometrical boundary conditions and the relative estimation errors for each case. We performed a total of 34 experiments; however, we report for comparison only those experiments for which: (1) no fluid overflow nor sensible evaporation was detected, and (2) the estimates of volumes and heights of the menisci were accurate (as discussed in Sect. 3.2.5). As an example, Fig. 3.8 shows both experimental and numerical data for the force-versus-lateral displacement curve relative to the second experiment of Table 3.2.

The experimental results confirm that lateral capillary forces arising from cylindrical menisci of lower height (which also means lower volumes in our case, because of the adopted procedures) are proportionally larger. Furthermore, the comparison of the results reveals that both numerical and analytical models tend to overestimate the meniscus stiffness, with the numerical estimate closer to the experimental values. This is evidenced in Fig. 3.9 by the deviation of the model data's trend lines from the ideal trend. The relative errors of analytical estimates (up to 32 %, with an average of 18.6 %) are larger than those of the numerical model (up to 14 %, with an average of 7.2 %), as well. We attribute these facts to several factors and sources of errors, discussed in the next Section.

3.4.1 Sources of Error

The relative error on the reconstructed value of pad diameters D (shown in the second column of Table 3.2) can be assumed as an indication of the error on the estimates of volumes and heights of the menisci. These errors in turn directly affect the geometry of the reconstructed menisci, and thus both models' estimates of their lateral stiffnesses. Given the relative errors $\delta h/h$ and $\delta R/R$ on the estimates of h

and R, respectively, the relative error on the volume of a cylinder is:

$$\frac{\delta V}{V} = \frac{\delta h}{h} + 2 \cdot \frac{\delta R}{R} \tag{3.11}$$

which equals $3 \cdot \delta R/R$ assuming equal relative errors for h and R. Considering e.g. the case of experiment 2, with reference parameter and error values shown in Table 3.2, the effects of a 2.45 % relative error on h, D and V give a relative error on the simulated versus experimental values of meniscus stiffness of 8.6 % (for the case of errors in excess) and 7.62 % (in the opposite case), respectively. This and similar error propagation analyses, as well as the little error due to mesh refinement in the model, can partly explain the relative error between SE data and real data. We also remark that the relative uncertainty on the experimental value of the stiffness of the spring (~6 %, see 3.5) is by itself very-close to the average relative error of the SE model compared to experiments (7.2 %).

Relative errors on estimates of D do also affect, though to a lesser extent, the estimates for K_m^L resulting from the analytical model. However, with this regard we could deduce *a posteriori* that the axysimmetric meniscus geometries we tried to enforce by visual inspection at the beginning of each experiment were not always perfectly cylindrical; we think this may be the main reason of the systematic analytical overestimates of the stiffnesses of the menisci.

Finally, a change in γ during the experiments, as induced e.g. by adsorbtion of contaminants from air, would proportionally affect the restoring forces, thus the stiffnesses of the menisci. While this is probably the case for high surface tension liquids like water, in our contextual experiments we could not detect any significant dispersion of the surface tension of silicone oils.

In view of all these plausible sources of errors, we consider the matching between our models and experiments to be well satisfying.

3.4.2 Hysteresis

Figure 3.10 shows the results of one early experiment, where relatively-large perturbations were imposed on an oil meniscus. Comparing forward and backward measurement curves and initial and final meniscus profiles, a clearly hysteretic behaviour was detected. This was due to the overflow of the fluid beyond the edge of the bottom pad (Fig. 3.11). This happened when the advancing edge angle of the meniscus on the bottom pad exceded the limit ζ_{max} set by the *Gibbs' inequality* [34]:

$$\zeta_{max} = \theta^a + 180° - \phi \tag{3.12}$$

θ^a being the advancing CA of the liquid, and ϕ the angle between the top and the side surface of the bottom pad (approximately equal to 90° in our case). Beyond this value of that angle, the fluid was no longer confined by the rim of the pad and it wet the

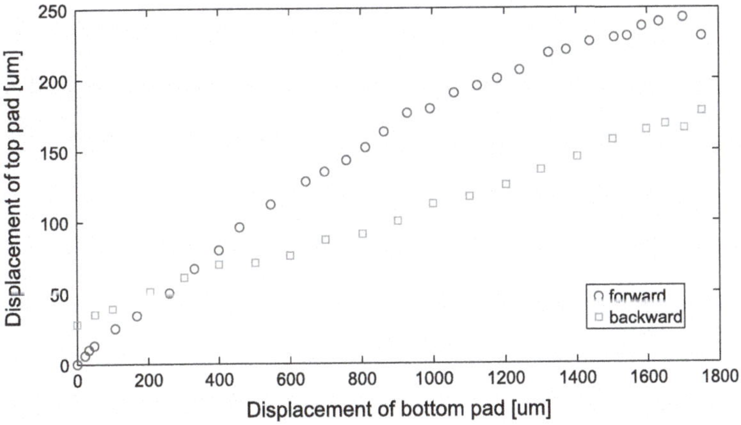

Fig. 3.10 Experimental *top* pad-versus-*bottom* pad (i.e. x_d^t vs. x_d^b) displacements during forward and backward scans in presence of hysteresis. The final large *top* pad displacement induced the oil overflow over the edge of the *bottom* pad, changing the volume and the profile of the meniscus and consequently the magnitude of the restoring force

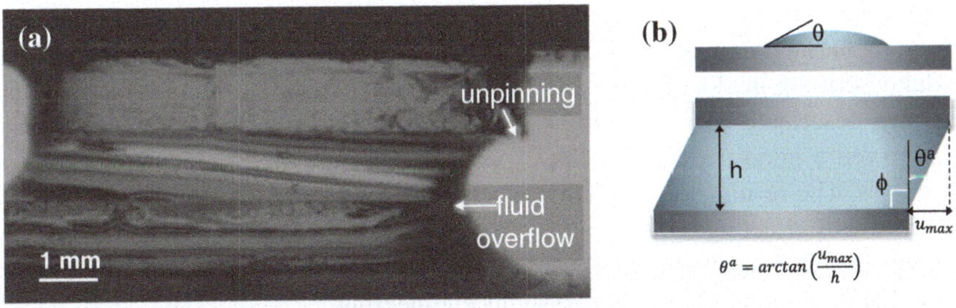

Fig. 3.11 Hysteresis by large displacement of the top relative to the bottom pad (**a**). Beyond a geometrically-predictable value of displacement, i.e. for an advancing angle $\theta^a + 180 - \phi$ between the fluid and the bottom pad (**b**), the fluid is no longer confined on top of the bottom pad surface, and it overflows. The volume of the residual fluid confined between the pads decreases, changing the profile of the meniscus. Triple contact-line unpinning along the edge of the top pad is also seen

vertical side of it, overflowing (Fig. 3.11a). The overflow changed the residual amount of fluid in the meniscus. The profile of the meniscus during the backward movement was thus different as compared to the forward movement; hence the difference in the resulting restoring force. Depending on θ^a and the actual height h of the meniscus, fluid overflow could be avoided by ensuring $u < u_{max} = h \cdot \tan(\theta^a)$ (Fig. 3.11b).

We incidentally observe the following. Thanks to: (1) the aforementioned geometrical relation between u_{max}, h and θ^a, and (2) the possibility of tracking in real-time the relative displacement between top and bottom pad, our setup may enable a method (which we call the *overflow* method) to experimentally determine the contact angle of liquids. This novel method may be suitable to measure all CAs, and particularly useful for estimating very-small θ^a's, since they are hereby magnified by the additional value 180° − ϕ due to the constrained geometry of the setup. An actual implementation of the method would require: (i) pads with non-circular shape and dimensions much larger than the fluid drops to avoid stresses on the fluid, and (ii) ideal control of the solid edges of the confining surfaces. Conversely, no thorough calibration of the sensing apparatus would be necessary—apart from a proper synchronization of laser and visual tracking systems.

Hysteretical curves were also measured when using water as fluid, even applying only small displacements. In this case the source of hysteresis was the rapid evaporation of water—that affected the volume and thus the curvature of the meniscus—and/or the higher susceptibility of the water surface to adsoption of surfactants from air, which sensibly affect its surface tension.

3.5 Conclusions

This chapter presented a comprehensive quasi-static study of lateral capillary forces arising from small perturbations of cylindrical fluid menisci. The lateral restoring forces were experimentally measured by means of a dedicated sensing apparatus exploiting the lateral force balance between capillary and elastic forces. An analytical model describing in closed-form the lateral stiffness of a cylindrical fluid meniscus was also described. The predictions of this model and those of a finite-element one showed a good match with the experimentally-measured values of meniscus stiffnesses. The results are satisfying, when considering the sources of noise, error and hysteresis affecting our sensing apparatus or intrinsic to this quasi-static type of experiments. The results provide a solid foundation for the widespread application of capillarity, particularly to microsystems engineering.

Our apparatus' capacity of tracking in real-time and with micrometric accuracy the relative displacements of top and bottom pads may enable, given proper boundary conditions, the direct measurement of the advancing contact angle of liquids. This novel method would exploit the sudden liquid overflow over the edge of the confining bottom surface induced by lateral meniscus perturbations exceeding the limit set by the Gibbs' inequality . The method may be useful to assess the quality of high-energy surfaces as an alternative to interferometric methods; and robust, being based on an isolated and dramatic event.

Finally, our experimental apparatus could constitute a reference testbed to further investigate lateral capillary forces arising from fluid menisci of arbitrary profiles and shapes. The meniscus geometry can be defined by the geometry of the confining pads—besides the other parameters discussed earlier. Scaling properties may also be studied, as well as the self-alignment dynamics [23] enabled by capillary forces. This knowledge is fundamental to many present-day microsystem integration technologies, from hybrid microhandling [38] to capillary part-to-substrate self-assembly [30].

Table 3.3 Summary of the estimates of the spring's bending stiffness K

Estimation method	K [Nm^{-1}]
Auxiliary cantilever—analytical	1.1506
Auxiliary cantilever—dynamic	0.9323
Auxiliary cantilever—weighting	1.0563
Dynamic	0.9375
Analytic	0.9036

Acknowledgments This work was supported by the EU Hydromel Project. The authors thank Bruno Tartini and Jean-Salvatore Mele for their valuable help in manufacturing the mechanical components of the experimental setup, Jean-Baptiste Valsamis for computational support, and IOP Publishing Ltd. for content reproduction.

Appendix

Bending Stiffness of the Spring

We estimated the bending stiffness K of our double-cantilever sliding spring in 3 different ways, obtaining 5 estimates. All geometrical parameters of the spring were known: cantilever length $L = 282$ mm, width $b = 12.7$ mm, thickness $t = 0.102$ mm; total spring mass (including both beams, shuttle and top pad) $M = 18.448\,g$. Multiple alternative estimates were motivated by the uncertainty on the actual Young Modulus E (assumed: 210 GPa) and density ρ (assumed: 7800 Kgm^{-3}) of our steel cantilevers, directly affecting our analytic estimates. A good agreement between all estimates was obtained—as summarized in Table 3.3 and detailed below. Nonetheless, we attributed the highest confidence to the 2 fully-experimental estimates of K (defined below as K_3 and K_4), both of which avoided the use of E and ρ. Therefore, we assumed for the spring a bending stiffness equal to the average of K_3 and K_4, i.e. $K = 0.9969\,Nm^{-1}$, with a relative uncertainty of 5.96 %.

We remark that such high sensitivity enabled the spring's desired force resolution (see requirement (iii) of Sect. 3.2.1) but also its high susceptibility to environmental perturbations (see Sect. 3.4).

Auxiliary Cantilever Method

First estimates of K involved an auxiliary steel cantilever of known dimensions (Precision Brand, length $l = 86.4$ mm, thickness $t = 0.102$ mm, width $b = 12.7$ mm). The measurement exploited the lateral balance between the elastic forces of the perturbed cantilever and spring (Fig. 3.12). Starting from the rest position, common to

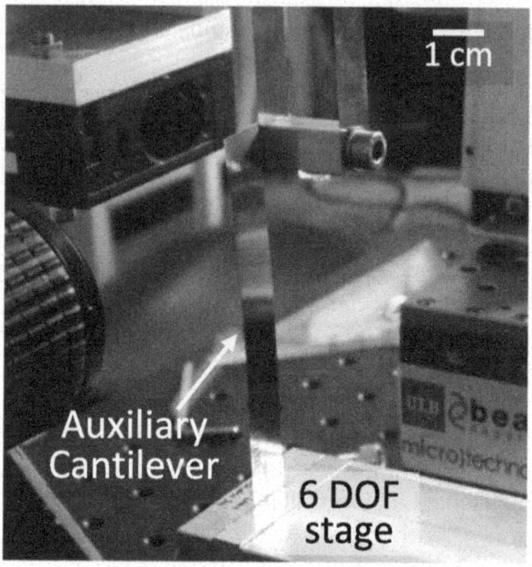

Fig. 3.12 The auxiliary cantilever used to estimate the spring stiffness

both cantilever and spring, a laser-tracked lateral displacement imposed on the cantilever induced a laser-tracked lateral displacement on the spring. After determining the bending stiffness of the cantilever, the stiffness of the spring was obtained from its force-versus-displacement curve by first-order polynomial fitting.

We estimated the bending stiffness k of the auxiliary cantilever in 3 ways. Assuming the standard stainless steel's Young's Modulus E and density ρ, the bending stiffness of a cantilever k for small deformations is given analytically by (see [49]):

$$k = \frac{3EI}{L^3} \tag{3.13}$$

where I is the cantilever's second moment of the area. We estimated I:

1. Analytically, as $I = \frac{bt^3}{12}$. Inserting this in (3.13) leads to $k_1 = 1.097\,\mathrm{Nm}^{-1}$.
2. From the knowledge of the first resonance f_1 of the cantilever, as obtained by solving Euler's beam equation ([49], p. 273):

$$I = \frac{2\pi f_1 \rho}{E\beta_1^4} \tag{3.14}$$

where $\beta_1 = 1.875$. We measured the vibration period $t = 97$ ms of the cantilever analyzing its laser-tracked oscillations on a digital oscilloscope. Hence, from (3.14) and (3.13) we got $k_2 = 0.8889\,\mathrm{Nm}^{-1}$.

We also estimated the bending stiffness of the cantilever by measuring and fitting numerically its tip load-versus-tip displacement curve. This gave us a value of $k_3 = 1.0071\,\mathrm{Nm}^{-1}$. We considered this the most reliable of our estimates of k.

We consequently obtained one value of the bending stiffness $K_\#$ from each $k_\#$ value: $K_1 = 1.1506\,\mathrm{N/m}$, $K_2 = 0.9323\,\mathrm{Nm}^{-1}$ and $K_3 = 1.0563\,\mathrm{Nm}^{-1}$.

Dynamic Method

From the natural oscillation frequency f_1 of the double-cantilever spring ($f_1 = 1.266\,\text{Hz}$ in our case, as measured by oscilloscope), its stiffness K_4 can be directly estimated according to:

$$K_4 = 4\pi^2 f_1^2 M_{\text{eff}} \tag{3.15}$$

where M_{eff} is the effective spring mass, including the mass of the shuttle and the kinetic energy-averaged mass of the cantilevers (according to Rayleigh method; see [49], p. 23, and 3.5 for details). We obtained $K_4 = 0.9375\,\text{Nm}^{-1}$.

Analytic Method

We also calculated K fully-analytically. We assumed that K_5 had 2 components: (1) the mechanical stiffness of 2 parallel, coupled cantilevers—with their unclamped extremities constrained by the shuttle to slide along a direction perpendicular to the cantilevers—given by material strength theory [49]; and (2) a component due to the gravitational potential energy, converted into a *gravitational stiffness.*

The mechanical component was obtained from:

$$K_{\text{mech}} = 2 \cdot \frac{12EI}{L^3} \tag{3.16}$$

We estimated the gravitational stiffness as (see 3.5 for details):

$$K_{\text{grav}} = \frac{6g}{5L}\left(m_{\text{sh}} + \frac{m_{\text{b}}}{2}\right) \tag{3.17}$$

where g is the acceleration of gravity. All parameters being known, we got a value of $K_5 = 0.9036\,\text{Nm}^{-1}$.

Effective Spring Mass

The mass of the spring's 2 cantilever ($m_{\text{b}} = 5.7\,g$) was not negligible compared to that of the shuttle and top pad ($m_{\text{s}} = 12.778\,g$). Therefore, in the dynamic estimation of the spring's stiffness we introduced an equivalent mass for both beams m_{eq}, which would have the same kinetic energy as the actual cantilevers for the same shuttle velocity v according to:

$$\frac{1}{2}m_{\text{eq}}v^2 = 2 \cdot \frac{1}{2}\int_0^L v^2(z)dm' = 2 \cdot \frac{1}{2}\lambda\int_0^L v^2(\xi)d\xi \tag{3.18}$$

where $dm' = \lambda d\xi$, and λ has the dimension of mass per unit length. The velocity $v(z)$ of each cantilever element located at a distance z from the clamped extremity was assumed to be proportional to its displacement computed by material strength theory:

$$v(z) = \frac{q(z)}{u}v \tag{3.19}$$

where the element $q(z)$ is given by:

$$q(z) = \frac{F}{EI}\left(\frac{Lz^2}{4} - \frac{z^3}{6}\right) \tag{3.20}$$

and $u = q(L)$. Using (3.20), (3.19), and (3.16) we get:

$$v^2(z) = v^2\left(\frac{9z^4}{L^4} - \frac{12z^5}{L^5} + \frac{4z^6}{L^6}\right) \tag{3.21}$$

which, inserted in (3.18), leads to:

$$m_{\text{eq}} = \frac{13}{35}m_{\text{b}} \tag{3.22}$$

Finally, the effective spring mass M_{eff} is given by:

$$M_{\text{eff}} = m_{\text{sh}} + m_{\text{eq}} = 14.8171g \tag{3.23}$$

Gravitational Stiffness

The gravitational component of the spring's stiffness arises from the fact that an horizontal displacement u of the shuttle is concurrent to a vertical parasitic motion p given by Henein ([18], formula 5.13) as[3]:

$$p \approx \frac{3u^2}{5L} \tag{3.24}$$

Considering that the shuttle undergoes a p upward displacement while each beam's mass center undergoes a $p/2$ vertical displacement, the gravitational stiffness K_{grav} is defined as follows (m_b is the mass of the 2 cantilevers):

$$\frac{1}{2}K_{\text{grav}}u^2 = m_{\text{sh}}gp + m_{\text{b}}g\frac{p}{2} \tag{3.25}$$

which together with (3.24) leads to (3.17).

---

[3] Using (3.24), we can estimate p for the top pad displacement induced by lateral force measurements. For e.g. experiment 2 of Table 3.2, a $u_{max} = 273$ µm led to a negligible $p_{max} = 0.16$ µm.

References

1. A.W. Adamson, *Physical Chemistry of Surfaces* (Wiley-Interscience, New York, 1997)
2. B. Berge, J. Peseux, Variable focal lens controlled by an external voltage: an application of electrowetting. Eur. Phys. J. E. **3**, 159–163 (2000)
3. R.W. Bernstein, X. Zhang, S. Zappe, M. Fish, M. Scott, O. Solgaard, Characterization of fluidic microassembly for immobilization and positioning of drosophila embryos in 2-d arrays. Sens. Actuators A **114**, 191–6 (2004)
4. K.F. Böhringer, U. Srinivasan, R.T. Howe, Modelling of capillary forces and binding sites for fluidic self-assembly, in *IEEE 14th International Conference on Micro Electro Mechanical Systems* (Interlaken (CH), 2001), pp. 369–374
5. M. Boncheva, D.A. Bruzewicz, G.M. Whitesides, Millimeter-scale self-assembly and its applications. Pure Appl. Chem. **75**, 621–630 (2003)
6. M. Boncheva, G.M. Whitesides, Making things by self-assembly. MRS Bull. **30**, 736–742 (2005)
7. K.A. Brakke, The surface evolver. Exp. Math. **1**, 141–165 (1992)
8. J. Chung, W. Zheng, T.J. Hatch, H.O. Jacobs, Programmable reconfigurable self-assembly: parallel heterogeneous integration of chip-scale components on planar and nonplanar surfaces. J. Microelectromech. S. **15**, 457–464 (2006)
9. C. Clasen, B.P. Gearing, G.H. McKinley, The flexure-based microgap rheometer (fmr). J. Rheol. **50**, 883–905 (2006)
10. C.D. Cooper, F.C. Alley, *Air Pollution Control: a Design Approach* (Waveland, Long Grove (IL, USA), 1994)
11. M. De Volder, J. Peirs, D. Reynaerts, J. Coosemans, R. Puers, O. Smal, B. Raucent, A novel hydraulic microactuator sealed by surface tension. Sens. Actuators A **123**(4), 547–554 (2005)
12. P.-G. de Gennes, F. Brochard-Wyart, D. Quere, *Capillarity and Wetting Phenomena: Drops, Bubbles, Pearls, Waves* (Springer, New York, 2004)
13. T. Fukushima, H. Kikuchi, Y. Yamada, T. Konno, J. Liang, K. Sasaki, K. Inamura, T. Tanaka, M. Koyanagi, New three-dimensional integration technology based on reconfigured wafer-on-wafer bonding technique, in *International Electron Devices Meeting* (Washington, D.C. (USA), 2007), pp. 985–8
14. T. Fukushima, Y. Yamada, H. Kikuchi, T. Tanaka, M. Koyanagi, Self-assembly process for chip-to-wafer three-dimensional integration, in *Electronic Components and Technology Conference (ECTC)* (Reno (NV, USA), 2007), pp. 836–841
15. L.S. Goodman, Geometrical optimization of controlled collapse interconnections. IBM J. Res. Dev. **13**, 251–265 (1969)
16. M. Gad el Hak, (ed.), *The MEMS Handbook* (CRC Press, Boca Raton, 1998)
17. X. Guo, H. Lic, B.Y. Ahn, E.B. Duoss, K.J. Hsia, J.A. Lewis, R.G. Nuzzo, Two- and three-dimensional folding of thin film single-crystalline silicon for photovoltaic power applications. Proc. Natl. Acc. Sci. **106**, 20149–20154 (2009)
18. S. Henein, Conception des structures articulées à guidages flexibles de haute précision. Ph.D. thesis, Ecole Polytechnique Federale de Lousanne, 2000
19. G. Humpston, Flip chip solder bonding for microsystems. In: IEE Colloquium on Assembly and Connections in Microsystems, (London (UK), 1997), pp. 1–3
20. J. Israelachvili, *Intermolecular and Surface Forces* (Academic Press, New York , 1994)
21. C. Kallmayer, H. Oppermann, G. Engelmann, E. Zakel, H. Reichl, Self-aligning flip-chip assembly using eutectic gold/tin solder in different atmospheres, in *IEEE/CPMT International Electronics Manufacturing* (Austin, (TX, USA), 1996), pp. 18–25
22. P. Lambert, *Capillary Forces in Microassembly* (Springer, New York, 2007)
23. P. Lambert, M. Mastrangeli, J.-B. Valsamis, G. Degrez, Spectral analysis and experimental study of lateral capillary dynamics for flip-chip applications. Microfluid. Nanofluid. **9**, 797–807 (2010)

24. S.-H. Liang, X. Xiong, K.F. Bohringer, Toward optimal designs for self-alignment in surface tension driven micro-assembly, in *IEEE 17th International Conference on Micro Electro Mechanical Systems* (Maastricht (NL), 2004), pp. 9–12
25. J. Lienemann, A. Greiner, J.G. Korvink, X. Xiong, Y. Hanein, K.F. Bohringer, Modelling, simulation and experiment of a promising new packaging technology parallel fluidic self-assembly of microdevices. Sens. Update **13**, 3–43 (2003)
26. C. Lin, F. Tseng, C.-C. Chieng, Studies on size and lubricant effects for fluidic self-assembly of microparts on patterned substrates using capillary effect. J. Electron. Pack. **130**, 021005–1(2008)
27. W. Lin, S.K. Patra, Y.C. Lee, Design of solder joints for self-aligned optoelectronics assembly. IEEE T. Compon. Pack. B **18**, 543–551 (1995)
28. P.M. Martino, G.M. Freedman, L.M. Racz, J. Szekely, Predicting solder joint shape by computer modeling, in *Electronic Components and Technology Conference (ECTC)* (Washington, D.C. (USA), 1994), pp. 1071–8
29. M. Mastrangeli, Enabling capillary self-assembly for microsystem integration. Ph.D. thesis, Arenberg Doctoral School, Katholieke Universiteit Leuven, 2010
30. M. Mastrangeli, S. Abbasi, C. Varel, C. van Hoof, J.-P. Celis, K.F. Bohringer, Self-assembly from milli- to nanoscale: methods and applications. J. Micromech. Microeng. **19**, 083001 (2009)
31. M. Mastrangeli, W. Ruythooren, C. Van Hoof, J.-P. Celis, Conformal dip-coating of patterned surfaces for capillary die-to-substrate self-assembly. J. Micromech. Microeng. **19**, 045015 (2009)
32. M. Mastrangeli, J.-B. Valsamis, C. Van Hoof, J.-P. Celis, P. Lambert, Lateral capillary forces of cylindrical fluid menisci: a comprehensive quasi-static study. J. Micromech. Microeng. **20**, 075041 (2010)
33. C.J. Morris, B.A. Parviz, Micro-scale metal contacts for capillary force-driven self-assembly. J. Micromech. Microeng. **18**:015022 (10pp) (2008)
34. J.F. Oliver, C. Huh, S.G. Mason, Resistance to spreading of liquids by sharp edges. J. Colloid Interf. Sci. **59**, 568–581 (1977)
35. S.K. Patra, Y.C. Lee, Modeling of self-alignment mechanism in flip-chip soldering—part ii: multichip solder joints, in *Electronic Components and Technology Conference (ECTC)* (1991), pp. 783–788
36. S.K. Patra, Y.C. Lee, Quasi-static modeling of the self-alignment mechanism in flip-chip soldering—part 1: single solder joint. J. Electron. Pack. **113**, 337–342 (1991)
37. E. Saeedi, S. Abbasi, K.F. Bohringer, B.A. Parviz, Molten-alloy driven self-assembly for nano and micro scale system integration. Fluid Dyn. Mater. Process. **2**, 221–246 (2007)
38. V. Sariola, Q. Zhou, H.N. Koivo, Hybrid microhandling: a unified view of robotic handling and self-assembly. J. Micro-Nano Mech. **4**, 5–16 (2008)
39. K. Sato, K. Ito, S. Hata, A. Shimokohbe, Self-alignment of microparts using liquid surface tension: behaviour of micropart and alignment characteristics. Precis. Eng. **27**, 42–50 (2003)
40. K. Sato, K. Lee, M. Nishimura, K. Okutsu, Self-alignment and bonding of microparts using adhesive droplets. Int. J. Precis. Eng. Man. **8**, 75–79 (2007)
41. K.L. Scott, T. Hirano, H. Yang, H. Singh, R.T. Howe, A.N. Niknejadk, High-performance inductors using capillary based fluidic self-assembly. J. Microelectromech. S. **13**, 300–9 (2004)
42. U. Srinivasan, D. Liepmann, R.T. Howe, Microstructure to substrate self-assembly using capillary forces. J. Microelectromech. S. **10**, 17–24 (2001)
43. S.A. Stauth, B.A. Parviz, Self-assembled single-crystal silicon circuits on plastic. Proc. Natl. Acc. Sci. **103**, 13922–7 (2006)
44. R.R.A. Syms, E.M. Yeatman, V.M. Bright, G.M. Whitesides, Surface-tension powered self-assembly of microstructures the state of the art. J. Microelectromech. S. **12**, 387–417 (2003)
45. A. Takei, K. Matsumoto, I. Shimoyama, Capillary torque caused by a liquid droplet sandwiched between two plates. Langmuir **26**, 2497–2504 (2010)
46. Q. Tan, Y.C. Lee, Soldering technology for optoelectronic packaging, in *Electronic Components and Technology Conference* (Orlando (FL, USA), 1996), pp. 26–36

47. A. Terfort, N. Bowden, G.M. Whitesides, Three-dimensional self-assembly of millimetre-scale components. Nature **386**, 162–4 (1997)
48. A. Terfort, G.M. Whitesides, Self-assembly of an operating electrical circuit based on shape complementarity and the hydrophobic effect. Ad. Mater. **10**, 470–3 (1998)
49. W.T. Thomson, M.D. Dahleh, *Theory of Vibration with Applications* (Prentice Hall, Upper Saddle River, 1998)
50. N. Van Veen, Analytical derivation of the self-alignment motion of flip-chip soldered components. Trans. ASME **121**, 116–121 (1999)
51. R.J.K. Wassink, *Soldering in Electronics* (Electrochemical Publications, UK, 2005)
52. C.E. Weatherburn, *Differential Geometry of Three Dimensions* (Cambridge University Press, Cambridge, 1955)
53. X. Xiong, Y. Hanein, J. Fang, Y. Wang, W. Wang, D.T. Schwartz, K.F. Bohringer, Controlled multibatch self-assembly of microdevices. J. Microelectromech. S **12**, 117–127 (2003)
54. X. Xiong, S.-H. Liang, K.F. Bohringer, Geometric binding site design for surface-tension driven self-assembly, in *IEEE International Conference on Robotics and Automation* (Barcelona (ES), 2004), pp. 1141–8
55. B. Yost, J. McGroarty, P. Borgesen, C.-Y. Li, Shape of a nonaxisymmetric liquid solder drop constrained by parallel plates. IEEE T. Compon. Hybr. **16**, 523–6 (1993)
56. X. Zhang, C.-C. Chen, R.W. Bernstein, S. Zappe, M.P. Scott, O. Solgaard, Microoptical characterization and modeling of positioning forces on drosophila embryos self-assembled in two-dimensional arrays. J. Microelectromech. S. **14**, 1187–1197 (2005)
57. W. Zheng, H.O. Jacobs, Self-assembly process to integrate and interconnect semiconductor dies on surfaces with single-angular orientation and contact-pad registration. Adv. Mater. **18**, 1387–1392 (2006)

Chapter 4
Capillary Torque

Atsushi Takei

Abstract Capillary forces make a liquid droplet's surface have the minimum area. If the droplet is sandwiched between two plates, it exerts capillary force on the plates. The magnitude of the force depends on the shape of the sandwiched droplet, which is in turn determined by the shape of the plates and the volume of the liquid. The liquid shape, however, is hard to determine analytically. In this chapter, the torque caused by a droplet sandwiched between two non-circular plates is experimentally and theoretically analyzed. We patterned a magnetic material on the surface of the plates and used it to apply a magnetic force to the plates. The torque on the plates was measured. The torque caused by capillary force was calculated by observing the equilibrium between the capillary force and magnetic force. We obtained approximate theoretical solutions for the liquid's shape and torque and verified that they were in accordance with the experimental results. The experimental and theoretical results presented in this chapter are useful for designing micro-devices or self-assemblies actuated by capillary force.

4.1 Introduction

As seen in Chap. 2, the surface energy of a pinned meniscus is proportional to its surface area and the capillary force makes the liquid's surface area take on its minimum value. The shape with the minimum surface area can be derived by solving the Young-Laplace equation [1], and a wide variety of liquid shapes have already been studied in the literature. Previous studies dealt with the shape of a droplet or liquid surface in contact with fiber or walls [2–6]. However, the shapes dealt with were circular; non-circular liquid shapes are still hard to treat analytically. On the other

A. Takei (✉)
Institute of Industrial Science, The University of Tokyo,
4-6-1 Komaba, Meguro-ku, Tokyo 153-8505, Japan
e-mail: atsushi.takei@espci.fr

P. Lambert (ed.), *Surface Tension in Microsystems*, Microtechnology and MEMS,
DOI: 10.1007/978-3-642-37552-1_4, © Springer-Verlag Berlin Heidelberg 2013

hand, there is a need to know the shapes conforming to a wide range of boundary conditions and liquid volumes, because capillary force becomes dominant on small scales (see Fig. 1.1) and it has been used for micro-manipulation and self-alignment [7–11]. Chapters 2 and 3 have already presented axial and lateral forces, and previous studies have elucidated the dynamics of liquid bridges between plates and spheres [12] (Dynamics will be studied separately in Part III of this book). Capillary force has also been used to fabricate three-dimensional structures [13–16]. The boundary conditions of the liquid can be controlled dynamically [17] and small objects in contact with the liquid can be manipulated [18] by using microelectro mechanical system (MEMS) technology. In reference [18], a small object was rotated by controlling the boundary conditions, and it could be used as a motor. The relationship between the boundary conditions and the capillary torque is required to estimate the object's dynamic characteristics.

In the following of this chapter, Sect. 4.2 will provide experimental evidence of capillary torque, before leading a thorough mathematical analysis. These results will be applied to a case study in Chap. 10.

4.2 Experiment

We measured and theoretically analyzed torques caused by a liquid droplet sandwiched between two structures. As shown in Fig. 4.1, a plate floating on the droplet stays at a particular position depending on its shape and the shape of the base plate. We varied the shapes of the base plate and floating plate and the liquid volume and

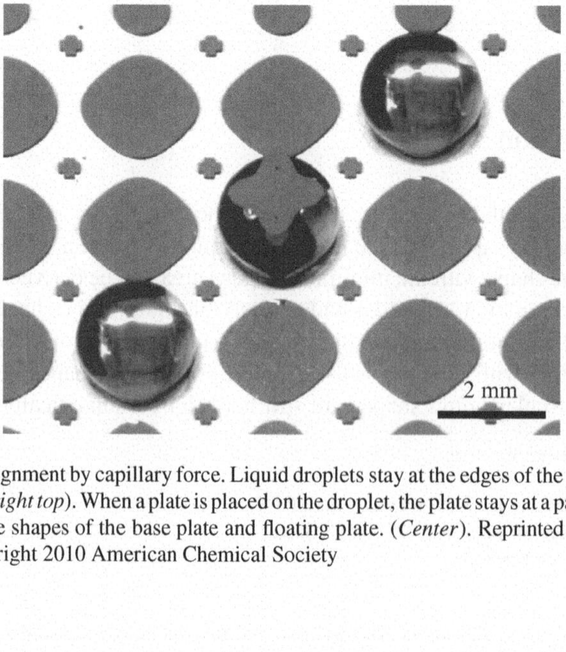

Fig. 4.1 Self-alignment by capillary force. Liquid droplets stay at the edges of the patterned silicon (*left bottom* and *right top*). When a plate is placed on the droplet, the plate stays at a particular position depending on the shapes of the base plate and floating plate. (*Center*). Reprinted with permission from [19]. Copyright 2010 American Chemical Society

examined the relationship between these parameters and torque around the axis perpendicular to both plates. In previous studies, a cantilever was used to measure the adhesive force of a capillary bridge [7, 8]. However, it is hard to provide only a rotational moment with cantilevers. For this reason, we patterned a magnetic material on the plate. Upon applying a magnetic field to the plate, the torque caused by the capillary force is measured in a non-contact condition. We also present calculations to derive the liquid's shape and the torque applied to the plates. In our theory, we considered a non-circular plate to be a circular plate with sinusoidal perturbations. The liquid's shape and the torque are derived by summing the effects of the circular part and the perturbation parts. The torque caused by capillary force is then analyzed experimentally and theoretically, and both results accord with each other.

4.2.1 Materials

We measured the torque caused by the capillary force of a liquid sandwiched between the two plates. The base plates were made from 500 μm-thick silicon wafers. The shapes of the base plates were formed by etching silicon layers 100 μm deep. Figure 4.2 shows photos of the fabricated base plates and their shapes in polar

(a)

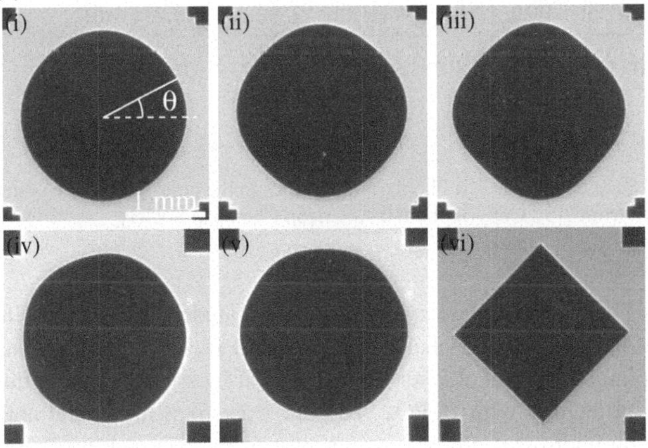

(b)

(i)	1000 + 12.5 cos4θ [μm]	(iv)	1000 + 12.5 cos5θ [μm]
(ii)	1000 + 25 cos4θ [μm]	(v)	1000 + 12.5 cos6θ [μm]
(iii)	1000 + 50 cos4θ [μm]	(vi)	1500 - μm - square

Fig. 4.2 Fabricated base plates. **a** Photos and **b** shapes in the polar coordinate system. Reprinted with permission from [19]. Copyright 2010 American Chemical Society

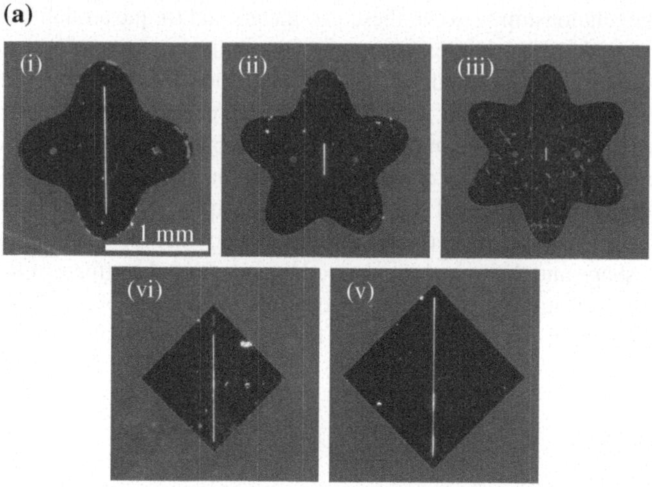

(b)

	Shape	Length of nickle
(i)	700 + 150 cos4θ [μm]	1200 [μm]
(ii)	700 + 150 cos5θ [μm]	300 [μm]
(iii)	700 + 150 cos6θ [μm]	120 [μm]
(iv)	1000 - μm - square	1200 [μm]
(v)	1250 - μm - square	1500 [μm]

Fig. 4.3 Fabricated upper plates. **a** Photos and **b** shapes in the polar coordinate system. Reprinted with permission from [19]. Copyright 2010 American Chemical Society

coordinates. The floating plates were also fabricated by etching a silicon wafer. Their thickness was 20 μm. A bar made of nickel was patterned on the surface of the floating plates. The nickel layer was deposited by sputtering, and the nickel bar was formed by etching with H_3PO_4. The bar was 2 μm thick, and approximately 5 μm wide (the width was not precise because of the wet etch process). Figure 4.3 shows photos of the fabricated floating plates and lists their shapes and the lengths of the nickel bars. When a droplet is sandwiched between the base plate and floating plate, the floating plate is stabilized at a particular position, as shown in the photos of Fig. 4.4. If the floating plate and the base plate have the same frequency of perturbations (in rad^{-1}), the directions of the patterns match those in Fig. 4.4. If the floating plate is displaced from the stabilization point, a restorative torque is generated. The torque is measured by applying a magnetic field varying with the rotational displacement. The measurement principle is depicted in Fig. 4.5. The torque T_m caused by the magnetic field H_{ext} (in Oe) is

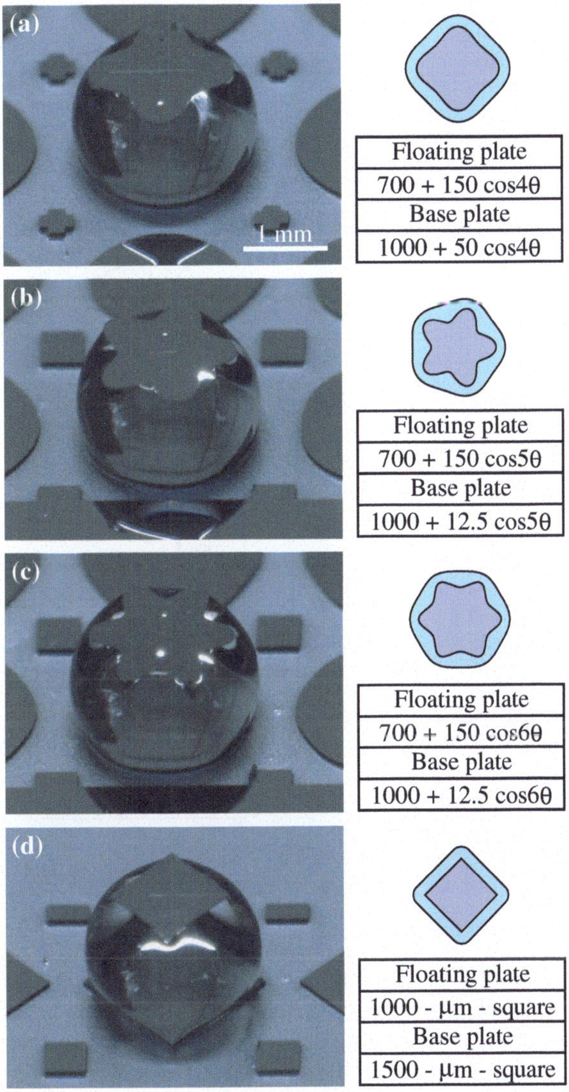

Fig. 4.4 Self-alignment by capillary force. Photos of alignment patterns are shown on the *left* side. The tables and schematic drawing of the floating plate and base plate pairs are shown on the *right* side. Reprinted with permission from [19]. Copyright 2010 American Chemical Society

$$T_{\mathrm{m}} = \mu_0 M H_{\mathrm{ext}} \cos\phi \tag{4.1}$$

where μ_0 is the magnetic permeability ($4\pi \times 10^{-7}\,\mathrm{N/A^2}$), M (in electromagnetic unit, emu) is the magnetic moment and ϕ the rotational displacement.[1] When the

[1] Please note that 1 Oe is $10^3/4\pi\,\mathrm{A/m}$ and 1 emu is $10^{-3}\,\mathrm{A\,m^2}$.

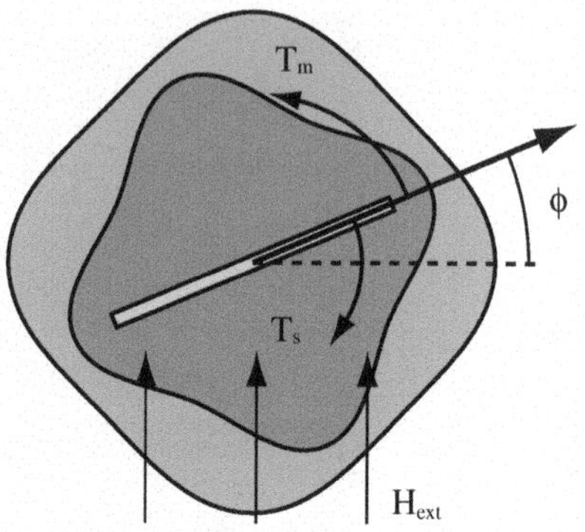

Fig. 4.5 Schematic drawing of torque measurement using magnetic force. Reprinted with permission from [19]. Copyright 2010 American Chemical Society

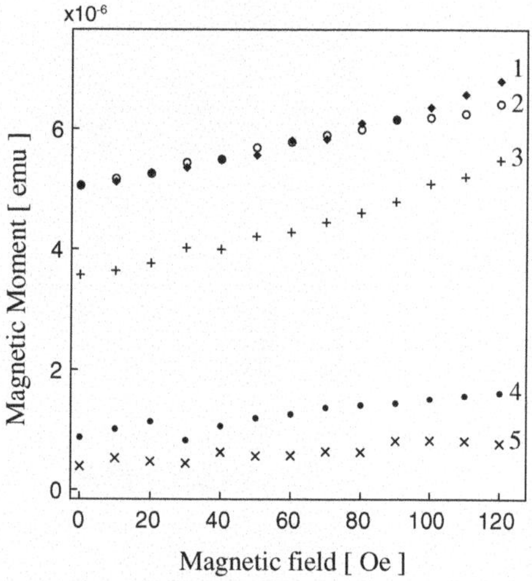

Fig. 4.6 Magnetic moment of long axis of the nickel pattern as a function of external magnetic field. The plot symbols stand for the following: 1 : 1250 μm square (*open circles*). 2 : 700 + 150 cos 4θ (*squares*). 3 : 1000 μm square (*crosses*). 4 : 700 + 150 cos 5θ (*filled circles*). 5 : 700 + 150 cos 6θ (*times*). Reprinted with permission from [19]. Copyright 2010 American Chemical Society

floating plate is at equilibrium, the torque T_s due to the capillary force is equal to T_m. By tuning the applied magnetic field, T_s can be measured as a function of ϕ. The magnetic moments of the floating plates were measured with a Superconducting Quantum Interference Device (SQUID). First, 500 Oe was applied to the long axis of the nickel bar, and the external magnetic field was turned down to 0 Oe. Then, the internal magnetization of the long axis of the nickel bar was measured while increasing the external magnetic field. The result is shown in Fig. 4.6. The internal magnetization of the minor axis was also measured. The internal magnetization was ten times less than that of the long axis. We assumed that the magnetization of the minor axis was negligible and that the direction of the magnetization matched the long axis. A pipette was used to place a droplet of pure water on a droplet, and the floating plate was placed on the droplet. In the initial state, the nickel bar and the external magnetic field were at right angles to each other.

4.2.2 Results

First, we measured the rotational displacement as a function of the applied magnetic field and calculated the torque for different perturbation amplitudes of the base plates. In this experiment, the base plates were of types (i), (ii), and (iii) shown in Fig. 4.2b, and the floating plate was of type (i) shown in Fig. 4.3b. The volume of the sandwiched droplet was 4.5 μL. To make the magnetization of the nickel bar the same as in the SQUID measurement shown in Fig. 4.6, we applied 500 Oe to the long axis of the nickel bar before the experiment. Figure 4.7b shows the relationship between the applied magnetic field and the rotational displacement. Figure 4.7c plots the relationship between the torque and rotational displacement. The calculation assumed the magnetic moment to be constant (5.0×10^{-6} emu) and used the mean values of the experimented rotational displacements. The angle between the nickel bar and the magnetic field was less than 22.5°. Since the magnetic field component along the long axis was small, the calculation used the magnetic moment at 0 Oe. The measured torque was proportional to the amplitude of the perturbations. The data was measured five times for each condition. In the experiment, the height of the droplet kept the same value in the rotational motion, and the effect of the evaporation was not observed.

Second, we observed the relationship between the torque and the frequency of the perturbations (in rad^{-1}) of the plates. The experiment tested the following pairs: type (i) base plate and type (i) floating plate, type (iv) base plate and type (ii) floating plate, and type (v) base plate and type (iii) floating plate. A 4.5 μL droplet was sandwiched between the plates. Figure 4.8b shows the relationship between the applied magnetic field and the rotational displacement, and Fig. 4.8c shows the relationship between the rotational displacement and torque. We assumed that the magnetic moments of the type (i), (ii), and (iii) floating plates were 5.0×10^{-6}, 0.9×10^{-6}, and 4.0×10^{-7} emu, respectively. The plots show that the torque fell dramatically as the frequency increased.

Third, we varied the volume of the sandwiched droplet and measured the torque. The experiment used the type (i) base plate and type (i) floating plate and droplets of 3.0, 3.5, 4.0, and 4.5 μL. The experimental results are shown in Fig. 4.9. The droplets with a smaller volume caused a larger torque. The larger droplets enlarged the gap between the two plates. The perturbations of the plates do not affect each other when there is a large gap, and the generated torque is small. Observation with a microscope revealed that the gap increased by approximately 0.1 mm each time the liquid volume was increased by 0.5 μL. This means the torque was sensitive to changing the gap between the plates.

Finally, we sandwiched a liquid droplet between square plates [type (vi) base plate and types (iv) and (v) floating plates; 4.5 μL of liquid.]. The results are shown in Fig. 4.10. The torque on the larger floating plate was four times larger than that on the smaller plate. The magnetic moments of the type (iv) and (v) floating plates were 3.6×10^{-6}emu and 5.0×10^{-6}.

(a)

	Upper plate	Base plate	Volume	
×	700 + 150 cos4θ	1000 + 12.5 cos4θ	4.5 μL	——
○	700 + 150 cos4θ	1000 + 25 cos4θ	4.5 μL	- - -
◆	700 + 150 cos4θ	1000 + 50 cos4θ	4.5 μL	······

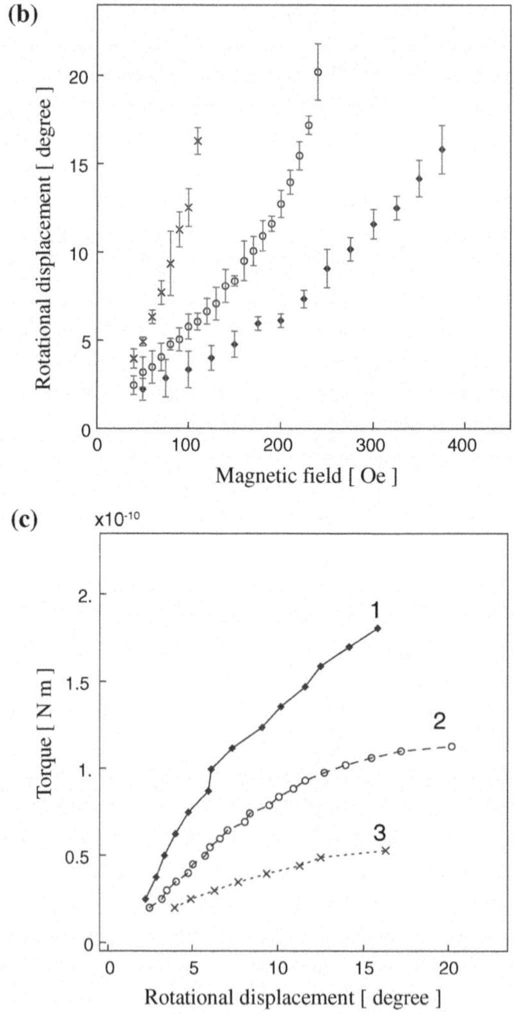

Fig. 4.7 Relationship between amplitude of perturbations and torque. **a** Table showing the experimental conditions and plot symbols for the graphs. The amplitudes of the perturbations were as follows: 12.5 μm (*times*), 25 μm (*open circles*), and 50 μm (*filled circles*). **b** Rotational displacement as a function of magnetic field. **c** Torque as a function of rotational displacement. Numbering is done for the comparison with the mathematical analysis in Sect. 4.3. Reprinted with permission from [19]. Copyright 2010 American Chemical Society

(a)

	Upper plate	Base plate	Volume	
×	700 + 150 cos4θ	1000 + 12.5 cos4θ	4.5 μL	——
◆	700 + 150 cos5θ	1000 + 12.5 cos5θ	4.5 μL	- - -
○	700 + 150 cos6θ	1000 + 12.5 cos6θ	4.5 μL	......

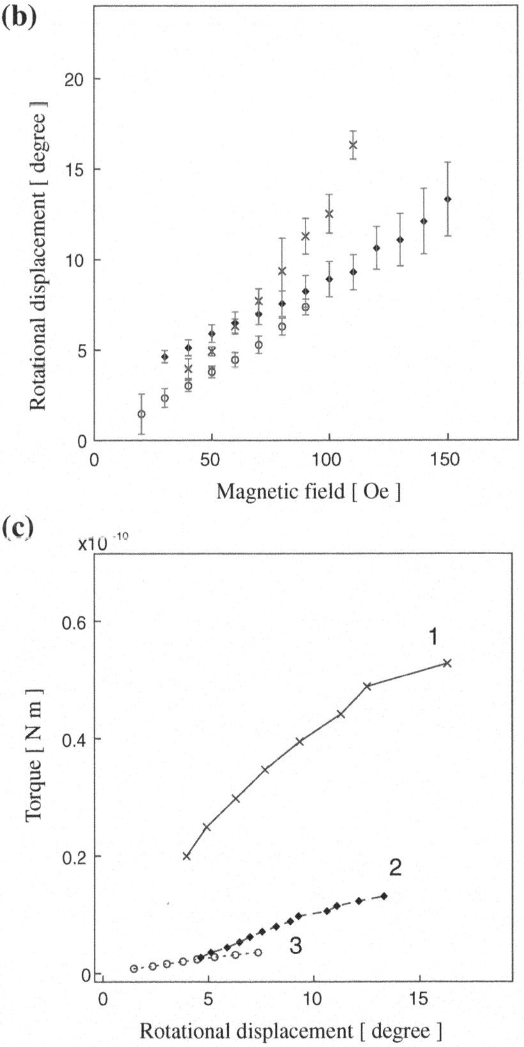

Fig. 4.8 Relationship between frequency of perturbations and torque. **a** Table showing the experimental conditions and plot symbols for the graphs. The frequencies of the perturbations were as follows: 12.5 μm cos 4θ (×, curve 1, 90° periodicity), 12.5 μm cos 5θ (◆, curve 2, 72° periodicity), and 12.5 μm cos 6θ (○, curve 3, 60° periodicity). **b** Magnetic field versus rotational displacement. Note that all of the plates are tilted by around 10° at around 100 Oe, but the masses of the patterned nickel on the plates are different. **c** Generated torque as a function of rotational displacement [same legend as in insight (b)] Reprinted with permission from [19]. Copyright 2010 American Chemical Society

(a)

	Floating plate	Base plate	Volume	
×	700 + 150 cos4θ	1000 + 12.5 cos4θ	4.5 μL	——
•	700 + 150 cos4θ	1000 + 12.5 cos4θ	4.0 μL	— -
○	700 + 150 cos4θ	1000 + 12.5 cos4θ	3.5 μL	- - -
♦	700 + 150 cos4θ	1000 + 12.5 cos4θ	3.0 μL	······

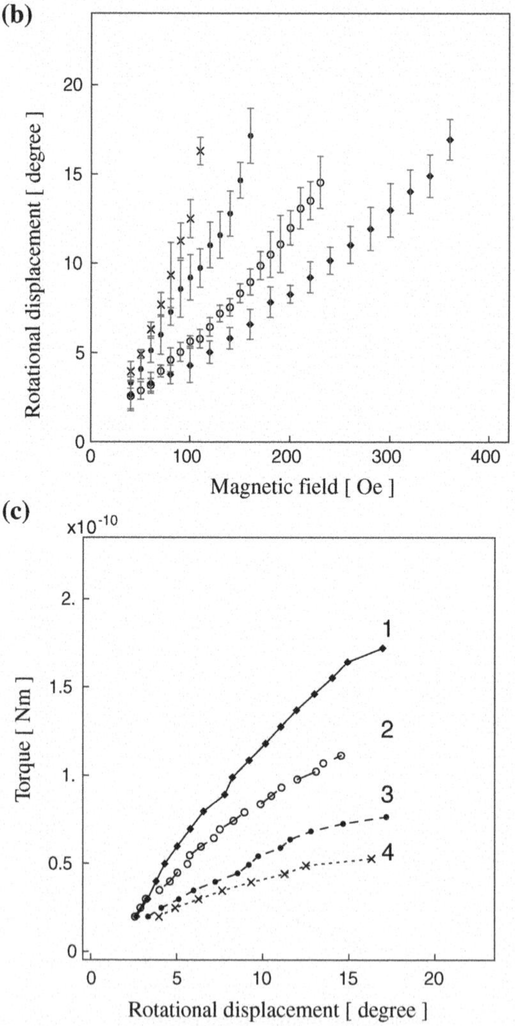

Fig. 4.9 Relationship between liquid volumes and torque. **a** Table showing the experimental conditions and plot symbols for the graphs. The volumes were as follows: 3.0 μL (♦), 3.5 μL (○), 4.0 μL (•), and 4.5 μL (×). **b** Magnetic field versus rotational displacement. **c** Generated torque as a function of rotational displacement. Reprinted with permission from [19]. Copyright 2010 American Chemical Society

(a)

	Floating plate	Base plate	Volume	
×	1000- μ m-square	1500- μm-square	4.5 μL	——
○	1250- μm-square	1500- μm-square	4.5 μL	- - -

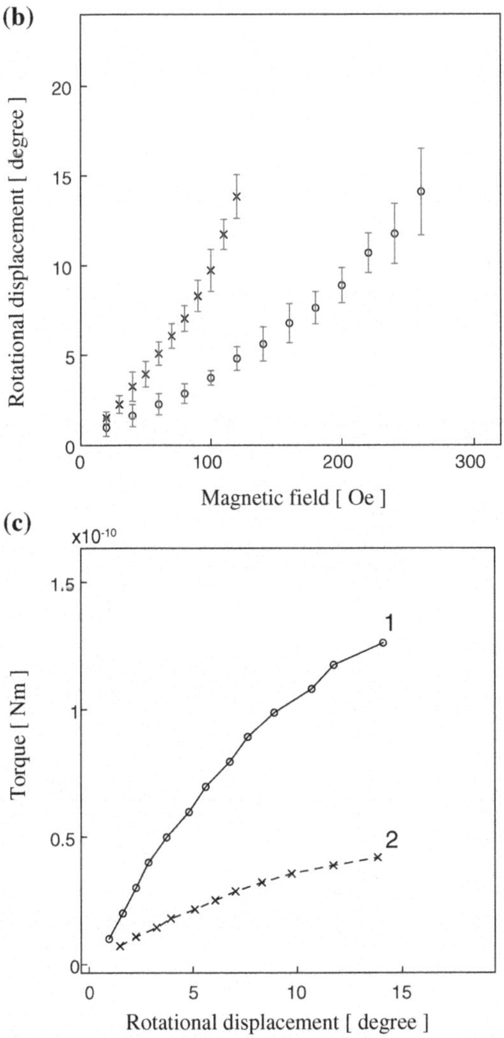

Fig. 4.10 Torque with square plates. **a** Table showing plot symbols and experimental conditions. **b** Magnetic field and rotational displacement. **c** Generated torque as a function of rotational displacement. (○ : 1250 μm, × : 1000 μm). Reprinted with permission from [19]. Copyright 2010 American Chemical Society

4.3 Mathematical Analysis

We analyzed how the shape of the plates and the liquid volume affect the generated torque. The torque can be calculated only if the surface area of the liquid is known, since the potential energy of the liquid is proportional to its surface area. The torque T_s applied to the floating plate is given by,

$$T_s = -\gamma \frac{\partial A}{\partial \phi} \tag{4.2}$$

where γ is the surface free energy of the liquid, A is the surface area, and ϕ the rotational displacement of the floating plate. First, the difference in surface areas between the $\phi = 0°$ and $\phi = 15°$ cases was roughly estimated from the experimental results shown in Fig. 4.7. The torque (indicated by ($\times$) in Fig. 4.7) at a rotational displacement of 15° was around 0.5×10^{-10} Nm. Since no torque is generated when there is no rotational displacement, the order of the difference in surface area can be roughly estimated by multiplying the torque by the rotational displacement and then dividing the result by the surface energy. Pure water was used in the experiment (surface energy is $72 \times 10^{-3}\text{J/m}^2$). Accordingly, the difference in surface areas between both cases is approximately $2 \times 10^{-10}\,\text{m}^2$. Since the surface area of the 4.5 µL sphere droplet is $13.2 \times 10^{-6}\,\text{m}^2$, the difference should be less than 0.002 % of the entire surface. Such a minute difference is hard to distinguish with a finite element method; another method is required. To determine the liquid shape, we set the coordinate system as illustrated in Fig. 4.11. The boundary conditions at the top and bottom plane are given by:

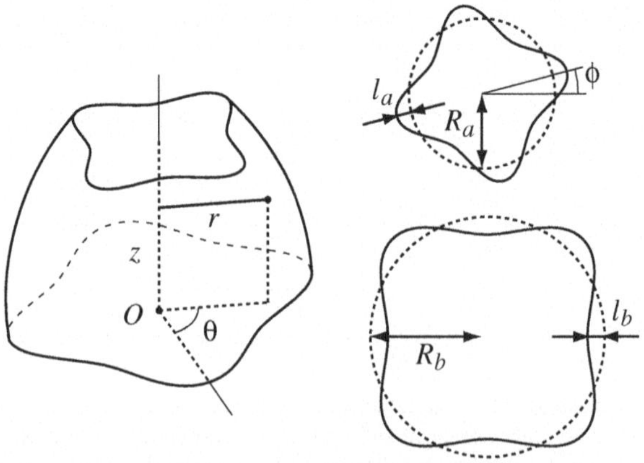

Fig. 4.11 Schematic drawing of the model and its coordinates and parameters. Reprinted with permission from [19]. Copyright 2010 American Chemical Society

$$r_a(\phi) = R_a + l_a \cos(n(\theta - \phi)) \text{ when } z = h \tag{4.3}$$

$$r_b(\phi) = R_b + l_b \cos(n\theta) \text{ when } z = 0 \tag{4.4}$$

where R_a and R_b are the diameters of the plates, l_a and l_b are the amplitudes of the perturbations, n is the frequency of the perturbations, and h is the gap of the plates. The schematic drawing shows the case of $n = 4$. We considered the liquid's shape to be a function of the rotational displacement ϕ and described it in terms of $r(\theta, z, \phi)$. As explained in [1] The surface $r(\theta, z, \phi)$ is obtained by solving the Young-Laplace equation. The surface area A is given by:

$$A = \int_0^{2\pi} \int_0^h d\theta dz r \sqrt{1 + \left(\frac{\partial r}{r\partial \theta}\right)^2 + \left(\frac{\partial r}{\partial z}\right)^2} \tag{4.5}$$

However, this equation can be analytically solved only in limited cases; the shape in this case can not be solved analytically. For this reason, we tried to obtain an approximate solution for the given boundary conditions. In general, free droplets are spherical, and a sphere satisfies the Young-Laplace equation. If gravity is negligible, a droplet sandwiched between circular plates can be considered to be part of a sphere. Thus, we can solve for the shape of liquid sandwiched between circular plates of radii R_a and R_b; then, we can add the effect of the deformation due to boundary conditions (4.3) and (4.4). The relationship between liquid volume V and the diameters of the circular plates is

$$V = \pi \left(R_m{}^2 \sqrt{R_m{}^2 - R_a{}^2} - \frac{1}{3} \sqrt{R_m{}^2 - R_a{}^2}^3 \right.$$
$$\left. + R_m{}^2 \sqrt{R_m{}^2 - R_b{}^2} - \frac{1}{3} \sqrt{R_m{}^2 - R_b{}^2}^3 \right) \tag{4.6}$$

where R_m is the radius of the droplet. Equation (4.6) can be solved numerically. The height of the sandwiched droplet h is

$$h = \sqrt{R_m{}^2 - R_a{}^2} + \sqrt{R_m{}^2 - R_b{}^2} \tag{4.7}$$

in this case, liquid's shape is described as

$$r'(\theta, z) = \sqrt{R_m{}^2 - \left(z - \sqrt{R_m{}^2 - R_b{}^2}\right)^2} \tag{4.8}$$

de Gennes studied the effect of perturbations at the bottom plane to the surface [20, 21]. His theory dealt with a flat surface (not cylindrical) in the x-z plane. Accordingly, if a perturbation, described as $\varepsilon \times \sin(qx)$ ($\varepsilon \ll 1$), is applied to the bottom plane ($z = 0$) of the flat surface, the surface can be described as

$\varepsilon \times \exp(-qz) \times \sin(qx)$. In another words, the effect of the perturbation passing in the height direction is proportional to $\exp(-qz)$, where q is the wavelength of the perturbation. We hypothesized that the perturbation from the circular contact line is also proportional to $\exp(-qz)$ in order to estimate the effect of boundary conditions (4.3) and (4.4). The wavelengths of the perturbation of the top plane q_a and the bottom plane q_b are given by

$$q_a = \frac{n}{R_a}, \quad q_b = \frac{n}{R_b} \tag{4.9}$$

Next, we add the effect of the perturbation to (4.8) and obtain the liquid's shape $r(\theta, z, \phi)$ as

$$\begin{aligned} r(\theta, z, \phi) = & \sqrt{{R_m}^2 - \left(z - \sqrt{{R_m}^2 - {R_b}^2}\right)^2} + a \exp\left(\frac{n}{R_a} z\right) \cos(n(\theta - \phi)) \\ & + b \exp\left(-\frac{n}{R_a} z\right) \cos(n(\theta - \phi)) + c \exp\left(\frac{n}{R_b} z\right) \cos(n\theta) \\ & + d \exp\left(-\frac{n}{R_b} z\right) \cos(n\theta) \end{aligned} \tag{4.10}$$

Here a, b, c and d are constants to satisfy boundary conditions (4.3) and (4.4) at $z = 0$ and $z = h$. They are

$$\begin{aligned} a &= \frac{1}{2} \frac{l_a}{\sinh(q_a h)}, & b &= -\frac{1}{2} \frac{l_a}{\sinh(q_a h)} \\ c &= -\frac{1}{2} \frac{l_b \exp(-q_b h)}{\sinh(q_b h)}, & d &= \frac{1}{2} \frac{l_b \exp(q_b h)}{\sinh(q_b h)} \end{aligned} \tag{4.11}$$

To verify that the shape derived from (4.10) and (4.11) is adequate, we compared it with the surface calculated by finite element software (Surface Evolver [22] already presented in Sect. 2.2.2). The parameters of the calculation are as follows: $R_a = 0.9\,\text{mm}$, $R_b = 1.0\,\text{mm}$, $l_a = 0.1\,\text{mm}$, $l_b = 0.1\,\text{mm}$, $V = 4.5\,\mu\text{L}$, $\phi = 0$, $n = 4$. Figure 4.12a, b is the surfaces calculated from (4.10) and Surface Evolver, respectively. Figure 4.12c shows cross-sectional views of the derived surfaces. The surfaces coincide. Using theoretical $r(\theta, z, \phi)$ given in (4.10), we also calculated the volume of the liquid with the following equation:

$$V = \int_0^{2\pi} \int_0^{2\pi} d\theta dz \frac{1}{2} r^2 \tag{4.12}$$

The parameters used to derive the liquid surface $r(\theta, z, \phi)$ give a volume of $4.50387\,\mu\text{L}$. The surface satisfies the boundary conditions of the top and bottom plane and liquid volume. Consequently, the method of deriving the liquid surface is adequate. Figure 4.13 plots the surface area as a function of rotational displacement.

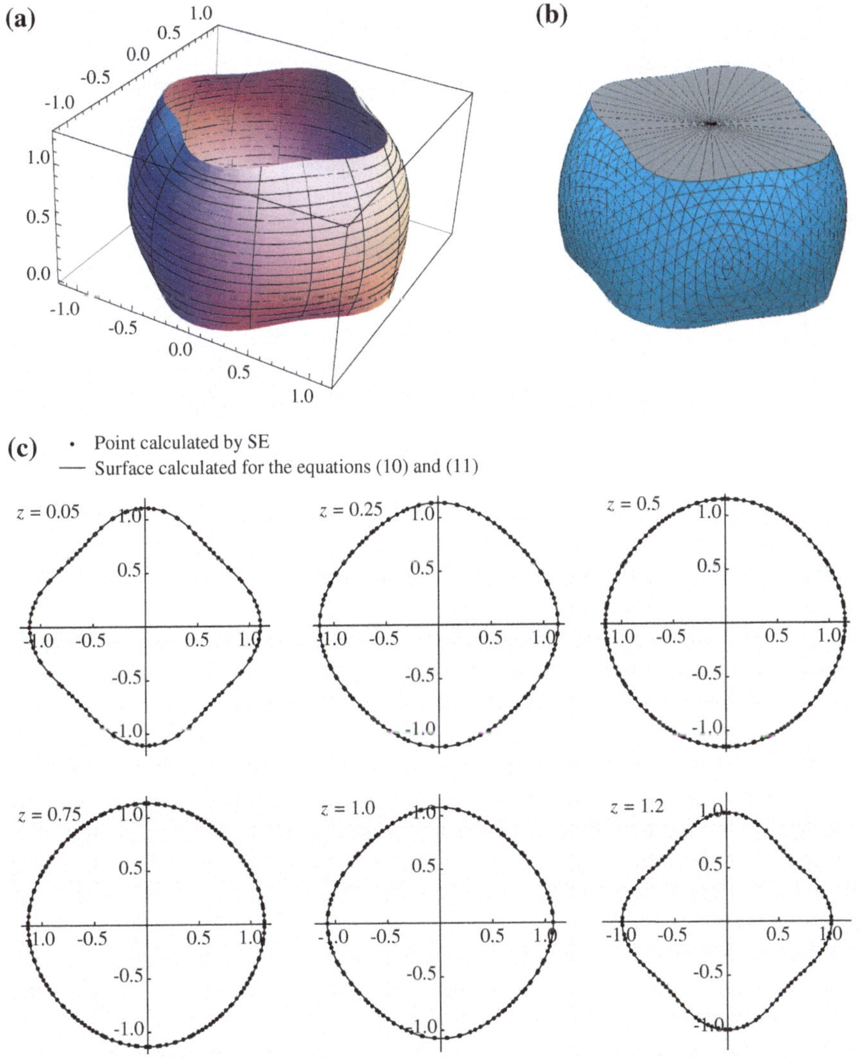

Fig. 4.12 Surface derived from **a** theory (4.10 and 4.11) and **b** Surface Evolver. **c** Horizontal cross-sectional views of both surfaces. Reprinted with permission from [19]. Copyright 2010 American Chemical Society

As predicted, the difference in surface area is less than $2 \times 10^{-10}\,\mathrm{m}^2$. The parameters of the calculation are: $R_a = 0.7\,\mathrm{mm}$, $R_b = 1.0\,\mathrm{mm}$, $l_a = 0.15\,\mathrm{mm}$, $l_b = 0.0125\,\mathrm{mm}$, $V = 4.5\,\mu\mathrm{L}$, $n = 4$. We calculated the torque with various parameters and verified that the calculations were in accordance with the experimental results.

Figure 4.14 shows the simulated relationship between the amplitude of the perturbations and torque. In this simulation, the base plates were (i), (ii), and (iii) and

Fig. 4.13 Simulated relationship between surface area and rotational displacement. The parameters are as follows: $R_a = 0.7\,\text{mm}$, $R_b = 1.0\,\text{mm}$, $l_a = 0.15\,\text{mm}$, $l_b = 0.125\,\text{mm}$, $V = 4.5\,\mu\text{L}$, $n = 4$. Reprinted with permission from [19]. Copyright 2010 American Chemical Society

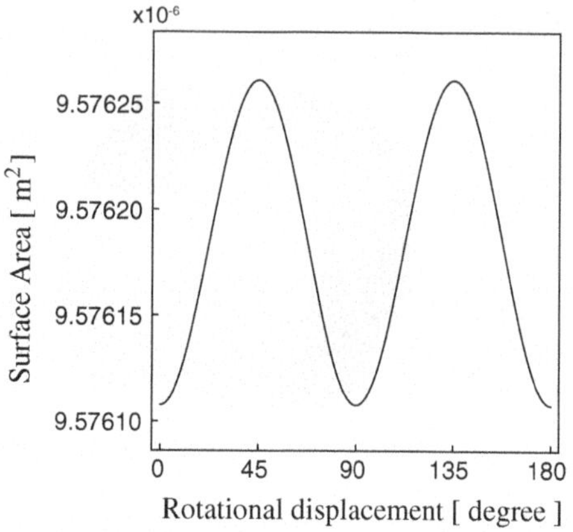

the floating plate was type (i). The liquid volume was $4.5\,\mu\text{L}$. The torque reaches a maximum at a rotational displacement of 22.5°. The torque is proportional to the amplitude of the perturbation. The experimental values were around two times larger than the simulated torques experimental values.

Figure 4.15 shows the simulated relationship between the frequency of the perturbation and torque. In this simulation, the base plate and floating plate pairs were type (i) and type (i), (iv) and (ii), and (v) and (iii). The liquid volume was $4.5\,\mu\text{L}$. The maximum torque occurs at 22.5, 18, and 15° for perturbation frequencies n of 4, 5 and 6, respectively. The torque falls to around one third for each increase in n. This tendency matches the measured one shown in Fig. 4.8. The reason behind this tendency is that the perturbation of the liquid passing in the height direction was reduced when the frequency increased, judging from (4.10). The data of the lines labeled 1 in Figs. 4.8 and 4.15 had the smallest frequencies and thus showed the largest torque.

Figure 4.16 shows the simulated relationship between liquid volume and torque. In this simulation, the base plate was type (i), the floating plate was type (i), and the liquid volumes were 3.0, 3.5, 4.0 and $4.5\,\mu\text{L}$. The $3.0\,\mu\text{L}$ droplet gave the highest torque. Moreover, the simulation matched the experimental data shown in Fig. 4.9. The above simulation only dealt with plates with sinusoidal perturbations. We also calculated the torque for a square plate. For instance, a square whose side lengths are $2L$ can be described in mathematical form as,

$$r = \begin{cases} \frac{L}{\cos(\theta-\pi/4)} (0 < \theta < \frac{\pi}{2}) \\ \frac{L}{\sin(\theta-\pi/4)} (\pi/2 < \theta < \pi) \\ -\frac{L}{\cos(\theta-\pi/4)} (\pi < \theta < \frac{3\pi}{2}) \\ -\frac{L}{\sin(\theta-\pi/4)} (\frac{3\pi}{2} < \theta < 2\pi) \end{cases} \tag{4.13}$$

Using the Fourier series expansion, it can be written as,

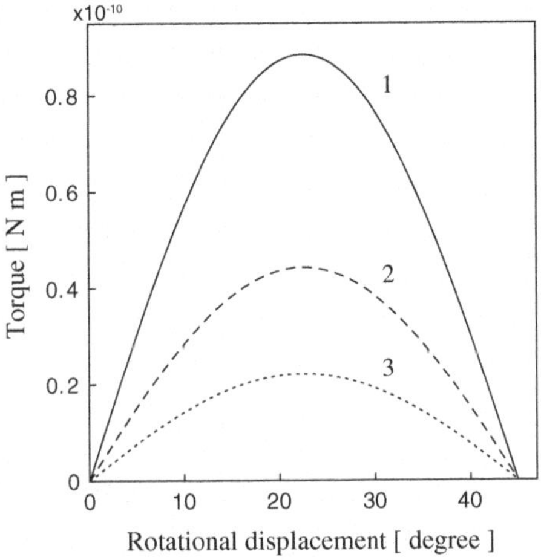

Fig. 4.14 Simulated relationship between amplitude of perturbations and generated torque. The base plates were as follows: $1 : 1000 + 50\cos 4\theta$. $2 : 1000 + 25\cos 4\theta$. $3 : 1000 + 12.5\cos 4\theta$. The floating plate was type (i): $700 + 150\cos 4\theta$. The volume of the liquid was 4.5 μL. The surface energy γ was $72 \times 10^{-3}\,\mathrm{Jm}^{-2}$. Reprinted with permission from [19]. Copyright 2010 American Chemical Society

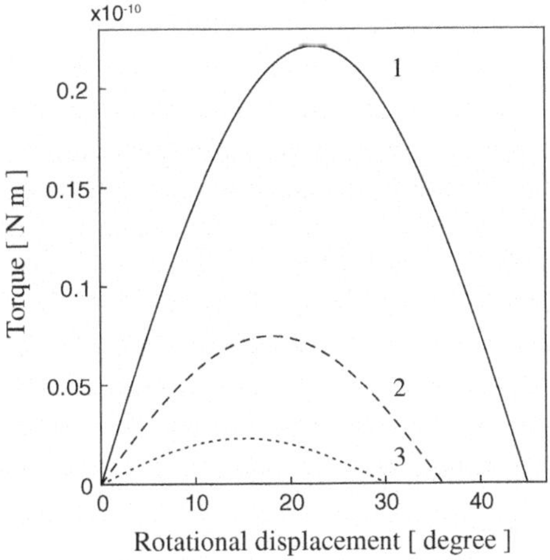

Fig. 4.15 Simulated relationship between frequency of perturbation and generated torque. The floating plate and base plate pairs were as follows: 1: $1000 + 12.5\cos 4\theta$ and $700 + 150\cos 4\theta$. $2 :$ $1000 + 12.5\cos 5\theta$ and $700 + 150\cos 5\theta$. $3 : 1000 + 12.5\cos 6\theta$ and $700 + 150\cos 6\theta$. The volume of the liquid was 4.5 μL. The surface energy γ was $72 \times 10^{-3}\,\mathrm{Jm}^{-2}$. Reprinted with permission from [19]. Copyright 2010 American Chemical Society

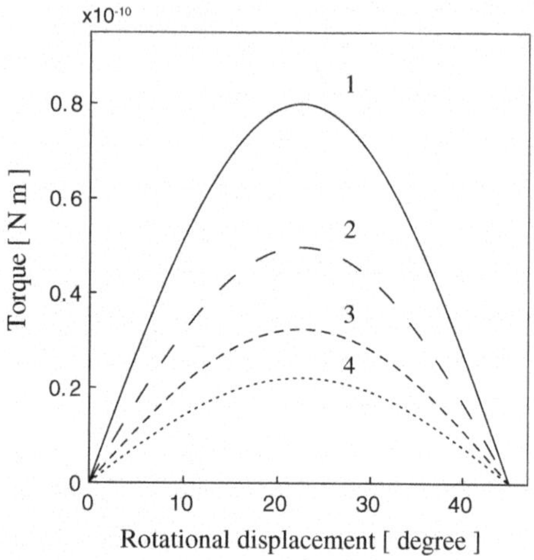

Fig. 4.16 Simulated relationship between liquid volume and generated torque. 1 : 3.0 μL 2 : 3.5 μL 3 : 4.0 μL 4 : 4.5 μL. The floating plate and base plate pair were $700 + 150 \cos 4\theta$ and $1000 + 12.5 \cos 4\theta$. The surface energy γ was $72 \times 10^{-3}\ \mathrm{Jm^{-2}}$. Reprinted with permission from [19]. Copyright 2010 American Chemical Society

$$r = (1.122 + 0.15644 \cos(4\theta) + 0.04934 \cos(8\theta) + 0.02341 \cos(12\theta) + \cdots)L \tag{4.14}$$

The torque was calculated with the same method as used to derive (4.10) and (4.11). Equation 4.14 means that a square gives the same torque as a circular plate with a radius of $1.122L$ and an infinite number of sinusoidal perturbations. The torque can be calculated by adding the effect of the second and subsequent terms of the right hand of (4.14) to that of the circular part. The effect of the fourth and subsequent terms is negligible. Figure 4.17 shows how square plates affect the torque. The base plate was of type (iv), and the floating plate was of type (iv) or (v). The liquid volume was 4.5 μL. The perturbations of the fifth and higher terms were neglected in the calculation. Even though the effect of the fourth term is very small, it was added for the sake of precision. The torque when using the 1250 μm-square plate was four times larger than when using the 1000 μm-square plate. This tendency and the order of the torque matched the experimental results shown in Fig. 4.10. We verified that the torque caused by a droplet sandwiched between plates with arbitrary shapes can be calculated by using a Fourier series expansion.

4.4 Conclusion

We measured and analyzed the torque caused by the capillary force of a liquid droplet sandwiched between microstructures by using a magnetic field technique. The diameters of the base plate and the floating plate were 1000 and 700 μm, respectively. The volumes of the sandwiched liquid were from 3.0 to 4.5 μL. The surface area changes when the floating plate is rotated. The change in surface area is around

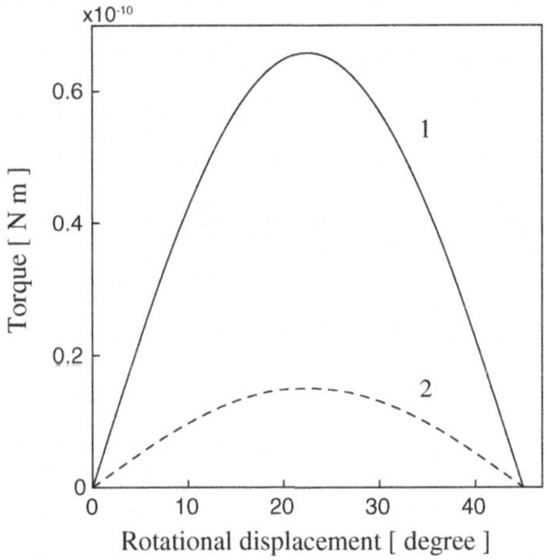

Fig. 4.17 Simulated torque using the square floating plates. 1 : 1250 μm − square2 : 1000 μm square. The volume of the liquid was 4.5 μL. The surface energy γ was 72×10^{-3} Jm^{-2}. Reprinted with permission from [19]. Copyright 2010 American Chemical Society

0.002 % of the entire surface. In the experiments, the maximum torque varied from 0.04×10^{-10} Nm to 1.7×10^{-10} Nm. In the experiment shown in Fig. 4.7 and simulation shown in Fig. 4.14, the torque is proportional to the amplitude of the plates shape perturbations. In the experiment shown in Fig. 4.8 and simulation shown in Fig. 4.14, the torque falls by three times for each unit increase in frequency of perturbation n. The maximum torque for $n = 4$, 5, and 6 were at 22.5, 18, and 15°. The liquid volume also affected the generated torque. Increasing the liquid volume by 0.5 μL, from 3.0 to 3.5 μL, almost doubled the torque. We verified that the torque can be calculated with the equations presented in this paper. By using a Fourier series expansion, the shapes of the plates can be separated into a circular part and perturbation parts. The torque can be calculated by summing the effects of all these parts. Although the experimentally obtained torque was two times larger than the calculated one, the order and relationship between the torque and the parameters were in accordance. The results and calculations presented in this chapter should thus be useful for designing micro-devices and self-assemblies actuated by capillary force.

4.5 Appendix

These development show to that the torque depends on 5 factors: liquid volume V, radius of floating plate R_a, radius of base plate R_b, perturbation of floating plate l_a, perturbation of base plate l_b, frequency of perturbation n. From the numerical results obtained in various conditions, the following scaling could be obtained:

$$T \sim \gamma \frac{l_a l_b R^9}{e^n V^3} \tag{4.15}$$

Table 4.1 In the simulation, we set $l_a = 0.1$ mm, $l_b = 0.1$ mm, $\gamma = 72 \times 10^{-3}$ J/m^2, $n = 4$

R_a (mm): R_b (mm)	0.5:1	0.6:1	0.7:1	0.8:1	0.9:1	1:1
T_s(N m) ($V = 3.0\,\mu$L)	5.7×10^{-11}	1.7×10^{-10}	4.0×10^{-10}	8.0×10^{-10}	1.4×10^{-9}	2.2×10^{-9}
T_s(N m) ($V = 4.0\,\mu$L)	2.2×10^{-11}	7.0×10^{-11}	1.6×10^{-10}	3.3×10^{-10}	5.4×10^{-10}	9.8×10^{-10}
T_s(N m) ($V = 5.0\,\mu$L)	1.0×10^{-11}	3.3×10^{-11}	8.0×10^{-11}	1.6×10^{-10}	2.9×10^{-10}	4.9×10^{-10}
T_s(N m) ($V = 6.0\,\mu$L)	5.4×10^{-12}	1.8×10^{-11}	4.3×10^{-11}	8.8×10^{-11}	1.6×10^{-10}	2.7×10^{-10}

The readers can estimate the torque according to the scaling law in (4.15). Please note that the scaling law is appropriate for the situation where the height h is the same order as R_a and R_b

where R is the representative scale of the system. We will add a table including calculation results, so that the readers can roughly estimate the torque for their own condition (Table 4.1).

Let us mention a recent result published by [23], providing an analytical estimate of the capillary torque T exerted by a liquid drop sandwiched between two square pads:

$$T(a) \approx \gamma L^2 \left(\frac{a}{\sqrt{1+a^2}} + \frac{1}{a+\sqrt{1+a^2}} \left(\frac{1}{a} + \frac{1}{\sqrt{1+a^2}} \right) - \frac{\log\left(a+\sqrt{1+a^2}\right)}{a^2} \right) \tag{4.16}$$

where L is the square edge, h is the liquid height, θ is the twist angle and $a = \frac{L\theta}{2h}$.

References

1. C. Isenberg, *The Science of Soap Films and Soap Bubbles* (Dover, New York, 1992)
2. M. Strani, F. Sabetta, J. Fluid Mech. **141**, 233 (1984)
3. C. Huh, L.E. Scriven, J. Colloid Interface Sci. **30**, 323 (1969)
4. B.J. Carroll, J. Colloid Interface Sci. **57**, 488 (1976)
5. R. Fodecave, F.B. Wyart, Macromolecules **31**, 9305 (1998)
6. D. Langbein, *Capillary Surfaces Shape-Stability-Dynamics in Particular under Weightlessness* (Springer, Berlin, 2002)
7. P. Lambert, *Capillary Force in Microassembly: Modeling, Simulation, Experiments, and Case Study; Microtechnology and MEMS* (Springer, New York, 2007)
8. P. Lambert, A. Delchambre, A. Chau, Langmuir **24**, 3157 (2008)
9. P. Lambert, A. Delchambre, Langmuir **21**, 9537 (2005)
10. X. Xiong, Y. Hanein, J. Fang, Y. Wang, W. Wang, D. Schwartz, K. B'ohringer, J. Microelectromech. Syst. **12**, 117 (2003)
11. R.R.A. Syms, E.M. Yeatman, V.M. Bright, G.M. Whitesides, J. Microelectromech. Syst. **12**, 387 (2003)

12. J.N. Israelachvili, *Intermolecular and Surface Forces*, 2nd edn. (Academic, San Diego, 1992)
13. R.R.A. Syms, E.M. Yeatman, Electron. Lett. **29**, 662 (1993)
14. R.R.A. Syms, E.M. Yeatman, V.M. Bright, G.M. Whitesides, J. Microelectromech. Syst. **4**, 177 (1995)
15. A. Takei, Y. Yoshihata, I. Shimoyama, J. Micromech. Microeng. **21**, 85009 (2011)
16. C. Py, P. Reverdy, L. Doppler, J. Bico, B. Roman, C.N. Baroud, Phys. Rev. Lett. **98**, 156103 (2007)
17. J. Lee, H. Moon, J. Fowler, T. Schoellhammer, C.J. Kim, Sens. Actuators A **95**, 259 (2002)
18. A. Takei, K. Matsumoto, I. Shimoyama, Lab Chip **10**, 1781–1786 (2010)
19. A. Takei, K. Matsumoto, I. Shimoyama, Langmuir **26**, 2497–2504 (2010)
20. J.F. Joanny, P.G. de Gennes, J. Chem. Phys. **81**, 552 (1984)
21. P.G. de Gennes, F. Brochard-Wyard, D. Quéré, *Capillarity and Wetting Phenomena: Drops, Bubbles, Pearls, Waves* (Springer, New York, 2004)
22. K. Brakke, Exp. Math. **1**, 141 (1992)
23. J. Berthier, K. Brakke, F. Grossi, L. Sanchez, L. di Cocio, Self-alignment of silicon chips on wafers: a capillary approach. J. Appl. Phys. **108**, 054905 (2010)

Chapter 5
Surface Tension Effects in Presence of Gas Compliance

Cyrille Lenders

Abstract In previous chapters, surface tension effects have been extensively detailed. This chapter extends these to take into account gas menisci instead of liquid menisci. Fundamental difference relies in the fact that the meniscus volume is not constant anymore, but now obeys the gas law. This chapter presents how to adapt the models, and discusses also issues about stability of a gas bubble generation. A dimensionless study is presented to predict instability. The experimental validation of the proposed model is also given. Finally, we also present a microrobotic application of this concept.

5.1 Introduction

In the previous chapter, the effect of a liquid bridge has been deeply investigated. Here, we extend these concepts for gaseous bridges in a liquid. Practically, we state the existence of surface tension effects due to the presence of bubbles in a liquid media. We explain the consequences of having a compressible media in the capillary bridge, compared to the model of a liquid bridge. We also mention some applications from the literature that make use of microbubbles.

In Sect. 5.2, we first present how to take gas properties into the model of surface tension. We illustrate these concept with the case of a bubble generation, and show there may be an instability in the bubble growth due to the gas compressibility. Then we propose in Sect. 5.3 a dimensionless analysis to determine a threshold of the instability. We present some experimental results to confirm the validity of the model in Sect. 5.4. Then we finally present an application of this theory to bubble robotics in Sect. 5.5.

C. Lenders (✉)
Université libre de Bruxelles, avenue F.D. Roosevelt 50, Bruxelles 1050, Belgium
e-mail: clenders@ulb.ac.be

P. Lambert (ed.), *Surface Tension in Microsystems*, Microtechnology and MEMS,
DOI: 10.1007/978-3-642-37552-1_5,

5.2 Modeling

In this section we present how to take account of the compressibility of the media in the model of surface tension effect. We apply this to two examples. We also define some limitations of the model in case of bubbles with a too small diameter.

5.2.1 Laplace Law

Let us first remind some basic relations that will be useful through this chapter.

Laplace law is a relation linking surface tension to mean curvature of an interface and pressure drop across this interface.

$$\Delta p = p_{\text{in}} - p_{\text{out}} = 2\gamma H \tag{5.1}$$

Note that Δp is here considered positive if the pressure on the side of the center of curvature (p_{in}) is larger than the pressure on the opposite side (p_{out}). This will indeed be the case most of time with bubbles.

This law is valid no matter if the medium is compressible or incompressible.

5.2.2 Surface Tension and Compressibility

A bubble differs from a liquid drop by its compressibility. Therefore, it is not correct to assume the volume of the capillary bridge to be constant. One could question the importance of gas compressibility in the modeling of surface tension. To show the impact of size on the interaction of surface tension and compressibility, let us consider a single bubble in a liquid medium.

We will always assume bubbles are small enough to be considered as spherical. This hypothesis is valid when the pressure around the bubble may be assumed constant. The variation of hydrostatic pressure should hence be smaller than the pressure drop across the interface. This is true when Bond number associated with the radius R of the bubble is smaller than 1.

$$\text{Bo} = \frac{\rho g R^2}{\gamma} << 1 \tag{5.2}$$

For an air bubble in water, this condition is fulfilled if the radius is smaller than 2.7 mm. In this case, the mean curvature H is simply equal to R^{-1}. Due to surface tension, the pressure inside the bubble will be larger than the pressure in the surrounding liquid, and therefore the volume will be smaller than without surface tension. This increase of pressure is larger if γ is larger, and if the radius R of the bubble is smaller.

The gas law is:

$$p_{in} V = n R_g T \tag{5.3}$$

where n is the amount of gas moles present in the bubble, T is the absolute temperature, $R_g = 8.314\,\mathrm{J\,mol^{-1}\,K^{-1}}$ is the gas constant, p_{in} is the absolute pressure in the bubble and V is the volume of the bubble. We cannot assume that the volume of the bubble is constant, but in order to compare the influence of surface tension, we will compare different bubbles with the same number of molecules n.

If p_{out} is the pressure in the liquid around the bubble, i.e. the sum of hydrostatic pressure and atmospheric pressure, the combination of (5.1) and (5.3) leads to:

$$\left(p_{out} + \frac{2\gamma}{R}\right)\frac{4\pi R^3}{3} = n R_g T \tag{5.4}$$

Figure 5.1 shows the influence of surface tension on a free immersed gas bubble. It is clear that the impact of surface tension on the bubble radius through gas compressibility is not negligible in the microscale. For bubbles of 10^{-14} mol (bubbles of a few micrometers), the relative variation of diameter induced by a surface tension $\gamma = 72 \times 10^{-3}\,\mathrm{N\,m^{-1}}$ is about 26 %.

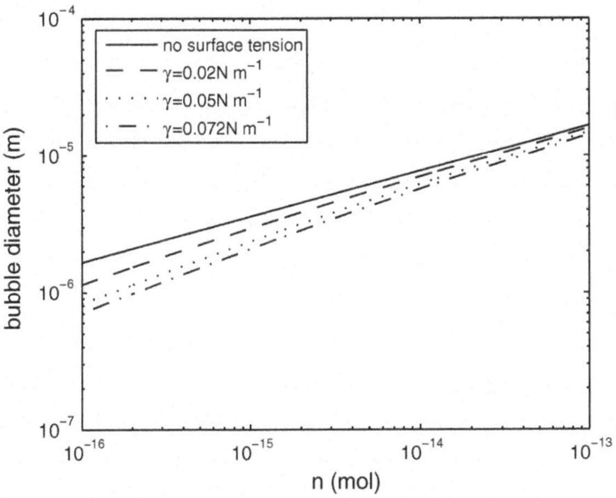

Fig. 5.1 Influence of surface tension and amount of gas mole on the diameter of a free gas bubble in a liquid. The influence of surface tension is more significant for small bubbles, and liquid-gas couples having a high surface tension ($T = 293.15$ K, $p_{out} = 101325$ Pa)

5.2.3 Effect of Compressibility When Generating a Bubble from a Hole

In order to illustrate the effects of gas compressibility on surface tension modeling, it is interesting to consider the problem of the generation of a single bubble gas, at a specific location, and with a control of its dimension. In addition, such a bubble generation is interesting to make use of surface tension in an actuator, where a bubble is placed between two solids (gaseous capillary bridge), as it will be described as application in Sect. 5.5.

One may find in the litterature numerous methods to generate gas bubbles in a liquid. But most of the time, these methods aim at generating bubbles continuously, in chemical reactors applications, or when studying the behavior of colloids [1, 8, 10, 12, 13].

To lead the study of the interaction of a gas bubble with a solid, we will focus on the means which allow to generate a single bubble, at a specific location, and which allow to control the dimensions of this bubble. We will briefly present some examples.

Electrolysis allows to neutralise ions contained in a liquid and, in that event, to form a gas [2, 3, 7]. Heat can also be used to grow a bubble [4], by heating a liquid above its boiling temperature. But electrolysis and heat based generation are difficult to implement practically, because it is necessary to create a specific nucleation area in order to control the position of the bubble generation [5, 11]. Moreover, heat based generation means that the liquid has been vaporized, so the gas will spontaneously return to its liquid state when the heat source is removed.

To ensure that the bubble grows where it is required to, we propose to flow gas out of a circular hole with a known diameter. The location of the bubble is therefore known for sure. The size of the bubble may be defined in two ways: either by controlling the pressure of the injected gas, or by controlling a volume change in a gas tank.

5.2.4 Pressure Control

The idea is here to basically control the pressure in the gas circuit. By increasing the pressure, the gas is pushed towards the outlet, and a bubble is generated.

The dimensions of the bubble, which are then defined by its diameter, are directly given by Laplace law (5.1). Indeed, in a quasi-static situation, there is no flow in the channel leading the gas to the outlet, hence the pressure may be assumed to be uniform. The pressure in the liquid being also uniform, the pressure drop across liquid-gas interface is given by the difference between the pressure in the gas circuit and the pressure in the liquid surrounding the bubble. This pressure drop directly defines the mean curvature H of the bubble, i.e. the radius of the spherical cap.

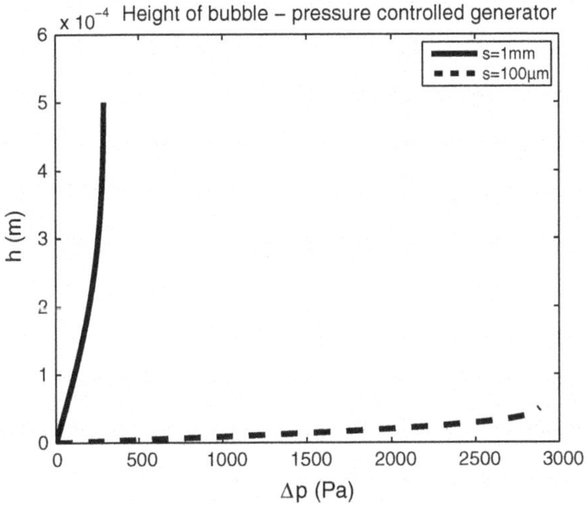

Fig. 5.2 Height of a bubble function of Δp, the pressure drop across the interface, in this case the increase of pressure in the gas circuit with respect to pressure outside the bubble. In this graph, $\gamma = 72 \times 10^{-3}\,\mathrm{Nm}^{-1}$. The pressure control is easier if outlet channel has a small diameter

It is obvious, and yet crucial, to note that the smallest possible radius of a spherical cap is the one of a half sphere, when the radius of the bubble equals the radius of the outlet hole s_1 (Fig. 5.3).

Raising the pressure from $\Delta p = 0$, the size of the bubble increases progressively, until the pressure is such as $\Delta p = 4\gamma/s_1$ where the bubble is a half sphere. If the pressure is increased beyond that value, bubbles will escape the hole in a continuous wisp.

It is therefore not possible, with this method, to generate single bubbles larger than a half sphere.

On Fig. 5.2, we see what is the height of the generated bubble for two different hole diameters, as a function of the pressure drop, for an air bubble in water ($\gamma = 72\,\mathrm{mN/m}$).

5.2.5 *Volume Control*

This is another version of gas injection based bubble generator. In this case, the volume of a finite gas tank is reduced, pushing the gas in the liquid and creating the bubble. This is illustrated in Fig. 5.3.

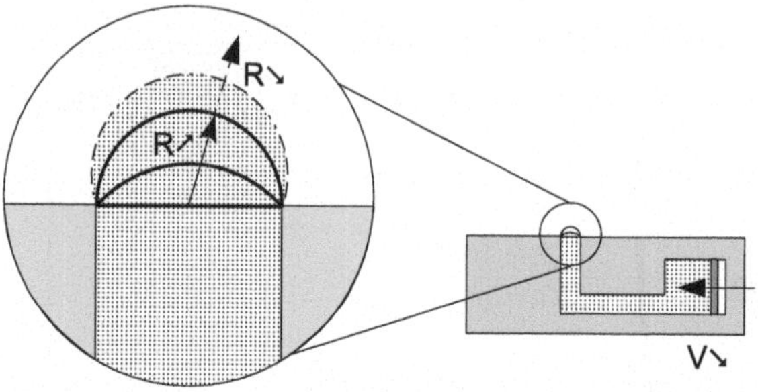

Fig. 5.3 Working principle of the volume controlled bubble generator. A finite volume of gas is confined in a chamber. By reducing this volume, the gas is expelled through a channel to the bubble growing location. By this mean, it is possible to inject any gas at a predefined location. The ability to grow any size of bubble will however be subjected to a further condition detailed in the next section

5.2.5.1 Basics

The device is like a syringe pump. However, it is not correct to assume that the volume swept by the piston in the syringe will be the volume of the bubble. Surface tension will play a considerable role.

In this case, control of bubble height will be based on the control of the volume in the gas container. In the case of a syringe, it is function of the position of the piston. Unlike in a pressure based control, there seems to be no limitation to the height of the bubble. The only size limitation appears when the bubble is so large that the adhesion forces on the edge of outlet channel are no longer large enough to balance the buoyancy force.

This principle has major advantages to other solutions: the size is easy to control, and any gas can be injected in any liquid (possible use of different liquids to vary surface tension). Finally, the position of the bubble is precisely defined.

5.2.5.2 Device Description

The device is like a syringe pump and is schematically represented on Fig. 5.4. The parameters are summarized in Table 5.1. The syringe is filled with the gas to inject, then the syringe tip is immersed in a liquid. The generation of a bubble is done by pushing the piston. The orientation of the tip (downwards or upwards) has no influence on the gas generation, since we are working with characteristic length below the capillary length. The gravity can henceforth be neglected.

The gas circuit includes the gas in the container and the gas in the bubble. It is clear that in the configuration shown in Fig. 5.4, the variation of volume ΔV in the gas circuit is given by:

$$\Delta V = \delta S \tag{5.5}$$

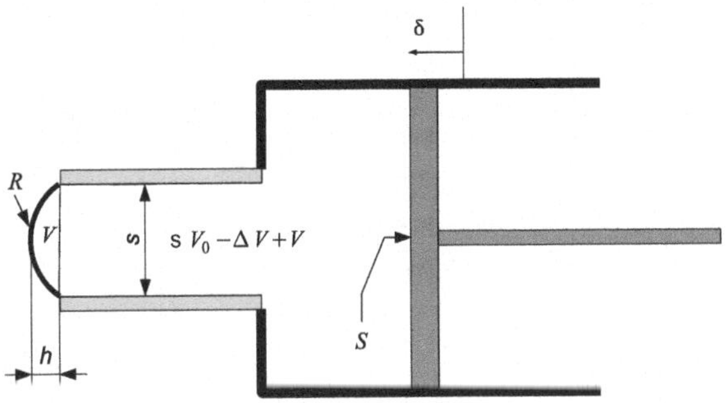

Fig. 5.4 Schematic representation of the volume controlled bubble generator. The device is essentially a syringe pump. The various parameters used in the model are shown

Table 5.1 Definition of model parameters

Parameter	Description
V_0 (m^3)	Initial volume in the syringe
V (m^3)	Volume of gas in the bubble
ΔV (m^3)	Variation of container volume (positive if volume decreases)
δ (m)	Movement of the piston
γ ($N\,m^{-1}$)	Surface tension at the liquid-gas interface
p_{atm} ($N\,m^{-2}$)	Atmospheric pressure
p ($N\,m^{-2}$)	Pressure in the gas container
T (K)	Temperature
R_g ($N\,m\,mol^{-1}\,K^{-1}$)	Universal gas constant
n (mol)	Total number of gas mole in the system
S (m^2)	Section of the piston
s (m)	Bubble channel diameter
R (m)	Radius of the bubble
h (m)	Height of the bubble

The next developments will be done using ΔV as it is a more general formulation. The model we will develop is to determine the bubble height h as a function of volume variation ΔV. We will see that it is possible to act on other parameters to modify the bubble height.

5.2.5.3 Model of Bubble Height

The purpose of the model developed in this section is illustrated on Fig. 5.5. The output of the model is the geometry of the generated bubble, which under the assumption we make has the shape of a portion of a sphere. Since we assume that the bubble is anchored to the border of the outlet channel having a diameter s, the geometry of the bubble can be deduced from its height h.

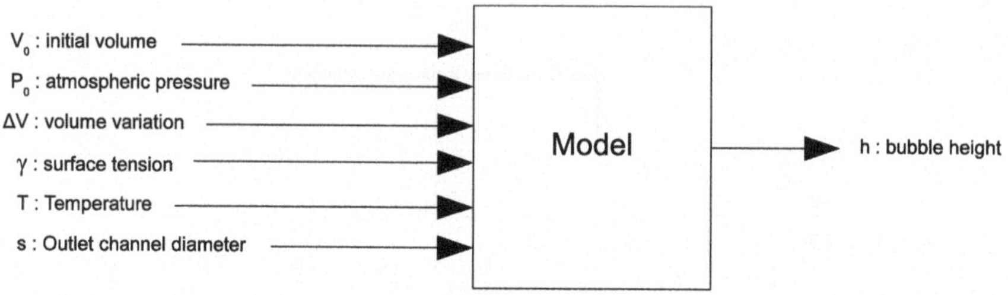

Fig. 5.5 Model of the volume controlled bubble generator. The output of the model is the geometry of the bubble, which we assume to be a portion of a sphere

5.2.5.4 Equations

To model the height of a bubble generated by a volume controlled generator, we base the reasoning on the next equations. We assume that the number of mole in the system is constant. Consequently, the total volume of gas is not constant due to the compressibility of the medium which is subject to surface tension force.

Laplace equation (5.1) links pressure drop across bubble interface, surface tension and the mean curvature of the interface. We assume the pressure inside the bubble to be constant, and since we suppose that the bubble is smaller than the capillary length, the mean curvature is constant too. Finally, assuming an axisymmetric configuration, the geometry of the bubble is a spherical cap (radius R):

$$\frac{2\gamma}{R} = p - p_{\text{atm}} \tag{5.6}$$

The gas law is used to take into account the volume variation induced by pressure variation:

$$p\,(V + V_0 - \Delta V) = n R_g T \tag{5.7}$$

It is possible to write equations linking the radius of the spherical cap R, the height of the bubble h, the syringe tip diameter s and the volume of the spherical cap V:

$$R = \frac{s^2}{8h} + \frac{h}{2} \tag{5.8}$$

$$V = \frac{\pi h}{6}\left(\frac{3s^2}{4} + h^2\right) \tag{5.9}$$

The combination of (5.6), (5.7), (5.8) and (5.9) leads to a fifth degree polynomial.

$$\frac{\pi}{12} p_{\text{atm}} h^5 + \frac{\pi}{3}\gamma h^4 + \frac{\pi}{12} p_{\text{atm}}\, s^2 h^3 + \left(-\frac{1}{2} n R_g T + \frac{1}{2} p_{\text{atm}}(V_0 - \Delta V) + \frac{\pi}{4}\gamma s^2\right) h^2$$
$$+ \left(\frac{\pi}{64} p_{\text{atm}} s^4 + 2\gamma (V_0 - \Delta V)\right) h + \frac{1}{8} p_{\text{atm}} s^2 (V_0 - \Delta V) - \frac{1}{8} n R_g T s^2 = 0 \tag{5.10}$$

Table 5.2 Sets of parameters for simulations

Parameters	First set	Second set
V_0 (m^3)	2×10^{-6}	5×10^{-9}
n (mol)	8.175×10^{-5}	2.044×10^{-7}
s (m)	2×10^{-3}	150×10^{-6}
ΔV (m^3)	$0 \rightarrow 2 \times 10^{-6}$	$0 \rightarrow 5 \times 10^{-9}$

Equation (5.10) is an implicit analytic function giving h, a function of 8 parameters. n could be seen as a dependent parameter if the initial filling of the container is made at atmospheric conditions ($p_{\text{atm}} V_0 = n R_g T$). We have chosen to leave it as an independent parameter to keep the ability to modify the value independently from the initial filling conditions, for example to consider supplementary gas injection or a gas leak. To the best of our knowledge, there is no analytical solution to explicit a fifth degree polynomial. However, it is easy to solve it numerically.

Our main interest is to find the height of the bubble with respect to container volume variation, but (5.10) allows us to check the influence of other parameters. Since it is a fifth order polynomial, we know there are 1, 3 or 5 real solutions h for sets of other parameters.

5.2.5.5 Shape of the Solution

To check the shape of the solution $h = h(\Delta V)$ and make it easier to interpret the multiplicity of solutions, we propose two sets of parameters. In each case, we suppose to inject air in water, and we assume $\gamma = 72 \times 10^{-3}\,\text{N}\,\text{m}^{-1}$. We also suppose to work at normal atmospheric pressure $p_{\text{atm}} = 101325\,\text{Pa}$, and normal temperature $T = 298.15\,\text{K}$. As a reminder, gas constant is $R_g = 8.314\,\text{J}\,\text{mol}^{-1}\,\text{K}^{-1}$. The value for other parameters are summarized in Table 5.2.

The first set corresponds to a larger syringe (13.3× larger outlet tip diameter and 400× larger volume) than the second one. In this table, the values of n are calculated using the gas law. We suppose the syringe to be filled with air under atmospheric conditions. In each case, we will compare the results with the height of a bubble of incompressible fluid:the height of a spherical cap having a volume equal to ΔV, the volume swept by the piston. The evolution of bubble height h with ΔV is shown in Fig. 5.6 for the first set of parameters and Fig. 5.7 for the second set of parameters.

The analysis of the results for the first set of parameters shows that there is only one solution for the whole range of ΔV. The combination of gas compressibility and surface tension clearly has an influence and leads to a smaller gas bubble volume that what could be expected. This phenomenon tends to disappear for large bubbles, since the effects of surface tension decrease with the size of the bubble. A large bubble has a smaller curvature, so the pressure drop is lower and the effect of gas compressibility disappears.

The results for the second set of parameters shows even more consequences of compressibility and surface tension. In this case, there is a range of ΔV values for

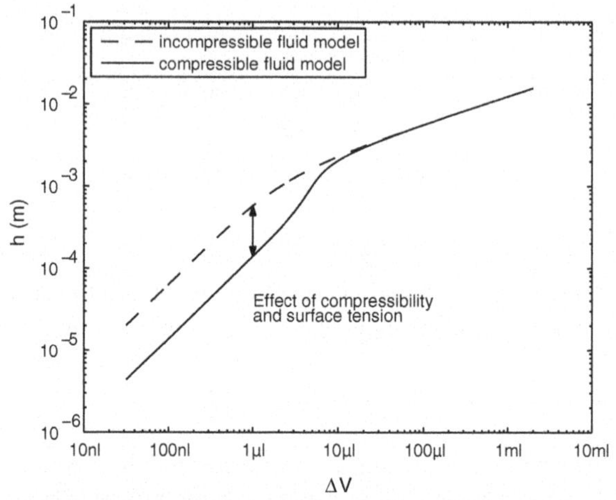

Fig. 5.6 Evolution of bubble height with volume variation (first set of parameters), i.e. with the volume swept by the piston in the syringe. The incompressible fluid model represents the height of a bubble having the same volume as ΔV, as if the fluid has a constant volume. The influence of gas compressibility and surface tension clearly have an influence on the volume of the bubble for small bubbles. In this configuration, there is a direct relation between ΔV and h

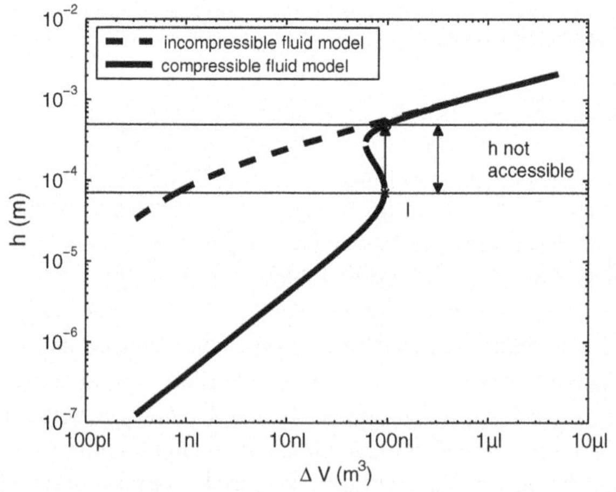

Fig. 5.7 Evolution of bubble height with the volume swept by the piston in the syringe (second set of parameters). In this configuration, the model gives three solutions for a range of ΔV. Physically, the bubble will grow all the way to point I, then its height will jump suddenly to the *upper curve*. This growing instability means there is a whole range of height that cannot be reached

which there are three solutions. If ΔV is increased progressively, the curve $h\,(\Delta V)$ reaches a point (I in Fig. 5.7) where its slope changes sign after being infinite. If ΔV increases beyond point I, the height of the bubble has no choice but to jump to the other solution (Arrow on Fig. 5.7), creating a sudden change in the bubble size. Of course, these results are for quasi-static situation, so the height change is not instantaneous, but this shows an instability in the bubble growing.

This means also that it is not possible to grow a bubble of any size with the second set of parameters. The area between point I and its corresponding solution on the upper curve is not accessible. One should also notice that the height variation at point I is significant: in this configuration, it increases by a factor 7. In this case, the buoyancy forces are generally large enough to lift the bubble. It is interesting to notice that point I is located approximately at $h = s/2$, i.e. when the bubble is a half sphere. The interpretation of this result is that in some cases, there is an accumulation of energy in the form of a pressure rise in the gas system. When the bubble grows from $h = 0$ to $h = s/2$, the pressure in the gas circuit increases, but when $h > s/2$, the pressure decreases. The accumulated pressure induces then a rapid growth of volume.

It should be interesting to find a criterion on the setup parameters to predict whether an instability is likely to occur (as for the second set of parameters) or if the growing will be continuous (as for the first set of parameters). This is done in Sect. 5.3, where we will translate our model in a dimensionless form.

5.2.5.6 Limit of the Model

Since the pressure inside a bubble increases for small bubbles, it is interesting to check whether the liquefaction of the gas could occur due to pressure rise.

It is not possible to liquefy a gas if its temperature is larger than its critical temperature. For example, critical temperature of nitrogen is $T_c = 126$ K, and for oxygen $T_c = 155$ K. Hence, liquefaction of air is not possible around ambient temperature, no matter how large the pressure is.

However, carbon dioxide has a critical temperature of 304 K, which is a bit larger than ambient temperature. Critical pressure of carbon dioxide is around 7.4×10^6 Pa. To reach such pressure inside the bubble, their diameter has to be around 10–40 nm at $T = 293$ K, which is very small. At this level, the gas law model is not correct anymore, because in this model, gas molecules are supposed not to interact with each other.

Other gases have a critical temperature reasonably larger than ambient temperature, like ammonia: $T_c = 405.5$ K. However, the pressure to liquefy the gas at ambient temperature is still in the range of 1 MPa, leading to bubble diameter of 90–320 nm.

To avoid model deviation due to the proximity of critical state of fluid, we shall consider this model is only valid for gas bubbles with a characteristic dimensions in the range of 1 μm to a few mm.

5.3 Dimensionless Analysis

Dimensionless numbers are useful to study a scaled model of a device, or to reduce the number of parameters in a model [9].

On the base of a dimensionless analysis of (5.10), we are showing in the following that the stability of bubble growth can be described by one dimensionless number only.

5.3.1 Buckingham Theorem

It is possible to construct a set of dimensionless numbers using Buckingham theorem [9]. In our case, the model can be expressed in the form:

$$\Psi\left(V_0, \Delta V, h, \gamma, p_{\text{atm}}, T, R_g, n, s\right) = 0 \tag{5.11}$$

There are nine parameters and five different dimensions are used (length, mass, time, temperature and mole). In the dimensional matrix hereafter, element i, j indicates the exponent of the ith dimension in the jth parameter:

$$\overline{\overline{D}} = \begin{array}{l|ccccccccc} & V_0 & \Delta V & h & \gamma & p_{\text{atm}} & T & n & s & R_g \\ \hline \text{mass} & 0 & 0 & 0 & 1 & 1 & 0 & 0 & 0 & 1 \\ \text{length} & 3 & 3 & 1 & 0 & -1 & 0 & 0 & 1 & 2 \\ \text{time} & 0 & 0 & 0 & -2 & -2 & 0 & 0 & 0 & -2 \\ \text{temperature} & 0 & 0 & 0 & 0 & 0 & 1 & 0 & 0 & -1 \\ \text{moles} & 0 & 0 & 0 & 0 & 0 & 0 & 1 & 0 & -1 \end{array} \tag{5.12}$$

The rank of $\overline{\overline{D}}$ is 4, so according to Buckingham's theorem, it is possible to find $9 - 4 = 5$ linearly independent relations between the exponents of each parameters. This leads to 5 independent dimensionless numbers: none of these 5 numbers can be found by a combination of the 4 others. To find such parameters, we only need to calculate the kern of $\overline{\overline{D}}$.

The problem of this method is that any combination of these terms will also produce a dimensionless number. It is not possible to predict which among all these possibilities will give the most relevant number to predict whether an instability will occur or not.

5.3.2 Dimensionless Equation

The solution to find the adequate dimensionless numbers is to formulate (5.10) in a dimensionless form. The idea is to norm every parameter of this equation. We propose to norm the parameters as follows:

$$\begin{aligned} &\widetilde{E} = \frac{n\,R_g\,T}{\gamma\,s^2} \quad \widetilde{\Delta V} = \frac{\Delta V}{s^3} \quad \widetilde{P} = \frac{p_{atm}}{\gamma\,/s} \\ &\widetilde{h} = \frac{2\,h}{s} \qquad \widetilde{V} = \frac{V_0}{s^3} \end{aligned} \tag{5.13}$$

Note that $\widetilde{h} = 1$ means the bubble is a half sphere. Equation (5.10) then becomes:

$$\begin{aligned} &\frac{\pi}{384}\widetilde{P}\widetilde{h}^5 + \frac{\pi}{48}\widetilde{h}^4 + \frac{\pi}{96}\widetilde{P}\widetilde{h}^3 + \left[-\frac{\widetilde{E}}{8} + \frac{\widetilde{P}}{8}\left(\widetilde{V} - \widetilde{\Delta V}\right) + \frac{\pi}{16}\right]\widetilde{h}^2 \\ &+\left[\frac{\pi}{128}\widetilde{P} + \left(\widetilde{V} - \widetilde{\Delta V}\right)\right]\widetilde{h} + \frac{\widetilde{P}}{8}\left(\widetilde{V} - \widetilde{\Delta V}\right) - \frac{\widetilde{E}}{8} = 0 \end{aligned} \tag{5.14}$$

As predicted by Buckingham's theorem, $\widetilde{h} = \widetilde{\mathrm{f}}\left(\widetilde{E}, \widetilde{P}, \widetilde{V}, \widetilde{\Delta V}\right)$ is now a function of 4 parameters instead of 8 in the dimensional formulation. We have reduced the number of parameters by 4, which was the rank of $\overline{\overline{D}}$.

Other simplifications can also be done in (5.14). The form of the equation shows that $\widetilde{V}$ and $\widetilde{\Delta V}$ are always together. It is possible in this case to reduce even more the number of parameters, posing:

$$\widehat{V} = \widetilde{V} - \widetilde{\Delta V} \tag{5.15}$$

$$\begin{aligned} &\frac{\pi}{384}\widetilde{P}\widetilde{h}^5 + \frac{\pi}{48}\widetilde{h}^4 + \frac{\pi}{96}\widetilde{P}\widetilde{h}^3 + \left(-\frac{\widetilde{E}}{8} + \frac{\widetilde{P}}{8}\widehat{V} + \frac{\pi}{16}\right)\widetilde{h}^2 \\ &+\left(\frac{\pi}{128}\widetilde{P} + \widehat{V}\right)\widetilde{h} + \frac{\widetilde{P}}{8}\widehat{V} - \frac{\widetilde{E}}{8} = 0 \end{aligned} \tag{5.16}$$

$\widehat{V}$ represents the normalized volume of gas remaining in the container after the movement the piston. $\widehat{V}$ cannot be relevant to predict any instability in the system. Indeed, this parameter is a function of ΔV, and the existence or absence of instability cannot depend on the piston position. The actual position of the piston does not have any influence on the fact that at a certain position, the bubble will grow suddenly. From a mathematical point of view, the function $h = h\,(\Delta V)$ does not depend on the abscissa point ΔV. As the instability is defined by the shape of this function, instability is not characterized by parameter ΔV. We can therefore state that the existence of an instability, if any, could be influenced by two dimensionless numbers: $\widetilde{E}$ and $\widetilde{P}$.

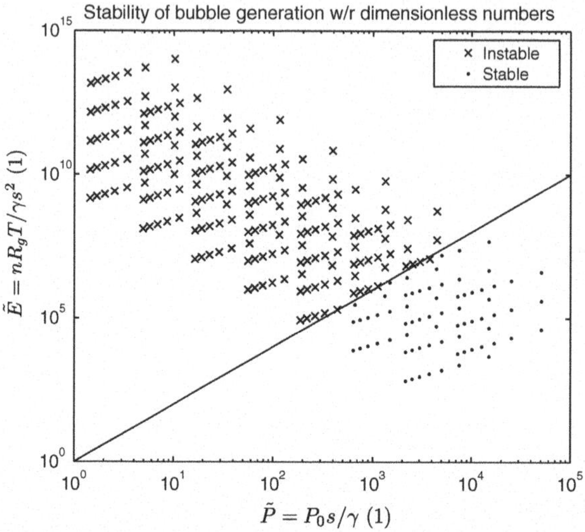

Fig. 5.8 We have run several simulations varying the input parameters. The *crosses* represent the configurations that led to an instability during bubble growing. The *dots* represent the configurations for which the growing was continuous. There is a clear border between both situations that can be approximated by a line

5.3.3 Numerical Approximation

In order to verify the existence of a threshold value for $\widetilde{E}$ and $\widetilde{P}$ beyond which an instability would occur, we made several simulations in which we varied the input parameters value. In each case, we checked if an instability occurs. The results are shown if Fig. 5.8. To identify the existence of an instability, we check the number of real solutions for (5.10): if there is more than one solution for a value of ΔV, we conclude the configuration lead to an unstable bubble growth.

In some cases, the combinations of parameters we used led to the same dimensionless values for $\widetilde{P}$ and $\widetilde{E}$. It is interesting to note that there is no $\widetilde{P}$, $\widetilde{E}$ couple leading to both stable and unstable growth. This confirms that only these two parameters are relevant to predict the existence of the instability.

There is a clear border between stable and unstable growth. This border can be approximated reasonably with a line in the bilogarithmic axes:

$$\log \widetilde{E} = 2 \log \widetilde{P} \tag{5.17}$$

Therefore, a stability condition regarding the experimental setup is:

$$\widetilde{P}^2 > \widetilde{E} \tag{5.18}$$

In its dimensional form, this criterion is:

$$\left(\frac{p_{\text{atm}}\, s}{\gamma}\right)^2 > \frac{n R_g T}{\gamma s^2} \tag{5.19}$$

This lead to a new dimensionless number, σ:

$$\sigma = \frac{p_{\text{atm}}^2 s^4}{\gamma\, n\, R_g\, T} \tag{5.20}$$

with the stability condition being simply:

$$\sigma > 1 \tag{5.21}$$

If the gas container is filled under atmospheric conditions,

$$n R_g T = p_{\text{atm}} V_0 \tag{5.22}$$

and σ can be rewritten in terms of easily measurable parameters:

$$\sigma = \frac{p_{\text{atm}} s^4}{\gamma V_0} \tag{5.23}$$

This criterion tells us that instability is more likely to occur if the gas container is large, the surface tension is high, the atmospheric pressure is low and the diameter of the injection point is small. A large surface tension and a small diameter will induce large surface tension effects, while a large gas container and a small atmospheric pressure will increase the effect of gas compressibility.

5.3.4 Analytical Approach

The results presented in Sect. 5.3.3 are only numeric. It is not possible to know if the criterion is correct for every single couple $\widetilde{P}, \widetilde{E}$ other than those presented in Fig. 5.8. Moreover, the line separating both regions is only approximative. Therefore we will also search for an analytic approach.

The method is based on the study of the partial derivative $\partial \widehat{V} / \partial \widetilde{h}$: in the case of an instability, the curve $\widetilde{h}\left(\widetilde{\Delta V}\right)$ has a vertical tangent corresponding to the condition:

$$\text{instability} \Leftrightarrow \exists \widetilde{h} \in \mathbb{R},\ \frac{\partial \widehat{V}}{\partial \widetilde{h}} = 0 \tag{5.24}$$

The partial derivative are:

$$\frac{\partial \widehat{V}}{\partial \widetilde{h}} = \frac{1}{16\left(\widetilde{h}^2 \widetilde{P} + 8\widetilde{h} + \widetilde{P}\right)^2} (-\pi\widetilde{P}^2\widetilde{h}^6 - 16\pi\widetilde{P}\widetilde{h}^5 - 3\pi\widetilde{P}^2\widetilde{h}^4 - 64\pi\widetilde{h}^4 - 32\pi\widetilde{P}\widetilde{h}^3$$
$$- 3\pi\widetilde{P}^2\widetilde{h}^2 - 64\widetilde{h}^2\pi - 16\widetilde{h}\pi\widetilde{P} + 128\widetilde{h}^2\widetilde{E} - \pi\widetilde{P}^2 - 128\widetilde{E}) \tag{5.25}$$

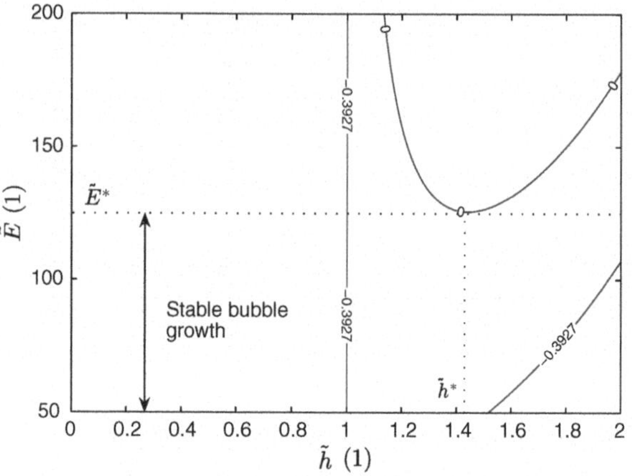

Fig. 5.9 Contour plot of $\partial\widehat{V}/\partial\widetilde{h}$ with respect to $\widetilde{h}$ and $\widetilde{E}$, in the case $\widetilde{P} = 10$. There is a range of $\widetilde{E}$ values for which the line $\widetilde{E}$ = constant has no intersection with the contour corresponding to $\partial\widehat{V}/\partial\widetilde{h} = 0$. The limit value is in this case between 100 and 150. The value of the partial derivative for $\widetilde{h} = 1$ is always $-\pi/8 \approx -0.3927$

It is remarkable that for $\widetilde{h} = 1$, $\partial\widehat{V}/\partial\widetilde{h} = -\pi/8$, whatever the values of $\widetilde{E}$ and $\widetilde{P}$. This means that an instability does not occur strictly at $h = s/2$ (the half-sphere configuration).

We consider a constant value of parameter $\widetilde{P}$. The condition (5.24) can be presented in a $\left(\widetilde{h}, \widetilde{E}\right)$ plane. Indeed, it is possible to plot the values of $\widetilde{h}$, $\widetilde{E}$ for which $\partial\widehat{V}/\partial\widetilde{h} = 0$. An example is shown in Fig. 5.9 for $\widetilde{P} = 10$.

The bubble growth will be stable if there is no intersection of the line $\widetilde{E}$ = constant with the contour $\partial\widehat{V}/\partial\widetilde{h} = 0$. We see in Fig. 5.9 that there is a range for $\widetilde{E}$ where this condition is satisfied. The limit value $\widetilde{E}^*$ is where the contour $\partial\widehat{V}/\partial\widetilde{h} = 0$ is minimal. This value corresponds to a value $\widetilde{h}^*$ that seems close to $\sqrt{2} \approx 1.41$. The condition (5.24) is equivalent to:

$$\forall\widetilde{P}, \text{instability} \Leftrightarrow \widetilde{E} > \widetilde{E}^*\left(\widetilde{P}\right) \tag{5.26}$$

This new criteria enables to determine an analytic expression of the criteria.

To find the value of $\widetilde{h}^*$, we consider the implicit equation $\partial\widehat{V}/\partial\widetilde{h} = 0$:

$$\begin{aligned}(&-\pi\widetilde{P}^2\widetilde{h}^6 - 16\pi\widetilde{P}\widetilde{h}^5 - 3\pi\widetilde{P}^2\widetilde{h}^4 - 64\pi\widetilde{h}^4 - 32\pi\widetilde{P}\widetilde{h}^3\\ &-3\pi\widetilde{P}^2\widetilde{h}^2 - 64\widetilde{h}^2\pi - 16\widetilde{h}\pi\widetilde{P} + 128\widetilde{h}^2\widetilde{E} - \pi\widetilde{P}^2 - 128\widetilde{E}) = 0\end{aligned} \tag{5.27}$$

from which we extract $\widetilde{E}_0 = \widetilde{E}_0\left(\widetilde{h}, \widetilde{P}\right)$, which is the equation of contour 0 in Fig. 5.9. We look for $\widetilde{h}^*$ such that $\partial\widetilde{E}_0/\partial\widetilde{h} = 0$:

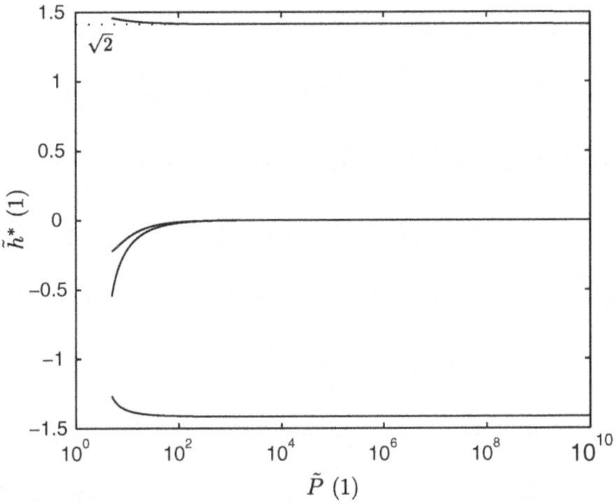

Fig. 5.10 $\widetilde{h}^*$ is the value at which the instability occurs (i.e. $\partial \widetilde{E}_0/\partial \widetilde{h}\;|_{\widetilde{h}=\widetilde{h}^*}= 0$). It is clear that $\widetilde{h}^*$ reaches quickly its asymptotic value $\sqrt{2}$. It is henceforth legitimate to state $\widetilde{h}^* = \sqrt{2}$ if $\widetilde{P} \geq 10$

$$\partial \widetilde{E}_0/\partial \widetilde{h} = \frac{\pi\left(\widetilde{h}^2\widetilde{P} + 8\,\widetilde{h} + \widetilde{P}\right)\left(\widetilde{P}\,\widetilde{h}^5 + 4\,\widetilde{h}^4 - \widetilde{P}\,\widetilde{h}^3 - 8\,\widetilde{h}^2 - 2\,\widetilde{P}\,\widetilde{h} - 4\right)}{32\left(\widetilde{h}-1\right)^2\left(\widetilde{h}+1\right)^2} \tag{5.28}$$

The first term of the numerator of (5.28) is equal to zero only for negative values of $\widetilde{h}$ which is not an acceptable solution. The second term will give us $\widetilde{h}^*$ with respect to $\widetilde{P}$. For large values of $\widetilde{P}$, the second term gives an asymptotic value of $\widetilde{h}^*$:

$$\begin{aligned}\lim_{\widetilde{P}\to\infty}\left(\widetilde{P}\,\widetilde{h}^5 + 4\,\widetilde{h}^4 - \widetilde{P}\,\widetilde{h}^3 - 8\,\widetilde{h}^2 - 2\,\widetilde{P}\,\widetilde{h} - 4\right) &= \widetilde{P}\left(\widetilde{h}^5 - \widetilde{h}^3 - 2\,\widetilde{h}\right)\\ &= \widetilde{P}\,\widetilde{h}\left(\widetilde{h}^2 - 2\right)\left(\widetilde{h}^2 + 1\right)\end{aligned} \tag{5.29}$$

so the only real positive value is $\widetilde{h}^* = \sqrt{2}$, as expected from Fig. 5.9. The evolution of $\widetilde{h}^*$ with respect to $\widetilde{P}$ is shown in Fig. 5.10. The asymptotic value is reached very quickly, and it is legitimate to assume $\widetilde{h}^* = \sqrt{2}$ if $\widetilde{P} \geq 10$.

Now if we return to (5.25) and replace in this partial derivative $\widetilde{h}$ by $\widetilde{h}^* = \sqrt{2}$, we find a relation between $\widetilde{E}$ and $\widetilde{P}$:

$$\frac{\partial \widehat{V}}{\partial \widetilde{h}}\;|_{\widetilde{h}=\widetilde{h}^*} = \frac{-384\,\pi + 128\,\widetilde{E} - 144\,\sqrt{2}\,\pi\widetilde{P} - 27\widetilde{P}^2}{16\left(8\,\sqrt{2} + 3\widetilde{P}\right)^2} = 0 \tag{5.30}$$

We can solve this equation with respect to $\widetilde{P}$ and find the relation that links $\widetilde{P}$ and $\widetilde{E}$ so that the instability exists. We note these values $\widetilde{P}^*$ and $\widetilde{E}^*$. The only real positive root is:

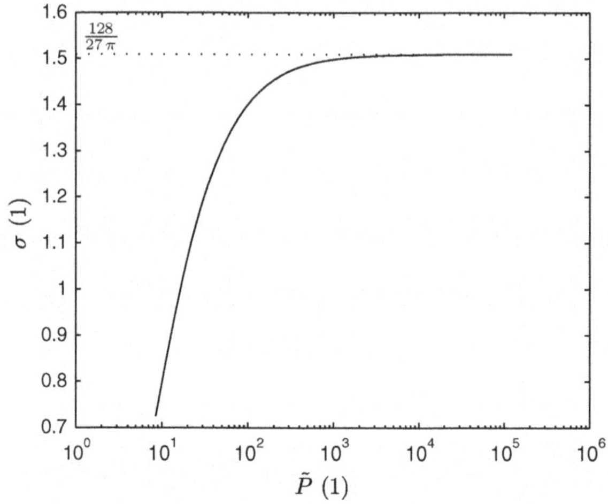

Fig. 5.11 This analytic study shows that the threshold value for the dimensionless number $\sigma = \widetilde{P}^2/\widetilde{E}$ depends on the dimensionless pressure $\widetilde{P}$. However, for values of $\widetilde{P}$ larger than 1000, σ tends to the asymptotic value $\frac{128}{27\pi}$

$$\widetilde{P}^* = \frac{8\left(-3\sqrt{2}\pi + \sqrt{6\pi\widetilde{E}^*}\right)}{9\pi} \tag{5.31}$$

Equation (5.31) is an analytical expression of the stability condition between $\widetilde{E}$ and $\widetilde{P}$. For large values of $\widetilde{E}^*$, it is possible to find an asymptotic value for the relation (see Fig. 5.11):

$$\widetilde{P_{lim}}^* = \frac{8\sqrt{6\pi\widetilde{E_{lim}}^*}}{9\pi} \tag{5.32}$$

We see that the criterion is close to the asymptotic value for $\widetilde{P} \geq 1000$. The stability criterion formulated from the numeric approximation is then:

$$\begin{aligned} \widetilde{P_{lim}}^{*2} &= \frac{128}{27\pi}\widetilde{E_{lim}}^* \\ \sigma &= \frac{128}{27\pi} \end{aligned} \tag{5.33}$$

It is slightly different from the value we found from the numerical approximation.

If we plot this criterion in Fig. 5.8, we see there are some values that cross the border. This is due to the numerical search of instability: if the configuration is almost stable, the range of h for which there are 3 solutions is short, and it is likely that the software jumps across this range, considering an unstable configuration to be stable (Fig. 5.12).

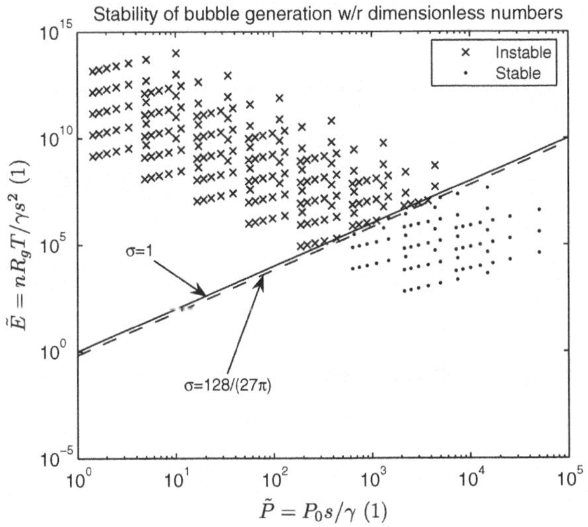

Fig. 5.12 Illustration of the analytical threshold between stable and unstable configurations. The error from numerical approximation obviously comes from the software instability detection, which considers stable an unstable configuration close to the limit

5.3.5 Stability Criterion

We have established a criterion σ allowing to predict if a volume controlled bubble generator will allow to grow bubbles continuously, or if the bubble will grow suddenly beyond a certain height.

The threshold of this criterion is constant for value of $\widetilde{P}$ larger than 1000: $\sigma = \frac{128}{27\pi}$.

5.4 Experimental Validation of the Volume Controlled Bubble Generation Model

To validate the model developed in Sect. 5.2.5.3 (Fig. 5.5), we made experiments on two different experimental setups.

5.4.1 First Experimental Setup

The first test bed is directly inspired from the syringe pump principle. It is represented schematically in Fig. 5.13.

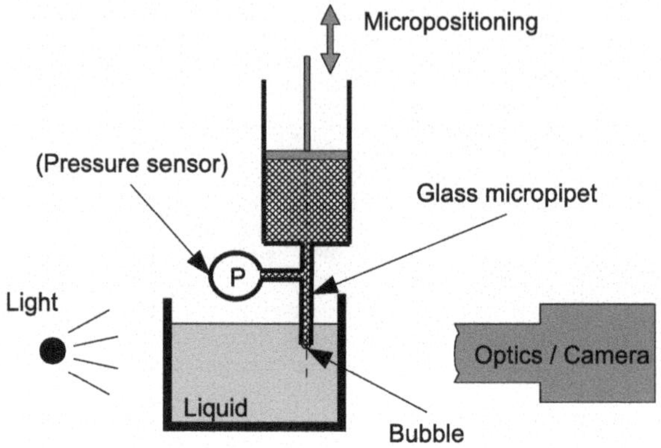

Fig. 5.13 Schematic representation of the first experimental setup for the validation of the volume controlled bubble generation model

The container is a syringe, and the volume variation is induced by the displacement of the piston. This displacement is operated manually using a micropositioning stage from *PI*. We used two different types of syringe:

- Syringe type "Exmire microsyringe MS-X05 ", volume 5 μL. Piston diameter is 0.35 mm (measured with a caliper)
- Syringe type "Narishige ITO", volume 800 μL. Piston diameter is 5 mm (measured with a caliper)

The outlet channel is a MDP-190-0 in vitro fertilization micropipette (Denuding micropipette), made of borosilicate glass. They have a nominal inner diameter of 190 μm.

The shape of the bubble is recorded with a magnifying optic mounted on a camera. The recorded images are analyzed on a computer with an own made Matlab code. The height of the bubble can therefore be correlated to the setup parameters and the position of the piston in the syringe.

A pressure sensor (*PMP-1400* from *Druck—GE*) has also been included later on the system, to correlate the curvature of the bubble to the pressure drop across its interface. The connection of the sensor to the rest of the system is presented in Fig. 5.14. The signal of the pressure transducer has been digitalised using a National Instrument USB acquisition system, with a sampling rate of 1 KHz. Then it has been filtered using a moving average on 25 samples.

As outlet channel, we have used *MDP-190-0* micropipettes. These pipettes should have a nominal inner diameter of 190 μm. However, when we measured the diameter with our calibrated vision system, it appears the diameter is only about 150 μm (measurement: $s = 155^{+3}_{-3}\,\mu\text{m}$). Only the measured value is used to compare the experimental results with the model. Examples of this pipette can be seen in Figs. 5.14 and 5.15.

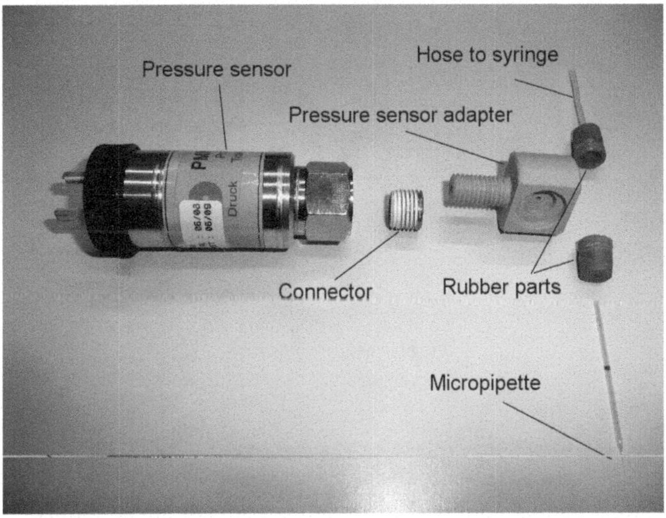

Fig. 5.14 View of the connection to the pressure sensor. This connection actually increases considerably the dead volume, mainly because of the rubber parts ensuring leak tightness with the micropipette and the flexible hose to the syringe, and because of the dead volume in the sensor itself

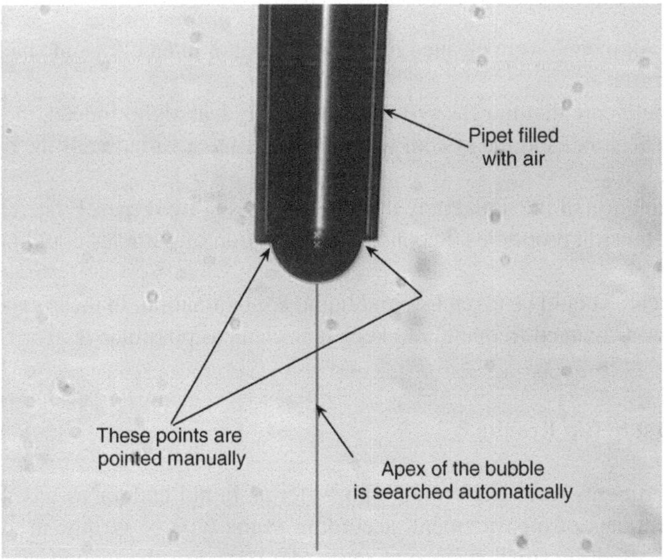

Fig. 5.15 Illustration of the image analysis software developed. The bubble grows at the end of the glass pipet. The user has to click on both ends of the bubble basis. In some case, the contrast is good enough to detect the apex of the bubble automatically. Otherwise, the user selects a third point of the bubble

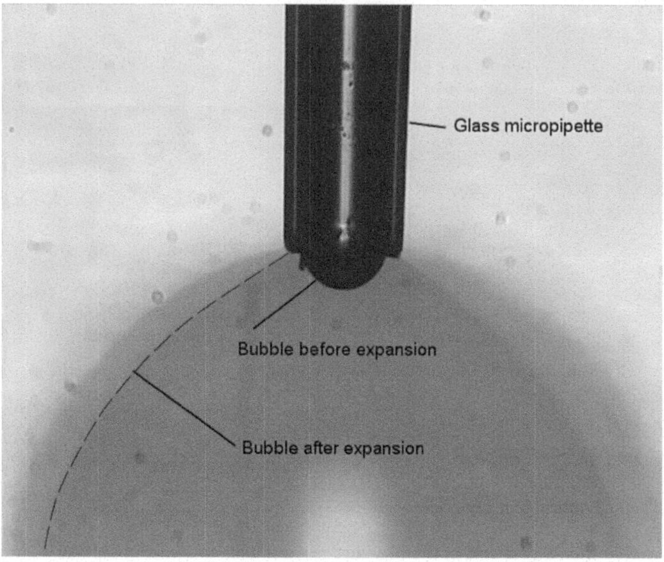

Fig. 5.16 View of a bubble growth instability. The video frame recorded the image of bubble before and after exploding

The micropipettes were cleaned in absolute ethanol in an ultrasonic bath for one minute before use.

One should care that the gas system is efficiently leak tight. Indeed, the volumes of gas may be small and the system will be sensitive to a variation of the number of gas molecules.

Contamination of the liquid may induce problems of two types. First, it may alter the surface tension properties. Second, any deposition of particles could jam up the gas channel.

Special care should be taken to avoid liquid contamination. In these experiments, the liquid was changed frequently to keep it as clean as possible.

5.4.1.1 First Setup Results

These experiments were performed with water as liquid and air as gas. We made two different sets of measurement, according to the type of syringe used. For the first set, we used the small syringe (*Exmire MS-X05*). For the second set, the large syringe (*Narishige ITO*) was used together with the pressure sensor, which makes the evaluation of the volume in the syringe less accurate due to the dead volume in the pressure sensor.

The parameters for both experimental setups are summarized in Table 5.3. The difference is essentially the initial volume V_0 and the section of the piston S, which induce a variation of the sensitivity of ΔV to the resolution of the positioning stage.

Table 5.3 Sets of parameters for experiments made on the first test bed

Parameters	Set 1	Set 2
V_0 (m^3)	47.6×10^{-9} (estimation including connections)	809×10^{-9} (estimation including connections)
S (m^2)	9.62×10^{-8}	1.96×10^{-5}
s (m)	150×10^{-6}	150×10^{-6}
σ	1.49×10^{-2}	1.28×10^{-3}

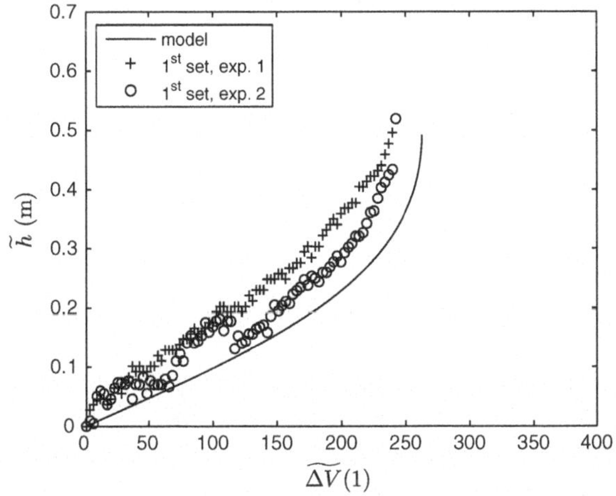

Fig. 5.17 Experimental validation of the volume controlled bubble generator, first setup, first set of parameters. The model fits well the experimentation points, except an horizontal offset whose origin is due to the manipulation

The other parameters are: $p_{\text{atm}} = 101325$ Pa (not measured), $\gamma = 72 \times 10^{-3}\,\text{Nm}^{-1}$, and the temperature which varied between 298 and 308 K (for theoretical curve, we used 298 K). The dimensionless number σ is also indicated in the table, and shows that these experiments should lead to unstable bubble growth (Fig. 5.16).

In Fig. 5.17 we have plotted the results of the first set of experiments, in dimensionless form. The experiment has been performed twice. The shape of the theoretical model shows good adequacy to the experimentation points, except that there seems to be an offset. This offset can be easily explained. The point at which we start the experimentation ($\Delta V = 0$) should correspond to a flat bubble ($h = 0$). But it is very difficult during the experiment to determine the exact zero height. An horizontal offset is henceforth acceptable.

In Fig. 5.18, we have plotted the results of the second set of experiments (first setup), in dimensionless form. The experiment has also been performed twice. In this setup, the pressure sensor has been included. We generated several bubbles successively. When the bubble expands due to instability, another bubble starts to grow.

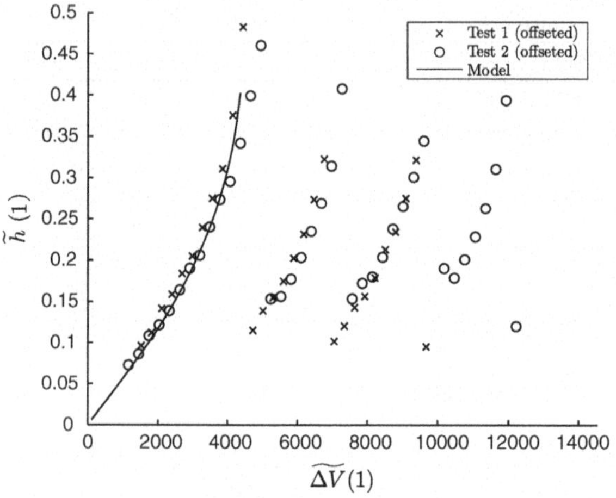

Fig. 5.18 Experimental validation of the volume controlled bubble generator, first setup, second set of parameters. For each of both repetitions, we regenerated other bubbles after each release due to the instability. The model fits well the experimentation points

To correct the offset error, we made a least squares research so that the model fits best the experimental points. We see that the model fits indeed well the experimentation points. The bubbles generated after the first one have similar behavior, but the results are not exactly the same. This is due to the variation of the parameters in the system, since some gas has been released.

Another result of this experimental setup is the pressure measurement. The results are shown in Fig. 5.19 for both experimentations. The pressure sensor was chosen to be used with smaller output channels as well. In this experimental setup, and because of technological limitations, we used only the large pipettes, and the pressure will rise only to around 2000 Pa, which is 2 % of the full scale range of the sensor. The signal was rater noisy. To filter it, we first made a moving average on 25 samples. Then we searched for the minimal and the maximal value on a 0.5 s time interval around the moment corresponding to the captured video frame.

The vertical bars represent the range of pressure value in this time interval of 0.5 s. The crosses represent the pressure deduced from image analysis, using Laplace equation (5.4). Pressure sensor measurements and image analysis measurements show good agreement for the first step of the experimentations.

We observe however that results from the pressure sensor tend to be larger than the results from image analysis for the next steps of the experimentations. Since the correspondence between both measurements relies only on Laplace equation there could be four explanations for this deviation. A modification of surface tension is possible, but the contamination of the liquid should lead to a lower value, which is the contrary of the measurement. Another explanation concerns an error in geometry measurement, such as image calibration error. But it should occur for every measurement, including the first one. Thirdly we have considered that the pressure measurement

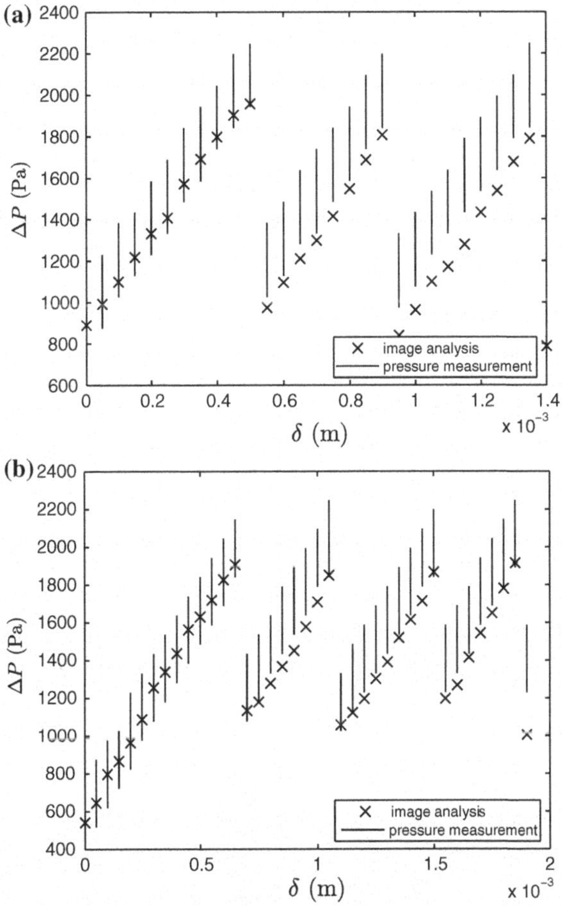

Fig. 5.19 Comparison of pressure measurement (**a** first repetition, **b** second repetition): crosses represent the pressure deduced from image analysis together with Laplace law, and the *vertical lines* show the results from pressure sensor measurement. The line stands for the range of value in a time interval around the video frame used for image analysis. The trend to overestimate pressure from pressure sensor measurement is most likely explained by a drift of the sensor, which is in this case used at only 2 % of its full scale

is not localized at the same place than the image measurement. Indeed the pressure sensor is connected to the gas system a few centimeters above the micropipette tip. Between these two locations, any free surface due to liquid or gas inclusion could modify the pressure. However in this case there is no reason why this problem should evolve during the experiment. Therefore the most likely explanation is a drift of the pressure sensor.

We can conclude that using an adequate pressure sensor, it is possible to measure the pressure, and henceforth the height of the bubble, from a distant location.

5.4.2 Second Experimental Setup

So far we have demonstrated the validity of the volume controlled bubble generator, but only in situations where an instability during the bubble growth occurs (i.e. $\sigma < 1$). In this case, we will demonstrate the continuous growing of a bubble in case of a small gas container and a larger outlet channel.

The second test bed is based on volume variation, but does not use syringe. It is schematically represented on Fig. 5.20. The bubble is generated from a platform made of aluminum which is immersed in the liquid (Fig. 5.21). This platform contains three through holes having a diameter of 1.23, 1.26, and 1.25 mm for hole 1, 2 and 3 respectively. These holes are each connected to a different flexible hose made of silicone and having an inner diameter of 0.5 mm. At the other end, the flexible hoses are connected to a large tank filled partially with liquid (water), partially with gas (air). The tank can be isolated with valves placed on the hoses.

The idea is that the gas container for the bubble generator is the flexible hose. The volume of gas is limited by pushing incompressible fluid from the large tank into the hoses until the gas remaining between the liquid from the tank and the outlet channel on the platform reaches the desired set point.

To vary the volume of the container and therefore push the gas out of the platform to generate bubbles, we only have to compress the hose where it contains liquid. To compress the hoses, we have placed them between a fixed wall and a micropositioning stage.

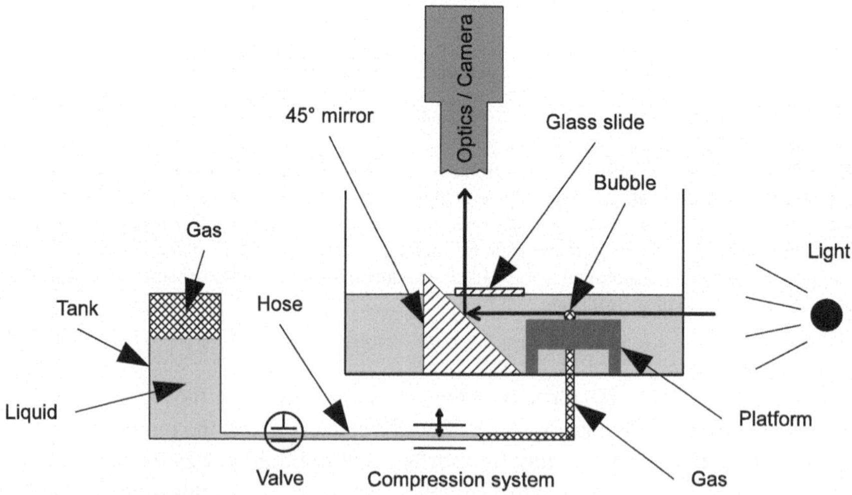

Fig. 5.20 Schematic view of the second experimental setup used to validate the volume controlled bubble generation model

Fig. 5.21 Stable bubble growth: when the bubble becomes larger than the capillary length, the hypothesis of spherical shape is not valid anymore. The pressure is larger at the *bottom*, creating the buoyancy force. The real bubble is taller than predicted by the model

5.4.2.1 Setup Issues

The second version of the syringe pump test bed has the advantage that the leakages should be reduced because of the few connections. It has also a rather small volume compared to some syringe, because of the low diameter of the flexible hose. However, it also has some issues, in addition to those of the first test bed.

Because of the actuation system, the allowed volume variations are small. It depends on the contact surface of the compression system. But of course, if we want to increase the volume that can be pushed, the sensitivity regarding the compression system displacement will increase. In our case, the hoses are compressed against the moving stage of a *Newport M-SDS40* micropositioning stage.

The calibration of the system is necessary since the ratio between the hose compression and the volume variation is non linear. Consequently we have monitored the displacement of the micropositioning stage with a *Keyence LC-2440* laser displacement sensor. We have filled the hose with an incompressible fluid (water), and we have captured image of the liquid droplet forming at the outlet channel on the platform. Hence the volume of the drop corresponds to the volume pushed by the compression system. To filter the results, we made a polynomial fitting to find a function giving ΔV with respect to the movement of the micropositioning stage. The correspondence between the table position and the volume pushed, together with the polynomial fittings, is shown in Fig. 5.22 for the first hole. The polynomials are degree 6 for first and second hole, and degree 9 for the third hole.

Because of the polynomial fitting, there are some ripples in the curves. But these ripples appear when the flexible is completely squeezed (the volume does not increase anymore), which is beyond the range of interest for our experiments.

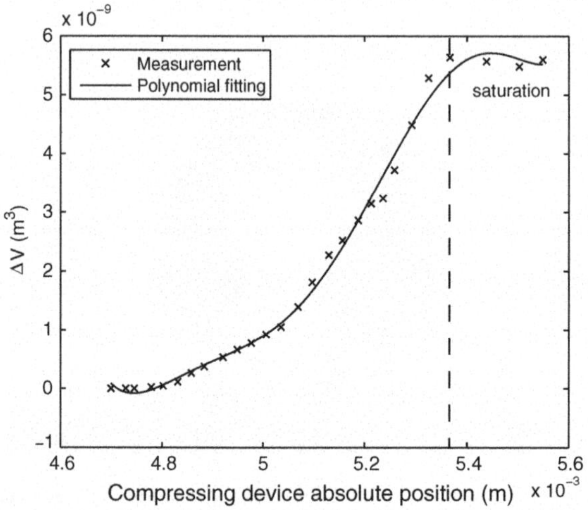

Fig. 5.22 Calibration of the compression system for volume variation and the polynomial fitting used during experimentations. The ripples are due to the polynomial fitting when the pushed volume does not increase anymore (saturation because the hose is fully squeezed). This is beyond the range of interest for our experiments

Table 5.4 Sets of parameters for experiments made on the second test bed

Parameters	Set 1	Set 2	Set 3
V_0	$98.2 \times 10^{-9}\,\text{m}^3$	$78.5 \times 10^{-9}\,\text{m}^3$	$58.9 \times 10^{-9}\,\text{m}^3$
s	$1.23 \times 10^{-3}\,\text{m}$	$1.23 \times 10^{-3}\,\text{m}$	$1.23 \times 10^{-3}\,\text{m}$
σ	32.9	41.2	54.9

A video camera is placed above the system, and a 45° mirror is used to view the bubbling area. This system is interesting if the area of interest is surrounded by other measurements devices, such as an atomic force microscope. However in that case, the optic path crosses the free surface of the liquid. We have seen experimentally that the images are more clear if we place a thin glass slide floating on the surface of the water, so that the optic path crosses no free liquid-air surface.

5.4.2.2 Second Setup Results

The experimental parameters of this setup have been summarized in Table 5.4. The three tests have been performed on the hole 1. The liquid we used is water and the gas is air. The only parameter changing from one experiment to the other is the initial volume. Pressure has been assumed to be $p_{\text{atm}} = 101325\,\text{Pa}$, temperature to be $T = 298.15\,\text{K}$, and surface tension to be $\gamma = 72 \times 10^{-3}\,\text{Nm}^{-1}$.

The results of the three experiments are shown on Fig. 5.23. It must be noted that when the bubble reaches a certain height larger around the capillary length, the hypothesis of part spherical shape is not correct anymore. This is why for large values of ΔV, the model underestimates the height of the bubble (Fig. 5.21). These reasons explain the deviation of the experimentation points and model for large ΔV values. For small values of ΔV, the bubble is pushed from the hole and is not immediately anchored on the edge of the border. This explains the deviations of the experimentation points and model for small values of ΔV.

5.4.3 Conclusion of Experimental Validation

The experiments performed on both experimental setups corroborate the model of bubble generation based on volume control, established in Sect. 5.2.5.3. Besides, we have seen that there exists, as predicted, two cases of bubble growth: one is discontinuous, as seen on the experimentations performed on the first test bed, the second is continuous as seen on the experimentations performed on the second test bed. Figure 5.24 shows a summary of two test results, each made on a different experimental setup. The dimensionless number established in Sect. 5.3.5 is useful to predict the continuous or discontinuous growth, based on the parameters of the setup. It shows, for example, that if a bubble of any size must be generated from a small hole, the gas volume must be lower than a value predicted by dimensionless criterion:

$$V_0 < \frac{27\,\pi\, p_{\text{atm}}\, s^4}{128\,\gamma} \tag{5.34}$$

We have also demonstrated that the pressure measurement is a good indicator of the bubble shape. This could allow to avoid the use of expensive and bulky image acquisition system. Besides, the pressure measurement can be localized anywhere in the gas system. This could be an advantage if for example the space around the bubble is restricted.

5.5 Application to Bubble Robotics

Bubble generation has been applied in order to generate three fluidic legs supporting a new microrobotic platform (Fig. 5.25).

This kind of platform actuated by capillary effects introduces compliance in microsystem, thanks to surface tension on the one hand and gas compressibility on the other hand. Combining this in a microrobotic approach, we currently work on a microrobotic platform with 6 degrees-of-freedom, trying to actively control 3 of them, as explained in the following.

Fig. 5.23 Experimental results on the second test bed. This validates the model for the continuous bubble generation. **a** Experimentation set 1. **b** Experimentation set 2. **c** Experimentation set 3

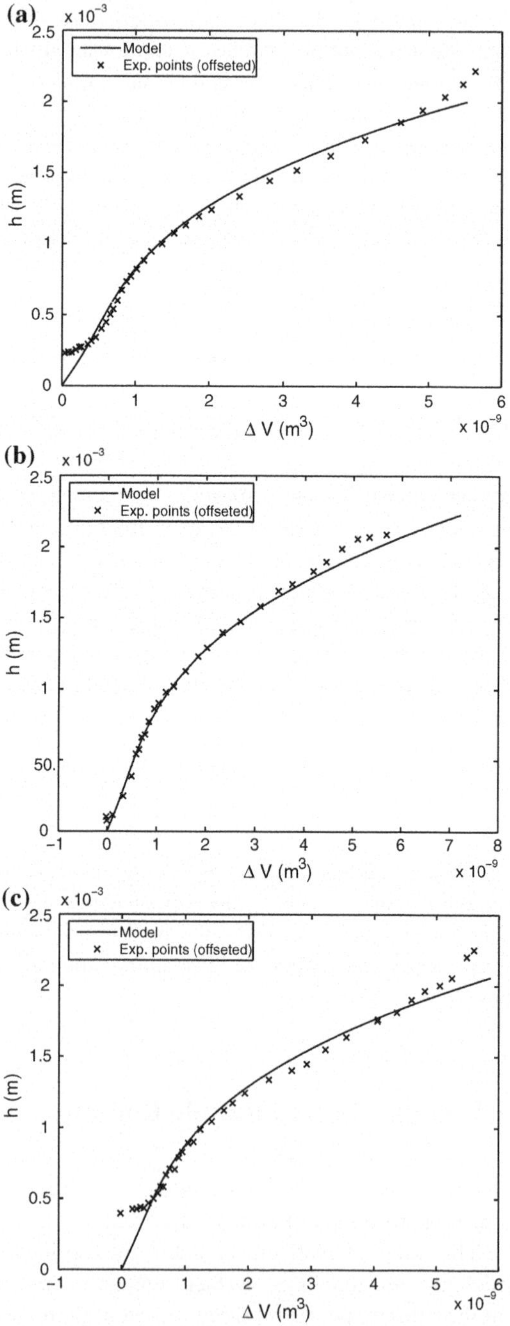

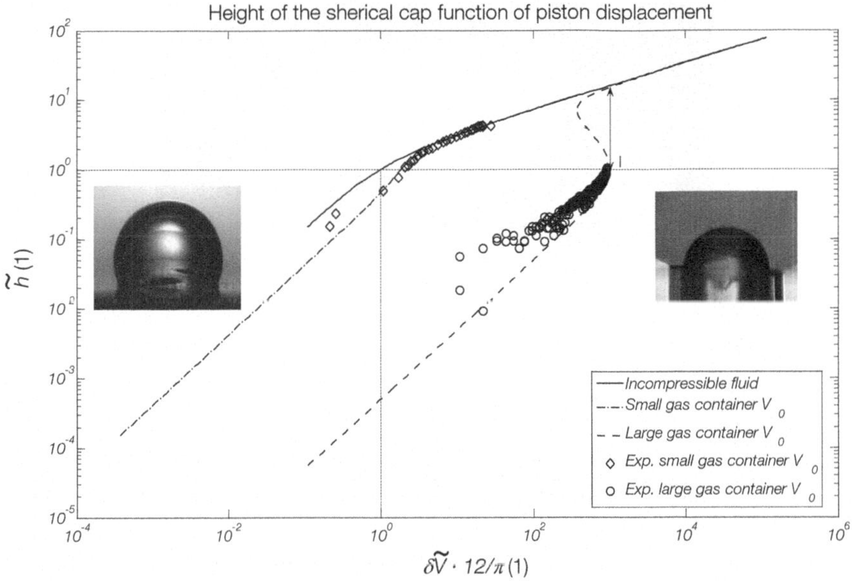

Fig. 5.24 Summary of experimental validation on both setup, allowing to validate both stable and unstable bubble growth. This figure combines the results of parameters set 1 in Table 5.3 and parameters set 2 in Table 5.4. The results have been plotted in a dimensionless form

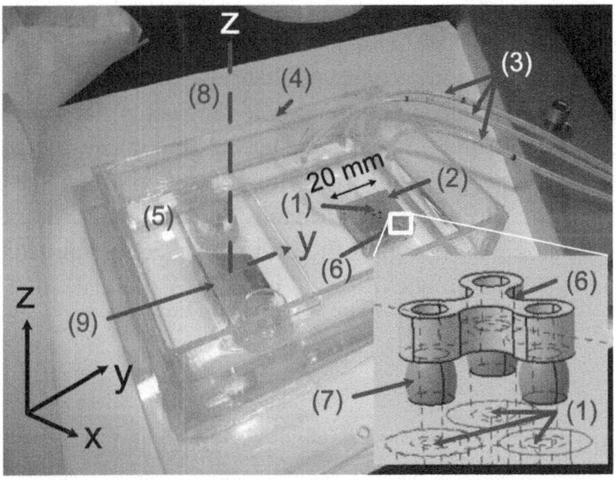

Fig. 5.25 General overview of the device: 3 holes (*1*) can be seen on the substrate (*2*), which are connected to the inlet tubes (*3*) which supply the gas. This substrate is immersed in a tank (*4*) containing silicone oil (*5*). Once the bubbles have been generated (they cannot be seen on the picture), the platform (*6*) is roughly positioned above the substrate. The fluidic legs (*7*) can be imaged with a camera whose optical axis is vertical (*8*), thanks to a mirror (*9*) tilted with 45°

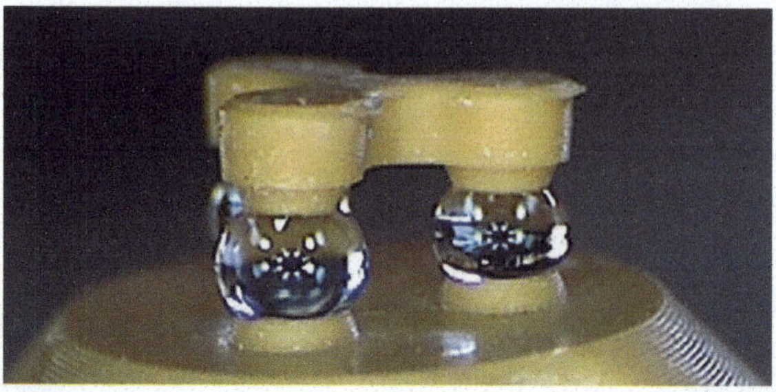

Fig. 5.26 Robulle, an example of Bubble Robotics: a millimetric platform is actuated along 3 degrees-of-freedom thanks to 3 liquid legs confined by surface tension (Courtesy Lycée Edgar Faure, Morteau, France)

Moreover, due to physics, we know that the proposed model allows to determine two unknowns among the position of the platform, the force exerted on it, the pressure in the fluidic legs and the number of gas molecules in the legs. If the latter may be known by calibration, this means that the knowledge of the pressure allows to compute the position and the force. The device acts in this case as a force-position sensor, combining surface tension and pressure effects. The device has 6 degrees of freedom (DOFs), among which, three are actuated: the z-axis translation having a stroke of a few hundreds of microns and θ_x and θ_y tilting up to about 15°. The platform is immersed in a liquid and placed on microbubbles whose shapes (e.g., height) are driven by fluidic parameters (pressure and volume). The modeling of this new type of compliant robot is fully described and compared with experimental measurements in a contribution recently published in Transactions on Robotics [6].

This development paves the way for an interesting actuation and robotic solution for submerged devices on the microscale. The fluidic leg can be seen as a natural miniaturization of macroscopic fluidic flexible actuators, in which the gas is confined by an elastic membrane. Here, this membrane is replaced by the liquid-gas interface ruled by surface tension. Our design is therefore intrinsically miniaturized.

Alternatively, legs can be made of liquid in air environment, such as depicted in Fig. 5.26.

5.6 Conclusions

In this chapter, we have extended the notion of capillary bridge to gas, and showed that it is important to take account of the compressibility of the gas. We have showed how to develop a model coupling gas law and Laplace law, while considering that the number of gas molecule is constant, not the volume.

This has been illustrated on the principle of generating a single bubble from a hole in a liquid, by pushing a volume of gas from a finite tank to this hole. Because of the gas compressibility, there is a chance that bubble growth will not be continuous and that an instability will occur.

We made a dimensionless analysis that demonstrated the existence of a threshold between continuous(stable) growth and instability. This criterion shows that a large initial volume and small size of hole are more likely to lead to an unstable growth, which confirms that this problem is especially related to microbubbles.

The existence of these two growing modes has been validated experimentaly, and demonstrate the need to take compressibility into account in the model of gaseous bridges.

Acknowledgments The author would like to thank the BRIC from Université libre de Bruxelles, and PHC-Tournesol funding (WBI-FNRS Egide) for their financial support.

References

1. C.N. Baroud, H. Willaime, Multiphase flows in microfluidics. C.R. Physique **5**, 547–555 (2004)
2. H. Bouazaze, S. Cattarin, F. Huet, M. Musiani, R.P. Nogueira, Electrochemical noise study of the effect of electrode surface wetting on the evolution of electrolytic hydrogen bubbles. J. Electroanal Chemi **597**, 60–68 (2006)
3. R. Darby, M.S. Haque, The dynamics of electrolytic hydrogen bubble evolution. Chem. Eng. Sci. **28**, 1129–1138 (1973)
4. D. Lapotko, E. Lukianova, Laser-induced micro-bubbles in cells. Int. J. Heat Mass Transf. **48**, 227–234 (2005)
5. S. Lee, W. Sutomo, C. Liu, E. Loth, Micro-fabricated electrolytic micro-bubblers. Int. J. Multiph. Flow **31**, 706–722 (2005)
6. C. Lenders, M. Gauthier, R. Cojan, P. Lambert, Three dof microrobotic platform based on capillary actuation. IEEE Trans Robot. **28**, 1157–1161 (2012)
7. S. Lubetkin, The motion of electrolytic gas bubbles near electrodes. Electrochim. Acta **48**, 357–375 (2002)
8. H.K. Nahra, Y. Kamotani, Prediction of bubble diameter at detachment from a wall orifice in liquid cross-flow under reduced and normal gravity conditions. Chem. Eng. Sci. **58**, 55–69 (2003)
9. T. Szirtes, *Applied Dimensional Analysis and Modeling* (MacGraw-Hill, New York, 1997)
10. R.B.H. Tan, W.B. Chen, K.H. Tan, A non-spherical model for bubble formation with liquid cross-flow. Chem. Eng. Sci. **55**, 6259–6267 (2000)
11. A. Volanschi, W. Olthuis, P. Bergveld, Gas bubbles electrolytically generated at microcavity electrodes used for the measurement of the dynamic surface tension in liquids. Sens. Actuators A **52**, 18–22 (1996)
12. Z. Xiao, R.B.H. Tan, An improved model for bubble formation using the boundary-integral method. Chem. Eng. Sci. **60**, 179–186 (2005)
13. Z.L. Yang, T.N. Dinh, R.R. Nourgaliev, B.R. Sehgal, Numerical investigation of bubble growth and detachment by the lattice-boltzmann method. Int. J. Heat Mass Transf. **44**, 195–206 (2001)

Part III
Dynamic Modelling of Capillary Forces

Chapter 6
Introduction

Jean-Baptiste Valsamis and Pierre Lambert

Abstract This chapter is an introduction to Part III devoted to dynamics. The dynamic study leads us to consider the liquid bridge as a mechanical joint presenting several degree of freedom. We here introduce the basis of a dynamical mechanical model and the governing equations of the fluid as well.

6.1 Liquid Bridge as Mechanical Joint

Liquid bridges generate capillary forces that can be repulsive or attractive according the numerous properties of the fluid and the solids (see Part II). These properties cover geometrical and physical aspects as well. From a mechanical point of view, a liquid bridge can be considered as a mechanical join between two solids (a substrate and a component). If the substrate is fixed, the component has six degrees of freedom with respect to the substrate: three translations x, y and z, and three rotations θ_x, θ_y and θ_z (Fig. 6.1).

We propose to study this joint as a second order system: Chap. 7 will be devoted to the study of the z-motion (axial dynamics) while Chap. 8 will study the radial motion.

Generally speaking, the equation governing an harmonic system excited by an input $x(t)$ is:

$$m\ddot{y}(t) + b\dot{y}(t) + ky(t) = x(t) \tag{6.1}$$

P. Lambert (✉) · J.-B. Valsamis
BEAMS Department, Université libre de Bruxelles, Brussels, Belgium
e-mail: pierre.lambert@ulb.ac.be

J.-B. Valsamis
e-mail: jvalsami@ulb.ac.be

P. Lambert (ed.), *Surface Tension in Microsystems*, Microtechnology and MEMS,
DOI: 10.1007/978-3-642-37552-1_6,

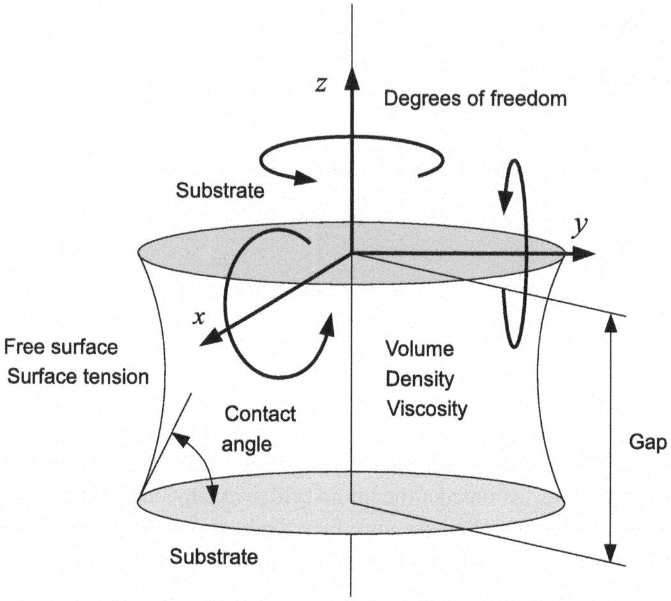

Fig. 6.1 Degrees of freedoms between the two solid interfaces. The substrates refer to solid interfaces, in general

With periodic input $x(t)$ of pulse $\omega = 2\pi f$, the system can be rewritten with phasors giving a transfer function of gain G and phase ϕ:

$$G(\omega) = \sqrt{\left(k - \omega^2 m\right)^2 + \omega^2 b^2} \tag{6.2}$$

$$\phi(\omega) = \arctan \frac{\omega b}{k - \omega^2 m} \tag{6.3}$$

The system is resonating when the mass m is higher than a critical mass m_c. We define the reduced mass $\hat{m}$:

$$m_c = \frac{b^2}{2k} \tag{6.4}$$

$$\hat{m} = \frac{m}{m_c} = \frac{2k}{b^2} m \tag{6.5}$$

- The system has low inertia if $m < m_c$ (or $\hat{m} < 1$). In this case (Fig. 6.2a, c), the system will have three steady states delimited by frequencies:

$$f_{c^-} = \frac{k}{2\pi b}\left(\frac{\sqrt{1+2\hat{m}} - 1}{\hat{m}}\right) \quad \rightarrow \frac{k}{2\pi b} \tag{6.6}$$

$$f_{c^+} = \frac{k}{2\pi b}\left(\frac{\sqrt{1+2\hat{m}} + 1}{\hat{m}}\right) \quad \rightarrow \infty \tag{6.7}$$

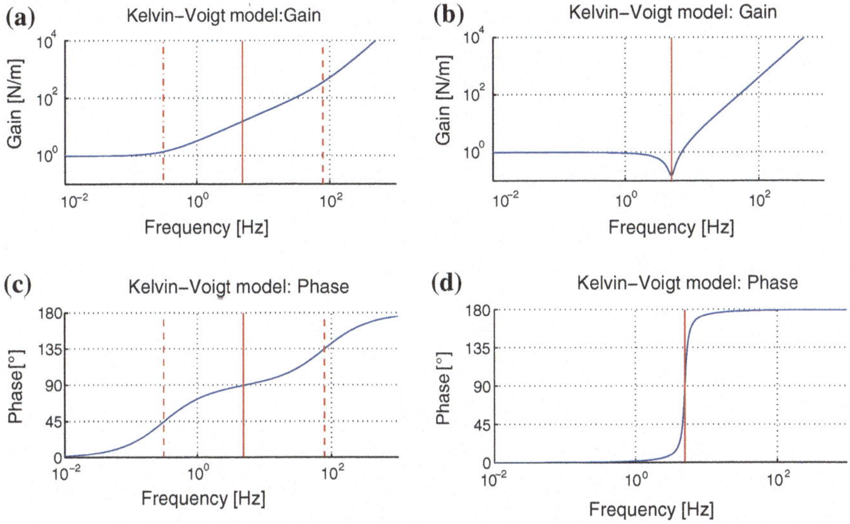

Fig. 6.2 Bode plots of the Kelvin-Voigt model for two configurations: $\hat{m} < 1$ (*left*) and $\hat{m} > 1$ (*right*). The *vertical line* is the resonant frequency and the *dashed lines* are the cut-off frequencies delimiting the three states. **a** Gain of transfer function with $k = 1\,\mathrm{N/m}$, $b = 5 \times 10^{-1}\,\mathrm{N\,s/m}$ and $m = 10^{-3}\,\mathrm{kg}$. **b** Gain of transfer function with $k = 1\,\mathrm{N/m}$, $b = 5 \times 10^{-3}\,\mathrm{N\,s/m}$ and $m = 10^{-3}\,\mathrm{kg}$. **c** Phase of transfer function with $k = 1\,\mathrm{N/m}$, $b = 5 \times 10^{-1}\,\mathrm{N\,s/m}$ and $m = 10^{-3}\,\mathrm{kg}$. **d** Phase of transfer function with $k = 1\,\mathrm{N/m}$, $b = 5 \times 10^{-3}\,\mathrm{N\,s/m}$ and $m = 10^{-3}\,\mathrm{kg}$

For each steady state, the gain can be approached by asymptotic lines:

$$G_k(\omega) \approx k \qquad \text{for} \quad \omega < \omega_{c^-} \tag{6.8}$$

$$\phi(\omega) \approx 0 \tag{6.9}$$

$$G_b(\omega) \approx \omega b \qquad \text{for} \quad \omega_{c^-} < \omega < \omega_{c^+} \tag{6.10}$$

$$\phi(\omega) \approx \frac{\pi}{2} \tag{6.11}$$

$$G_m(\omega) \approx \omega^2 m \quad \text{for} \quad \omega > \omega_{c^+} \tag{6.12}$$

$$\phi(\omega) \approx \pi \tag{6.13}$$

- For a high inertial system, $m > m_c$ (or $\hat{m} > 1$), the damping state is overridden by the two others states (Fig. 6.2b, d). The system will resonate at the frequency f_m when the gain is minimum. When $\hat{m}$ increases, this frequency f_m rapidly converges to f_r (the stiffness exactly compensates inertial effects, the phase is 90°).

$$f_m = \frac{k}{\sqrt{2}\pi b}\left(\frac{\sqrt{\hat{m}-1}}{\hat{m}}\right) \rightarrow f_r = \frac{k}{\sqrt{2}\pi b}\sqrt{\frac{1}{\hat{m}}} \tag{6.14}$$

At f_m, the gain is minimum and at f_r, the phase is 90°. The minimum of the gain is:

$$G(\omega_m) = \frac{k}{\hat{m}}\sqrt{2\hat{m} - 1} \approx \frac{\sqrt{2}k}{\sqrt{\hat{m}}} \tag{6.15}$$

For each steady state, the gain can be approached by asymptotic lines:

$$G_k(\omega) \approx k \qquad \text{for} \quad \omega < \omega_m \tag{6.16}$$

$$\phi(\omega) \approx 0 \tag{6.17}$$

$$G_m(\omega) \approx \omega^2 m \qquad \text{for} \quad \omega > \omega_m \tag{6.18}$$

$$\phi(\omega) \approx \pi \tag{6.19}$$

Any oscillating system is completely described by the knowledge of the coefficients k, b and m: the transfer function can be computed and the second order differential equation associated to the system can be easily solved (for example the behaviour of the system after a step input). There are several ways to determine these coefficients:

- **Analytically**: considering the physics, it is sometimes possible to simplify the governing equations with a proper geometry. For example, it is rather intuitive that the damping coefficient b can be associated to the viscosity of a fluid. Under certain assumptions, the Navier-Stokes equation can give an analytic expression.
- **Numerically**: the finite element method (for example) provides a way to solve the complete equation system and determine a specific solution of (6.1). Numerical simulations are powerful to validate approximations made in analytical approach.
- **Experimentally**: the system is excited physically (periodic input or step input). This way is used to validate the previous methods.

These coefficients can be determined by analysing the response of the system submitted to a specific input, matching the output with the theoretical one. In Chap. 7, the input is a periodic signal of variable frequency ((6.2) and (6.3) are used) to identify the coefficients k, b and m. In Chap. 8, the analysis of a step response is used and compared to the solution of (6.1) to identify the coefficient b.

6.2 Equations Governing Fluids

Fluids are governed by the Navier-Stokes equation and the mass conservation equation:

$$\rho\left(\frac{\partial \overline{u}}{\partial t} + (\overline{u}\cdot\overline{\nabla})\overline{u}\right) = \rho\overline{g} - \overline{\nabla}p + \overline{\nabla}\cdot\tau \tag{6.20}$$

$$\overline{\nabla}\cdot\overline{u} = 0 \tag{6.21}$$

Table 6.1 Terms of the Navier-Stokes equation

Term	Denomination	Description
$\rho\frac{\partial\overline{u}}{\partial t}$	Acceleration term	The force due to the temporal acceleration of a fluid particle
$\rho(\overline{u}\cdot\overline{\nabla})\overline{u}$	Convection term	The force due to the spatial variation of the velocity of a particle
$\rho\frac{\partial\overline{u}}{\partial t}+\rho(\overline{u}\cdot\overline{\nabla})\overline{u}$	Inertial term	The sum of above term constitutes the inertial forces
$-\overline{\nabla}p$	Pressure term	The force due to a variation of pressure
$\overline{\nabla}\cdot\tau$	Viscous term	The force due to internal friction
$\rho\overline{g}$	Gravity term	The force due to the gravity

Each term of this equation is a force per volume, the denomination is described in Table 6.1. Hence we will talk about viscous flow (or laminar), inertial flow, ...when the fluid is mainly governed by the viscous term, the inertial term, ...

The vector equation is developed in cartesian and cylindrical coordinates in Appendix A.

6.3 Fluid Dynamics Related Boundary Conditions

The Navier-Stokes equation and the flow conservation are applied on a domain Ω. To have a solution, the system must be accompanied by initial conditions (at a particular time usually $t = 0$) and boundary conditions on velocity and on pressure.[1] For these partial differential equations, there are three kinds of boundary conditions: the Dirichlet condition (that imposes a certain value on the boundary), the Neumann condition (that imposes a flux through the boundary) and a mix of both of them.

We summarise here all the boundary conditions usually encountered in fluid flow problems. In the next, p and $\overline{u}$ are the pressure and the velocity of the fluid, $\overline{1}_t$ and $\overline{1}_n$ are the normal and tangent vectors to the boundary, $\overline{v}_\Omega$ is the velocity of the boundary. An arbitrary value is denoted by *.

- **No slip**

$$\overline{u} = \overline{v}_\Omega \tag{6.22}$$

This condition means that both the velocities of the fluid and the boundary are identical.

[1] For steady-state (time is not a dimension of the problem), the initial conditions are useless.

- **Slipping**

$$\overline{u} \cdot \overline{1}_n = \overline{v}_\Omega \cdot \overline{1}_n \tag{6.23}$$

$$(\tau \cdot \overline{1}_n) \cdot \overline{1}_t = 0 \tag{6.24}$$

This condition means that the fluid is free to move along the boundary, but the normal velocities are equal. In addition, the tangential stresses are zero.

- **Viscous slipping**

$$\overline{u} \cdot \overline{1}_n = \overline{v}_\Omega \cdot \overline{1}_n \tag{6.25}$$

$$(\tau \cdot \overline{1}_n) \cdot \overline{1}_t = -\mu \frac{\overline{u} - \overline{v}_\Omega}{\beta} \cdot \overline{1}_t \tag{6.26}$$

When the slip goes along with friction, the friction force balances the shear stresses. The parameter β represents the slip length [1, 3]. As we mentioned earlier, this force is proportional to the relative tangential velocity between the fluid and the boundary. If this force is null, we recover the slipping case.

- **Inlet/outlet pressure**

$$p = p^* \tag{6.27}$$

$$\tau \cdot \overline{1}_n = 0 \tag{6.28}$$

This condition specifies a particular pressure at the boundary. The flow is free to flow across the boundary and the normal stress is zero.

- **Inlet/outlet velocity**

$$\overline{u} - \overline{v}_\Omega = \overline{u}^* \tag{6.29}$$

This condition specifies a particular relative velocity of the boundary. The no slip condition is a particular case for which $\overline{u}^* = 0$.

- **Surface force**

$$-p + (\tau \cdot \overline{1}_n) \cdot \overline{1}_n = f_n^* \tag{6.30}$$

$$(\tau \cdot \overline{1}_n) \cdot \overline{1}_t = f_t^* \tag{6.31}$$

This condition means that a force is applied on the boundary. The flow is free to flow across the boundary. Generally, the surface force is normal to the surface and $f_t^* = 0$. For a normal force exerted on a low viscous fluid, the force applied is equal to the pressure inside the fluid, as in inlet/outlet pressure condition. The reader will note the sign convention: the normal vector is outwards the domain. For example, $f_n^* < 0$ means the force is applied towards the fluid.

- **Surface tension interface**

$$-p + p_0 + (\tau \cdot \overline{1}_n) \cdot \overline{1}_n = -2H\gamma \tag{6.32}$$

$$(\tau \cdot \overline{1}_n) \cdot \overline{1}_t = 0 \tag{6.33}$$

$$\overline{v}_\Omega \cdot \overline{1}_n = \overline{u} \cdot \overline{1}_n \tag{6.34}$$

The particularity of microfluidics is the interface stress due to the surface tension. The description can be found in [2]. For a constant surface tension, we obtain the relations above. When the fluid is at rest, we recover the Laplace relation:

$$p - p_0 = 2H\gamma \tag{6.35}$$

There are however many cases where the surface tension is not constant in space. For example when there is gradient in the concentration of surfactants (such as soap) at the free interface, or when a gradient of temperature exists. The consequence is that the surface tension force is not only in the normal direction. The force due to a non constant surface tension is known to be the Marangoni force [3]:

$$\overline{F}_M = \overline{\nabla}\gamma \tag{6.36}$$

Given the free surface $S(e_1, e_2)$ defined by parameters e_1 and e_2, the interface stresses become:

$$(-p + p_0)\,\overline{1}_n + \tau \cdot \overline{1}_n = -2H\gamma\overline{1}_n + \frac{\partial \gamma}{\partial e_1}\overline{1}_{e_1} + \frac{\partial \gamma}{\partial e_2}\overline{1}_{e_2} \tag{6.37}$$

6.4 Overview of Next Chapters

Chapter 7 is devoted to the development of the harmonic model in the case of an axisymmetric liquid bridge excited by a vertical periodic input. The coefficients k, b, m are estimated analytically by simplifying the Navier-Stokes equation. The study is validated by experiments and numerical simulations of the system. We describe the experimental setup we designed to investigate vertical forces arising on the bottom pad from small periodic perturbations of the top pad confining the liquid meniscus. The setup allowed the accurate control of all physical and geometrical parameters relevant for the experiments. The parameters we investigated are both physical (viscosity and surface tension of the fluid) and geometrical (the edge angle between the meniscus and the pad, the height of the meniscus).

Chapter 8 is devoted to the study of lateral menisci oscillations, such as experienced in self-centering of components. The related results concern the expression of a characteristic time associated to the second order system, provided in a diagram as a function of only two governing non dimensional parameters (see Fig. 8.8).

Chapter 9 will draw the conclusion of this part on menisci dynamics.

Acknowledgments The work devoted to axial forces dynamics has been funded by a grant of the F.R.I.A.—*Fonds pour la Formation à la Recherche dans l'Industrie et l'Agriculture.*, while the study on lateral menisci dynamics was achieved among the framework of the European project.

References

1. COMSOL AB, *COMSOL Multiphysics V3.5a: User's Guide*, 2008
2. G.K. Batchelor, H.K. Moffatt, M.G. Worster, *Perspectives in Fluid Dynamics: A Collective Introduction to Current Research* (Cambridge University Press, Cambridge, 2003)
3. H. Bruus, *Theoretical Microfluidics* (Oxford University Press, Oxford, 2008)

Chapter 7
Axial Capillary Forces (Dynamics)

Jean-Baptiste Valsamis and Pierre Lambert

Abstract Up to now, we have only considered liquid bridges at equilibrium. Therefore, the forces applied by these menisci on solids they connect are only due to surface tension effects. If the meniscus is stretched or compressed with an important velocity or an important acceleration, viscous or inertial effects have to be added to capillary forces. Viscous effects will be governed by the viscosity of the liquid while inertial effects will be governed by density. Normally at small scales, density can be neglected, but attention must be paid in case of high accelerations, such those used in inertial micromanipulation [9]. From a mechanical point of view, these three terms form the well known Kelvin-Voigt model. After a brief state of the art, this chapter explains how to estimate the coefficients of this Kelvin-Voight model and compares the related results with numerical simulation and experiments.

7.1 Introduction

This chapter focuses on the behaviour of the vertical dynamic force generated by an axisymmetric liquid bridge. The next section describes the problem and the main assumptions on the meniscus. Section 7.3 gives the mathematical background of the Kelvin-Voigt Model. The analytic development of the coefficients k, b and m are given in Sects. 7.4, 7.5 and 7.6 respectively. Analytic relations are validated on a simple case in Sects. 7.7 and 7.8. An experimental setup (Sect. 7.9) has been performed to validate the model. Results are given in Sect. 7.10. A conclusion (Sect. 7.11) ends the chapter.

J.-B. Valsamis (✉) · P. Lambert
BEAMS Department, Université libre de Bruxelles, 50, av. F.D. Roosevelt, 1050 Brussels, Belgium
e-mail: jvalsami@ulb.ac.be

P. Lambert
e-mail: pierre.lambert@ulb.ac.be

P. Lambert (ed.), *Surface Tension in Microsystems*, Microtechnology and MEMS,
DOI: 10.1007/978-3-642-37552-1_7,

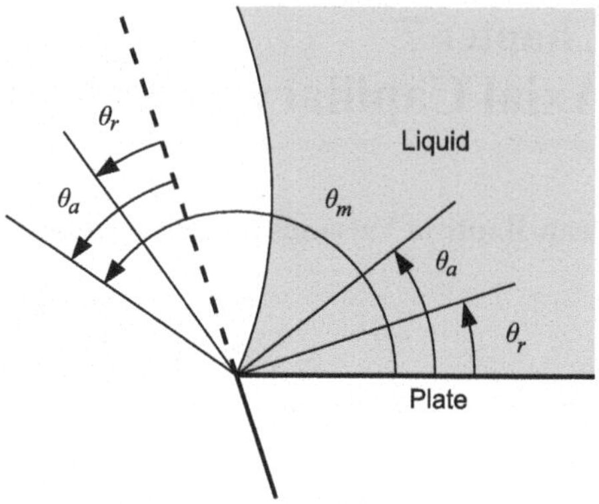

Fig. 7.1 Pinning contact angle. When a liquid is pinned, the contact angle can be higher than the advancing contact angle. This figure was published in [11]. Copyright © 2013 Elsevier Masson SAS. All rights reserved

7.2 Problem Description and Methodology

In this chapter, we will describe the liquid bridge as a mechanical join between two solids. The liquid bridge is assumed to be pinned by two circular and parallel interfaces providing an axial symmetry of the geometry around the z axis. The dimensions are submillimetric, 750 μm of radius and gap around 200 μm. The plate presents sharp edge to ensure the liquid to be pinned. This present the advantage to have a free contact angle. Indeed, the liquid will recede if the contact angle is below the receding angle θ_r and will advance if the contact angle is higher than the advancing angle θ_a (Fig. 7.1). Since the geometry present an edge, the advance of the liquid is done along the nearly vertical edge. The same argument can be done on this second edge. Thus, the liquid will spread out of the plate if the contact angle is higher than θ_m. In a certain way, the pinning increases the advancing contact angle.

For small displacements, the dynamic study of these degrees of freedom may be decoupled into 6 frequential responses. The latter can be described in several ways, depending on how the system is excited (the input) on one hand and on how the effect of the excitation is passed on (the output) on the other hand. Typical input/output are the position of an interface, its velocity, its acceleration and its force. The choice of position, velocity, acceleration of force is usually driven by the convenience of the way of experimental measurements. It is not necessary to characterise the transfer function for every pair of input/output: they can be retrieved through mathematical relationships (the velocity is the derivative of the position, the force is the product of the mass and the acceleration...).

We will consider the input as a displacement of the top interface and the output as the force exerted on the bottom interface. The liquid bridge is replaced by a Kelvin-Voigt model: a system made up of a spring, a damper dashpot and an equivalent mass connected in parallel (Fig. 7.3). According to liquid properties, it is expected to have a stiffness depending on the surface tension γ, a damping coefficient of the viscosity μ and an equivalent mass of the density ρ.

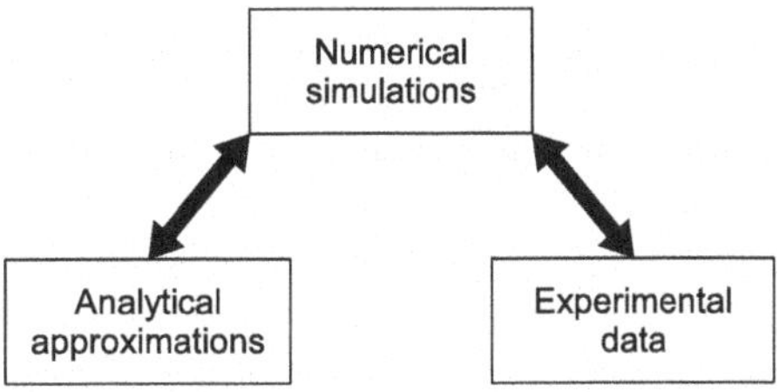

Fig. 7.2 Link between analytic, numeric and experimental results

The methodology adopted to validate the results consists of two steps (see Fig. 7.2): (i) a comparison between the analytical approximations of the mechanical parameters derived from the Navier-Stokes problem and numerical simulations, and (ii) a comparison between experimental and numerical data. That is, the numerical simulations act like a buffer between analytical expressions and experiments. The first comparison supports the simplifications made with the Navier-Stokes equation to obtain the analytic expressions of k, b and m. For the second comparison, we directly used experimental data in numerical simulations: due to experimental errors, the symmetry around the plane containing the neck ($z = 0$) was not exactly verified. The numerical simulations were performed with COMSOL MULTIPHYSICS 3.5a.

This problem has already been partially addressed: the literature review on vertical dynamics highlights the works of van Veen et al. [12], Cheneler et al. [1] and Pitois et al. [9].

In [12] van Veen has developed an analytical study of the motion of the flip chip soldered components. The model is based on the computation of the free surface energy for axisymmetric geometries delimited by two parallel and circular interfaces. The free interface is developed as a fourth order polynomial and its derivative is the origin of the motion equation. A second order equation governs the time evolution of the height of the bump. Cheneler et al. [1] propose a quite close analysis of liquid bridge with a totally different goal: they developed a micro-rheometer intended to determine the viscosity of liquids. However the device is not defined in the paper, the underlying idea is that the liquid added between two solids generates a dashpot in series with a calibrated spring/mass system. By measuring the phase shift between the position of the solids and the force exerted, the friction can be deduced. The method used is based on an estimation of the stiffness and the friction. This stiffness is deduced from an analytical approximation of the free interface, that is in this case a revolution of a part of a sphere (a piece of tore). However, the stiffness was calculated with a corrective multiplicative term of the capillary force, instead of the calculus of the derivative. The friction force is estimated from shear stresses inside the liquid. This friction and the stiffness have been linearized and added in the Newton equation governing the motion of the component wetted by the meniscus. Its validation is done thanks an numerical analysis. However, the numerical analysis does not solve the Navier-Stokes equation inside the meniscus. Finally, the authors do not support their study with any experimental data. In [9], Pitois et al. have studied the evolution of the force between two spheres moving aside at a given constant velocity. The analytical model involves also estimation of stiffness and friction. The stiffness is computed from the derivative of the capillary force, itself calculated with geometrical approximation of the different order of magnitude. The friction force is based on similar assumption but a corrective term is added. Contrarily to Cheneler, the liquid bridge is not pinned and the contact surface varies in time. The study is based on an analytical approach of the problem together with an experimental setup. As the main result, they highlighted different behaviours of the capillary forces (negative and positive, according the separating distance z), the influence of the velocity illustrating the effect of viscosity. Nevertheless, the authors do not expose some important aspects, such as the way to control the volume (volumes are small and the viscosity high, making the use of pipette very difficult), the variation of the contact angles during the separation and, as a corollary, the motion of the triple line, making the interpretation of their results difficult.

7.3 Mathematical Background of the Kelvin-Voigt Model

The mechanical model of the axial degree of freedom is presented in Fig. 7.3: a Kelvin-Voigt system made up of a spring (of stiffness k), a damper dashpot (of damping coefficient b) and an equivalent inertial force (of mass m) connected in parallel. Such a system can be described by its frequential response and is entirely defined when the coefficient k, b and m are known.

Typical input/output pairs are the position of an interface, its velocity, its acceleration and its force. The system can be completely described by the frequential response of a single input/output pair. Here, we will characterise the vertical translation considering as input **the displacement of the top interface** $x(t)$, and as output **the force exerted on the bottom interface** $F(t)$.

The gravitational effects can conversely be ignored because of the small dimension of the bridge. Furthermore, the gravitational force is completely static, without any effect on a dynamic study. The expression of the forces is:

$$\bar{f_k}(t) = -f_k(t)\bar{1}_x = -kx(t)\bar{1}_x \tag{7.1}$$

$$\bar{f_b}(t) = -f_b(t)\bar{1}_x = -b\dot{x}(t)\bar{1}_x \tag{7.2}$$

$$\bar{f_m}(t) = -f_m(t)\bar{1}_x = -m\ddot{x}(t)\bar{1}_x \tag{7.3}$$

The force exerted by the meniscus on the lower interface is thus:

$$\bar{F}(t) = F(t)\bar{1}_x = -\bar{f_k}(t) - \bar{f_b}(t) - \bar{f_m}(t) \tag{7.4}$$

In the next section, we will present a semi-analytical development to approximate the coefficients of the Kelvin-Voigt model describing the liquid bridge. The aim is to decouple the liquid properties, namely the surface tension γ, the viscosity μ and the

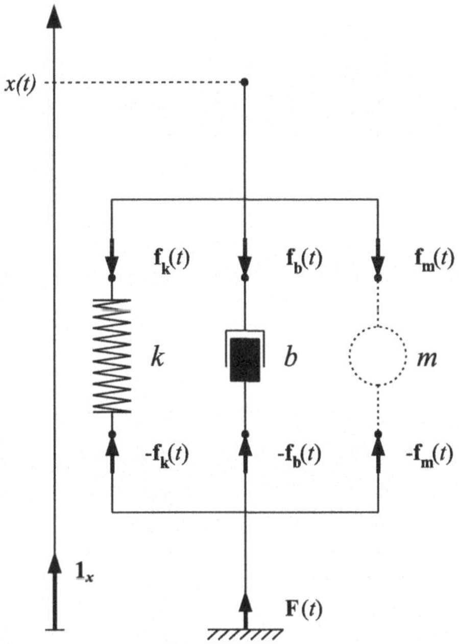

Fig. 7.3 The Kelvin-Voigt model. Forces $f_k(t)$ and $f_b(t)$ represent the force of the spring and of the *dashpot*, respectively, while $f(t)$ represents the total force exerted by liquid bridge on the system. Origin is assumed at the free length position. The direction of forces are represented for a stretched spring $x(t) > 0$, upwards velocity $\dot{x}(t) > 0$ and upwards acceleration $\ddot{x}(t) > 0$. This figure was published in [11]. Copyright © 2013 Elsevier Masson SAS. All rights reserved

density ρ, into the stiffness k (Sect. 7.4), the damping b (Sect. 7.5) and the inertial mass m (Sect. 7.6), respectively.

Briefly, the spring force considers the fluid at rest, the damping force considers only the viscous force driving the fluid, and the inertial force considers the force due to fluid acceleration. The Navier-Stokes equation is simplified consequently. The force applied by the fluid on the bottom interface is then computed, and the corresponding coefficients are derived. Gravitational effects are ignored, while inertial ones are contemplated where required.

7.4 Stiffness

7.4.1 Introduction

The force generated by a spring is independent of the speed at which its extremity moves. Therefore, to calculate the stiffness, it is appropriate to consider the liquid at rest. The pressure outside the liquid bridge p_{out} is assumed to be zero.

The geometry and the parameters are presented in Fig. 7.4. In addition to the axial symmetry, the geometry presents a symmetry with respect to the r axis, which further reduces the parameters in the study. The input parameters are the radius of

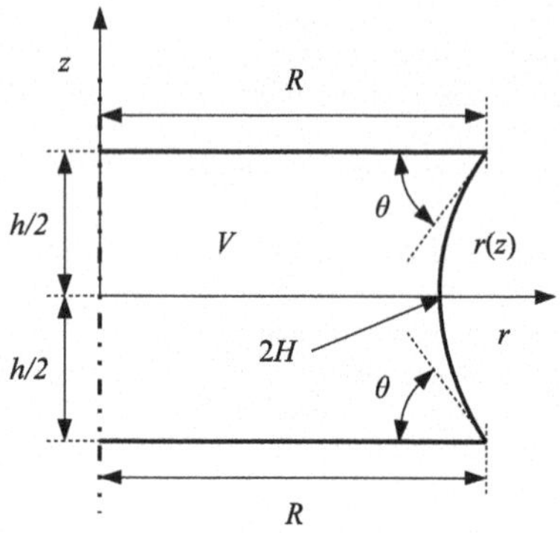

Fig. 7.4 Geometry and parameters for the analytic model of k. This figure was published in [11].

the plate R, the gap (or meniscus height) h, and the edge angle θ. By fixing these three parameters, the meniscus volume is automatically determined.

The meniscus stiffness k can be deduced from the derivative of the spring force f_k as follows:

$$k = -\frac{\mathrm{d}f_k}{\mathrm{d}h} = \frac{\mathrm{d}F}{\mathrm{d}h} \tag{7.5}$$

where h is the height of meniscus, $f_k = -kz$ and F is the total force exerted by the meniscus on the bottom pad.

Since the liquid is at rest, the inner pressure is only due to the curvature of the free interface:

$$p_{\text{in}} = 2H\gamma \tag{7.6}$$

As seen in Chap. 2, the force exerted by the liquid is thus the sum of the Laplace and the surface tension forces:

$$F = 2\pi\gamma R \sin\theta - 2H\gamma\pi R^2 \tag{7.7}$$

A positive force F means that the meniscus pull the substrate upwards while a negative force F means that the meniscus push the meniscus downwards. The stiffness k is given by:

$$k = \frac{\mathrm{d}F}{\mathrm{d}h} = 2\pi\gamma R \cos\theta \frac{\mathrm{d}\theta}{\mathrm{d}h} - \pi\gamma R^2 \frac{\mathrm{d}(2H)}{\mathrm{d}h} \tag{7.8}$$

The derivatives are done assuming the volume is constant. With the same convention,[1] the local curvature of an analytical axisymmetric shape $r(z)$ has already been given in (2.24):

[1] The sign of the curvature depends on the direction of the normal vector. The convention hereby adopted has the normal pointing outside of the liquid, giving a positive curvature for a sphere.

$$2H(z) = -\frac{r''(z)}{[1+r'^2(z)]^{\frac{3}{2}}} + \frac{1}{r(z)\sqrt{1+r'^2(z)}} \qquad (7.9)$$

Unfortunately, there is no analytical solution $r(z)$ and digital solving was requested.

7.4.2 Numeric Approach

The first method is the numerical integration of (7.9). The derivative of the curvature and the edge angle is computed by the finite difference method, leading to the results shown in Fig. 7.6.

Figure 7.5 shows the algorithm of the numerical approach. A first solution is computed with the boundary value problem `bvp4c` with the initial parameters (the radius R, the contact angle θ and the gap h). By giving an additional boundary value, the curvature $2H$ is handled as an unknown parameter to be determined by the solver. The volume V is derived from an integration of the shape (solution of `bvp4c`). The problem is solved again with new gap $(h + \Delta h)$ and an arbitrary starting value of the contact angle (the input value θ was used). For each contact angle, an output volume is computed. The correct contact angle is the one that keeps the volume constant. The search of this contact angle is done with `fzero`, with the contact angle as input and the volume difference to be zeroed. After reaching the solution, the derivatives of the curvature and the contact angles can be computed by finite difference.

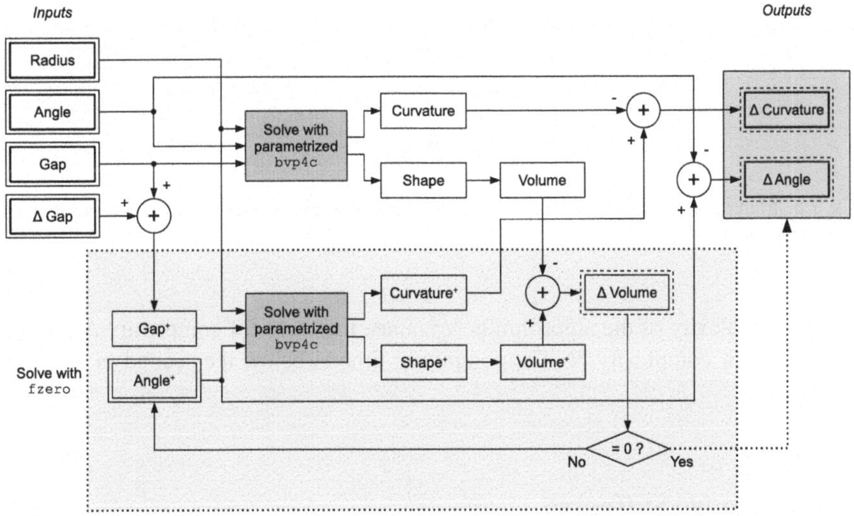

Fig. 7.5 Algorithm of numeric approach under MATLAB, using `bvp4c` and `fzero`

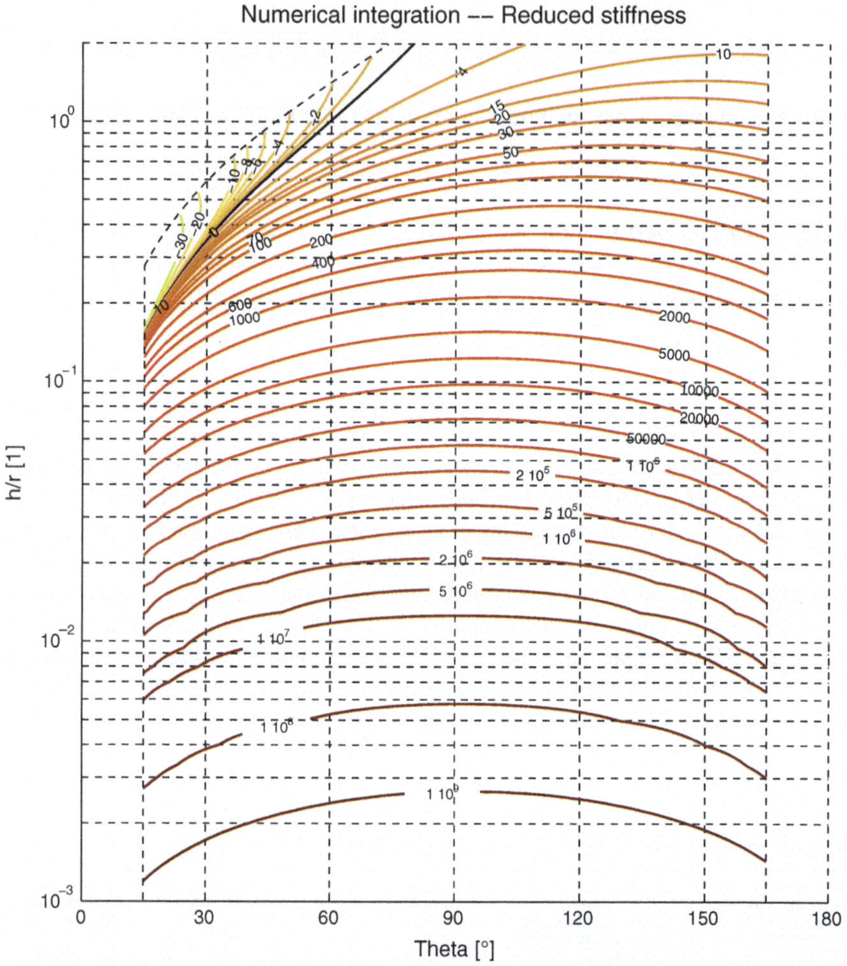

Fig. 7.6 Map of reduced stiffness $\hat{k} = k/\gamma$. This figure was published in [11]. Copyright © 2013 Elsevier Masson SAS. All rights reserved

The complexity of the algorithm is N^2, since the fzero (of complexity N) loops on `bvp4c` of complexity N. The computing time is below the second in most of cases.

7.4.3 Negative Stiffness

The reader will note that the stiffness coefficient is not always positive on Fig. 7.6. As illustrated in Fig. 7.7, when the gap increases, the meniscus curvature decreases, reducing the inner pressure. The second term of the derivative of the force (7.8) is always positive. On the contrary, the sign of the contribution of the surface tension force, expressed in the first term of (7.8), depend on the edge angle. The edge angle always decreases as the gap increases but the cosine is positive for angle below 90°. The contribution of the surface tension force is then negative when $\theta < 90°$ and can be bigger than the Laplace force for high gap.

Figure 7.8a represents a part of the map of the reduced force $\hat{F} = F/r\gamma$ according to the edge angle θ and the reduced height $\hat{h}$. The bold line is the evolution of the θ when the gap increases at constant volume. The evolution of the force along this curve is represented in Fig. 7.8b (plain line). As the gap increases, the force reaches a maximum and begins to decrease. The derivative (i.e., the stiffness) is therefore negative (dashed line).

Consequently, for certain configurations the meniscus can be considered as an *anti-spring*. Anti-springs are unstable because the force tends to deviate from the equilibrium state. However, if the anti-spring is mechanically constrained—as in the case of a meniscus—it is not unstable: the gap is fixed externally, whatever the force.

The effect of the negative stiffness on Bode plots is depicted in Fig. 7.9 for a low inertial and high inertial system (plain lines). In addition, the dashed lines recall the plots of Fig. 6.2 (positive stiffness).

For a low inertial system, the gains (Fig. 7.9a) are identical. The phase (Fig. 7.9c) starts at 180°, reaches 90° during the b-state and finishes at 180° in the m-state.

For a high inertial system, k-state and m-state interact because the b-state vanishes. Since the phase of the k and m-states are 180°, their effects are additive and the gain (Fig. 7.9b) does not show any minimum at the resonant frequency. The phase (Fig. 7.9d) is constant at 180° except near the resonant frequency.

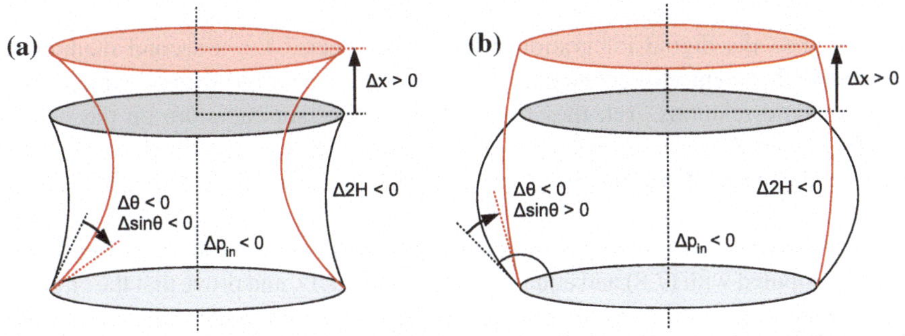

Fig. 7.7 Evolution of edge angle and inner pressure as the gap increases at constant volume. This figure was published in [11].

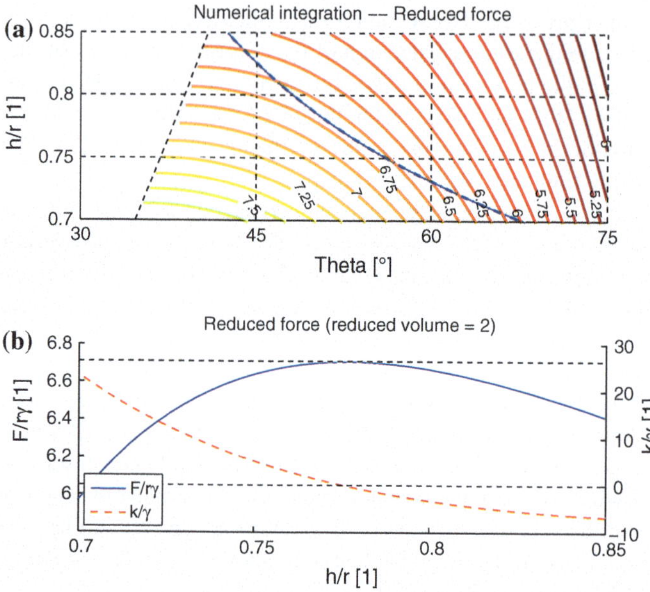

Fig. 7.8 Illustration of origin of negative stiffness. **a** Map of the reduced force $\hat{F} = F/r\gamma$ according to the edge angle θ and the reduced height $\hat{h} = h/r$. The *dashed line* is a constant volume line. **b** Evolution of the reduced force and reduced stiffness along the constant volume line, in the direction of increasing gap. This figure was published in [11]. Copyright © 2013 Elsevier Masson SAS. All rights reserved

7.4.4 Geometrical Assumptions

Compared to the digital integration presented in Sect. 7.4.2, a second method is to assume the meniscus geometry, i.e. parabolic or circular shapes, which provides analytic relations. Yet, they do not have a constant curvature on the whole interface since they are not solution of (7.9). The curvature and its derivative are taken at the neck of the meniscus ($z = 0$). These relations are summarised in Table 7.1.

The relative errors on the parabolic and circular model are shown on Fig. 7.10a and b (computed with (7.8) and equations from Table 7.1), and prove that the circular approximation is more accurate. Special care must be used in the evaluation of the relative error near the region where the stiffness is zero: by definition, any small difference of value produces an error tending to infinity. The error tends to zero when the shape approaches a cylinder. The circular model gives an error below 30 % for form factor $\hat{h} < 0.1$.

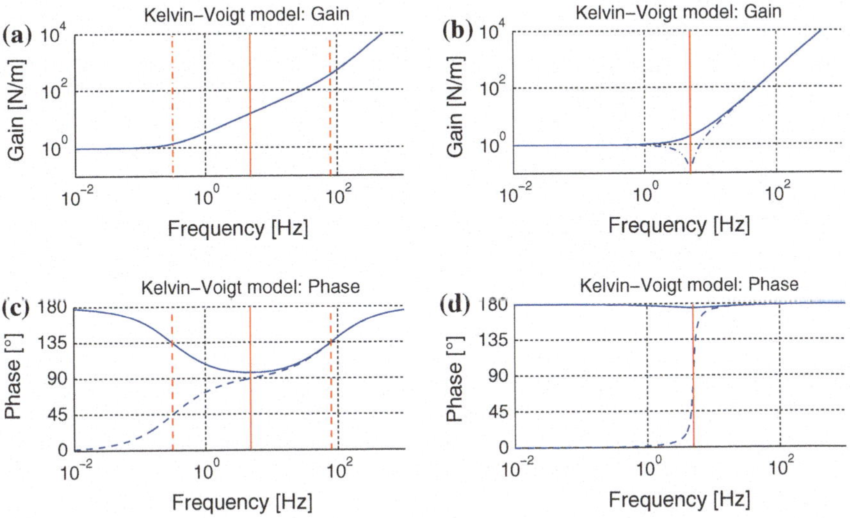

Fig. 7.9 Bode plots of the Kelvin-Voigt model within the stiffness is negative (*plain lines*) or positive (*dashed lines*), for two configurations: $\hat{m} < 1$ (*left*) and $\hat{m} > 1$ (*right*). The *vertical line* is the resonant frequency and the *dashed lines* are the cut-off frequencies delimiting the three states. **a** Gain of transfer function with $k = -1$ N/m, $b = 510^{-1}$ Ns/m and $m = 10^{-3}$ kg **b** Gain of transfer function with $k = -1$ N/m, $b = 510^{-3}$ Ns/m and $m = 10^{-3}$ kg. **c** Phase of transfer function with $k = -1$ N/m, $b = 5\,10^{-1}$ N s/m and $m = 10^{-3}$ kg. **d** Phase of transfer function with $k = -1$ N/m, $b = 5\,10^{-3}$ N s/m and $m = 10^{-3}$ kg

Table 7.1 Analytical expressions from the parabolic model and the circular model

Parabolic model	$\theta \to \frac{\pi}{2}$
$\frac{\mathrm{d}\theta}{\mathrm{d}h} = -\sin\theta \frac{30R^2\sin^2\theta - 20Rh\sin\theta\cos\theta + 3h^2\cos^2\theta}{10Rh^2\sin\theta - 2h^3\cos\theta}$	$\to -\frac{3R}{h^2}$
$\frac{\mathrm{d}2H}{\mathrm{d}h} = 2\cot\theta\left(\frac{1}{h^2} + \frac{2}{(4R - h\cot\theta)^2}\right) + \frac{2h}{\sin^2\theta}\left(\frac{1}{h^2} - \frac{2}{(4R - h\cot\theta)^2}\right)\frac{\mathrm{d}\theta}{\mathrm{d}h}$	$\to \frac{3}{4Rh} - \frac{6R}{4h^3}$
Circular model	
$\frac{\mathrm{d}\theta}{\mathrm{d}h} = -\frac{4R^2\cos^4\theta + 4Rh\sin\theta\cos^3\theta + 3h^2\cos^2\theta - h^2\cos^4\theta}{3h^3\sin\theta\cos\theta + 4Rh^2\cos^2\theta + \left(\frac{\pi}{2} - \theta\right)\left(2h^3\cos^2\theta - 3h^3 - 2Rh^2\sin 2\theta\right)}$ $- \frac{-\left(\frac{\pi}{2} - \theta\right)\left(3h^2\sin\theta + 4Rh\cos\theta\right)}{3h^3\sin\theta\cos\theta + 4Rh^2\cos^2\theta + \left(\frac{\pi}{2} - \theta\right)\left(2h^3\cos^2\theta - 3h^3 - 2Rh^2\sin 2\theta\right)}$	$\to -\frac{3R}{h^2}$
$\frac{\mathrm{d}2H}{\mathrm{d}h} = \frac{2\cos\theta}{h^2} + \frac{2\sin\theta}{h}\frac{\mathrm{d}\theta}{\mathrm{d}h} + \frac{1 - \sin\theta}{(2R\cos\theta - h(1 - \sin\theta))^2}\left(2\cos\theta - 2h\frac{\mathrm{d}\theta}{\mathrm{d}h}\right)$	$\to \frac{3}{4Rh} - \frac{6R}{4h^3}$

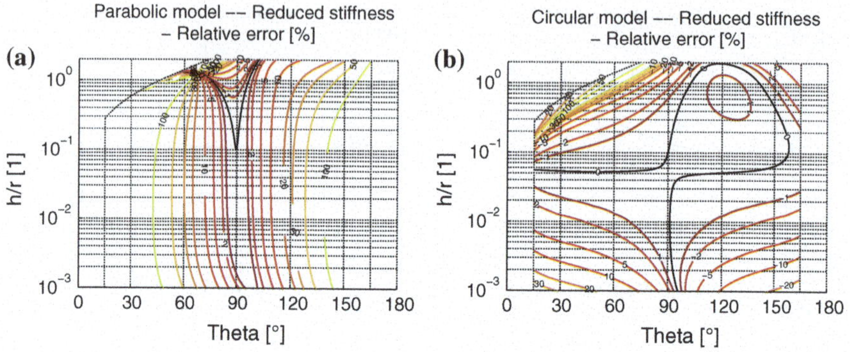

Fig. 7.10 **a** Relative error on the reduced stiffness computed from parabolic model $(\hat{k}_{pm} - \hat{k})/\hat{k}$. **b** Relative error on the reduced stiffness computed from circular model $(\hat{k}_{cm} - \hat{k})/\hat{k}$. Relative error of analytical models according to the edge angle θ and the reduced height $\hat{h} = h/r$, in logarithmic scale. This figure was published in [11].

7.5 Damping Coefficient

Physically, the damping coefficient is related to any friction effect. In our case, it highlights the effect of the viscosity of the fluid, as given in (7.40). We will assume the surface tension having an negligible impact of the viscous forces: as explained in Sect. 7.3 the spring force is defined by the position of the system, the damping force by its velocity and the inertial effect by its acceleration. For a periodic excitation $h = H\cos(\omega t)$, the maximal displacement is H, the maximal velocity is $H\omega$ and the maximal acceleration is $H\omega^2$. There is necessarily a pulse $\omega > \omega_{c^-}$ (see Chap. 6) for which the pressure inside the meniscus is driven only by viscous effects, vanishing the static effect of the surface tension. In addition, if $\omega < \omega_{c^+}$ the acceleration term can also be neglected, so that the Navier-Stokes equation contains only pressure and viscous terms. The 2D axisymmetric continuity and momentum equations read (full description of these equations are given in Appendix A):

$$\frac{1}{r}\frac{\partial(ru_r)}{\partial r} + \frac{\partial u_z}{\partial z} = 0 \tag{7.10}$$

$$-\frac{\partial p}{\partial r} + \mu\left(\frac{\partial^2 u_r}{\partial r^2} + \frac{\partial^2 u_r}{\partial z^2} + \frac{1}{r}\frac{\partial u_r}{\partial r} - \frac{u_r}{r^2}\right) = 0 \tag{7.11}$$

$$-\frac{\partial p}{\partial z} + \mu\left(\frac{\partial^2 u_z}{\partial r^2} + \frac{\partial^2 u_z}{\partial z^2} + \frac{1}{r}\frac{\partial u_z}{\partial r}\right) = 0 \tag{7.12}$$

The considered geometry is a cylinder whose parameters are depicted in Fig. 7.11. The contact angle fixed to 90°. Although a velocity field appears in the meniscus,

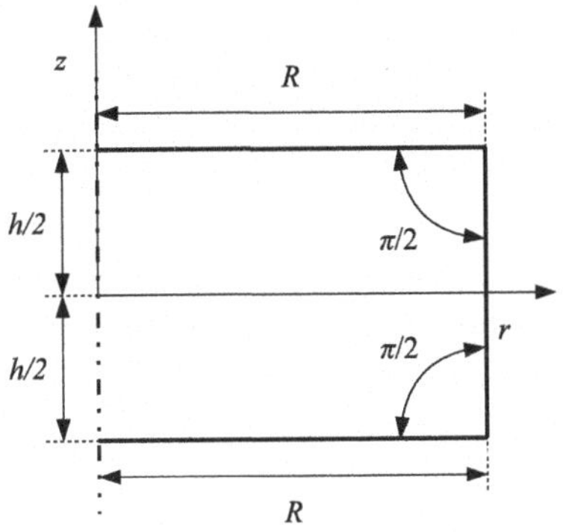

Fig. 7.11 Geometry and parameters for the analytic model of b and m

the amplitude is small and we will consider the deformation of the meniscus small enough to be ignored: the geometry is constant.

The force applied by the liquid on the bottom plate is defined by the sum of all the constrains on the plate:

$$F\bar{1}_z = \left[\int_\Gamma (-p\mathrm{I} + \tau)\cdot \mathrm{d}\bar{\mathrm{S}}\right]\cdot \bar{1}_z \tag{7.13}$$

$$= \left[\int_\Gamma \left(-p + 2\mu\frac{\partial u_z}{\partial z}\right) \mathrm{dS}\right]\bar{1}_z \tag{7.14}$$

In the complete review of Engmann et al. [5], the authors propose some assumptions on the velocity profile of a film of liquid squeezed by two parallel plates at constant velocity. The vertical velocity field (i.e. the z component) is assumed to be independent on r:

$$\bar{u} = u_r(t,r,z)\bar{1}_r + u_z(t,z)\bar{1}_z \tag{7.15}$$

Applying the continuity Eq. (7.10) (in the following, the $'$ will refer to the derivative with respect to z):

$$u_r = -\frac{r}{2}u_z'(t,z) \tag{7.16}$$

Using (7.11) and (7.12):

$$-\frac{\partial p}{\partial r} - \mu\frac{r}{2}u_z'''(t,z) = 0 \tag{7.17}$$

$$-\frac{\partial p}{\partial z} + \mu u_z''(t,z) = 0 \tag{7.18}$$

The last equation gives:

$$\frac{\partial^2 p}{\partial r \partial z} = 0 \tag{7.19}$$

so that (7.11) becomes finally:

$$u_z''''(t,z) = 0$$
$$\Rightarrow \quad u_z(t,z) = a(t)z^3 + b(t)z^2 + c(t)z + d(t) \tag{7.20}$$

We need 4 boundary conditions to determine a, b, c and d. They are provided by the velocity on upper and lower plates:

$$u_z(t, -\frac{h}{2}) = 0 \tag{7.21}$$
$$u_z(t, \frac{h}{2}) = \dot{h}(t) \tag{7.22}$$
$$u_r(t, r, -\frac{h}{2}) = 0 \quad \Rightarrow \quad u_z'(t, -\frac{h}{2}) = 0 \tag{7.23}$$
$$u_r(t, r, \frac{h}{2}) = 0 \Rightarrow u_z'(t, \frac{h}{2}) = 0 \tag{7.24}$$

where $\dot{h}(t)$ is the top plate velocity. The full velocity profile is finally given by:

$$\bar{u}(t,r,z) = \dot{h}(t)r\left(3\frac{z^2}{h^3} - \frac{3}{4h}\right)\bar{1}_r + \dot{h}(t)\left(-2\frac{z^3}{h^3} + \frac{3}{2}\frac{z}{h} + \frac{1}{2}\right)\bar{1}_z \tag{7.25}$$

To compute the force, pressure and velocity derivatives are required. The derivatives with respect to r and z are:

$$\frac{\partial \bar{u}}{\partial r} = \dot{h}(t)\left(3\frac{z^2}{h^3} - \frac{3}{4h}\right)\bar{1}_r \tag{7.26}$$
$$\frac{\partial \bar{u}}{\partial z} = 6\dot{h}(t)\frac{rz}{h^3}\bar{1}_r + \dot{h}(t)\left(-6\frac{z^2}{h^3} + \frac{3}{2h}\right)\bar{1}_z \tag{7.27}$$

The pressure is obtained by integrating (7.17):

$$p(r,t,z) = 3\mu\dot{h}(t)\frac{r^2}{h^3} + f(t,z)$$

With (7.18), $f'(t,z)$ is known. Consequently:

$$p(r,t,z) = 3\mu\dot{h}(t)\frac{r^2}{h^3} - 6\mu\dot{h}(t)\frac{z^2}{h^3} + C(t) \tag{7.28}$$

The last constant $C(t)$ (constant with respect to r and z) is found thanks to the stress equilibrium at the free interface:

$$-p + (\tau \cdot \bar{1}_r) \cdot \bar{1}_r = -p_0 = 0 \tag{7.29}$$

This expression is generally not satisfied. Indeed we assumed a particular vertical velocity profile that does not necessarily match with the condition at the interface [5]. However, several hypotheses can be found on the flow to define the constant C:

1. The total stress at the boundary is averaged and is equal to the outer pressure, giving C_1.
2. The total pressure at the boundary is averaged and is equal to the outer pressure, giving C_2.
3. The pressure is balanced at a particular point, that is the bottom triple line $(R, -\frac{h}{2})$ in this case, giving C_3.

The first condition leads to:

$$\int_{\Gamma} -p \, \mathrm{dS} + \int_{\Gamma} (\tau \cdot \bar{1}_r) \cdot \bar{1}_r \, \mathrm{dS} = 0 \tag{7.30}$$

$$\Leftrightarrow \quad C_1(t) = \mu \dot{h}(t)\Big(-3\frac{R^2}{h^3} - \frac{1}{2h}\Big) \tag{7.31}$$

With the second assumption, the constant $C_2(t)$ is:

$$C_2(t) = \mu \dot{h}(t)\Big(-3\frac{R^2}{h^3} + \frac{1}{2h}\Big) \tag{7.32}$$

For the third case, the viscous term is null at the triple line (at the top and the bottom plates as well). The balance of the pressure at the triple line is:

$$p(t, R, \pm\frac{h}{2}) = 0$$

$$\Leftrightarrow \quad C_3(t) = \mu \dot{h}(t)\Big(-3\frac{R^2}{h^3} + \frac{3}{2h}\Big) \tag{7.33}$$

If the ratio $R/h >> 1$, then $C_1(t) \approx C_2(t) \approx C_3(t)$ and the three assumptions give identical pressure field. Indeed, the height is small with respect to the radius, and the variation of the pressure due to the interface condition becomes negligible compared to the pressure induced by the friction forces inside the volume. According the boundary condition, at the bottom interface, the pressure is:

$$p_1(t, r, -\frac{h}{2}) = -3\frac{\mu \dot{h}(t)}{h^3}(R^2 - r^2) - 2\frac{\mu \dot{h}(t)}{h} \tag{7.34}$$

$$p_2(t, r, -\frac{h}{2}) = -3\frac{\mu \dot{h}(t)}{h^3}(R^2 - r^2) - \frac{\mu \dot{h}(t)}{h} \tag{7.35}$$

$$p_3(t, r, -\frac{h}{2}) = -3\frac{\mu \dot{h}(t)}{h^3}(R^2 - r^2) \tag{7.36}$$

The viscous stresses are null on the interface, the force is the integration of the pressure on the bottom pad:

$$F_n(t) = \frac{3\pi}{2}\frac{\mu\dot{h}(t)R^4}{h^3} + B_n\pi\frac{\mu\dot{h}(t)R^2}{h} \tag{7.37}$$

where n is the case number corresponding to the n$'h$ boundary condition and the constant B_n is defined by:

$$B_n = \begin{cases} 2 & \text{for the zero averaged stress, case (1)} \\ 1 & \text{for the zero averaged pressure, case (2)} \\ 0 & \text{for the zero pointwise pressure, case (3)} \end{cases} \tag{7.38}$$

The force applied by the liquid is a viscous force. In the Kelvin-Voigt developed applied to this configuration, the force due to the damper is $f_b(t) = b\dot{h}(t)$. The damping coefficient is consequently:

$$b_n = \frac{3\pi}{2}\frac{\mu R^4}{h^3} + B_n\pi\frac{\mu R^2}{h} \tag{7.39}$$

Here again, when $R/h >> 1$, the last term of (7.39) can be ignored and $b_1 \approx b_2 \approx b_3 \approx b$ whatever the boundary conditions:

$$b \approx \frac{3\pi}{2}\frac{\mu R^4}{h^3} \tag{7.40}$$

7.6 Inertial Effect

For the last coefficient of the Kelvin-Voigt model, m, we will assume the fluid to be governed by its density only. Following the argument developed in the previous section, there is a sufficiently high frequency ($\omega > \omega_{c^+}$) beyond which only inertial term (containing ω^2) will drive the pressure inside the meniscus. The 2D axisymmetric Navier-Stokes equation will contain only the inertial term and the pressure term. However, the convective term contains $(\bar{u} \cdot \bar{\nabla})\bar{u} \propto H^2\omega^2/h$ while the acceleration term contains $\dot{\bar{u}} \propto H\omega^2$. For a very small amplitude H, the convective term vanishes.

The geometry is identical to the one used to describe the damping coefficient (Fig. 7.11). The 2D axisymmetric continuity and momentum equations read:

$$\frac{1}{r}\frac{\partial(ru_r)}{\partial r} + \frac{\partial u_z}{\partial z} = 0 \tag{7.10}$$

$$\rho \frac{\partial u_r}{\partial t} = -\frac{\partial p}{\partial r} \tag{7.41}$$

$$\rho \frac{\partial u_z}{\partial t} = -\frac{\partial p}{\partial z} \tag{7.42}$$

When inertia is important, a laminar flow is characterised by a boundary layer on the wall.[2] The thickness of the layer tends to zero when the velocity increases, so we will neglect the boundary layer. The reader will note that the thinner the layer is, the stronger the viscous forces are. Fortunately, those viscous forces are tangent to the surface and are not considered in calculus of the force normal to the surface.

Similarly to Sect. 7.5, the vertical velocity field u_z is still assumed to be independent on r (7.15), the mass conservation Eq. (7.10) leads to:

$$\bar{u} = -\frac{r}{2} u_z'(t,z)\bar{1}_r + u_z(t,z)\bar{1}_z \tag{7.43}$$

Using (7.41) and (7.42):

$$\frac{\partial p}{\partial r} - \rho \frac{r}{2} \dot{u}_z'(t,z) = 0 \tag{7.44}$$

$$\frac{\partial p}{\partial z} + \rho \dot{u}_z(t,z) = 0 \tag{7.45}$$

The last equation gives:

$$\frac{\partial^2 p}{\partial r \partial z} = 0 \tag{7.46}$$

so that (7.41) becomes finally:

$$\begin{aligned} &\dot{u}_z''(t,z) = 0 \\ \Rightarrow \quad &\dot{u}_z(t,z) = a(t)z + b(t) \end{aligned} \tag{7.47}$$

We need two boundary conditions, given by the velocity on upper and lower plates. The acceleration of the top plate is fixed by $\ddot{h}(t)$:

$$\dot{u}_z\left(t, -\frac{h}{2}\right) = 0 \tag{7.48}$$

$$\dot{u}_z\left(t, \frac{h}{2}\right) = \ddot{h}(t) \tag{7.49}$$

The acceleration profile $\dot{\bar{u}}$ then writes:

---

[2] In a pipe submitted to a constant inlet pressure, the parabolic profile of velocity is the profile minimising the friction inside the flow. When the viscosity is low, the friction is too small compared to inertia, leading to a quasi-uniform profile with boundary layers. The thickness of these layers decreases with the Reynold number.

$$\dot{\bar{u}}(t,r,z) = -\frac{r}{2h}\ddot{h}(t)\bar{1}_r + \frac{z}{h}\ddot{h}(t)\bar{1}_z \tag{7.50}$$

The radial acceleration is constant on cylinder of radius r (surface normal to $\bar{1}_r$) and the axial acceleration is constant on plane normal to $\bar{1}_z$. The pressure is obtained by integrating (7.44):

$$p(t,r,z) = \rho\ddot{h}(t)\frac{r^2}{4h} + f(t,z) \tag{7.51}$$

With (7.45), the $f'(t,z)$ is known. Consequently:

$$p(t,r,z) = \rho\ddot{h}(t)\left(\frac{r^2}{4h} - \frac{z^2}{2h}\right) + C(t) \tag{7.52}$$

The last constant $C(t)$ (constant with respect to r and z) is found thanks to the stress equilibrium at the free interface:

$$-p = -p_{\text{out}} = 0 \tag{7.53}$$

As for the damping coefficient, this condition is generally not satisfied (we assumed a particular vertical velocity profile that does not necessarily match with the condition at the interface [5]). The hypotheses to define the constant C are:

1. The total pressure at the boundary is averaged and is equal to the outer pressure, giving a constant C_1.
2. The pressure is balanced at a particular point, that is the bottom triple line $(R, -\frac{h}{2})$ in this case, giving a constant C_2.

Constant C_1 is:

$$\int_\Gamma -p \, \mathrm{dS} = 0$$

$$\Leftrightarrow \quad C_1(t) = -\rho\ddot{h}(t)\left(\frac{R^2}{4h} - \frac{h}{24}\right) \tag{7.54}$$

And C_2 is:

$$p\left(t, R, -\frac{h}{2}\right) = 0$$

$$\Leftrightarrow \quad C_2(t) = -\rho\ddot{h}(t)\left(\frac{R^2}{4h} - \frac{h}{4}\right) \tag{7.55}$$

The force at the bottom is computed by considering only the pressure in (7.14). With constants $C_1(t)$ and $C_2(t)$, this reads:

$$F_1(t) = \pi R^2 h\rho\ddot{h}(t)\left(\frac{R^2}{8h^2} - \frac{1}{6}\right)$$

$$F_2(t) = \pi R^2 h\rho\ddot{h}(t)\left(\frac{R^2}{8h^2} + \frac{1}{8}\right)$$

If the equivalent mass is subject to the acceleration $\ddot{h}$, from 7.3 and 7.4 we have:

$$m_1 = \pi R^2 h\rho\left(\frac{R^2}{8h^2} - \frac{1}{6}\right) \tag{7.56}$$

$$m_2 = \pi R^2 h\rho\left(\frac{R^2}{8h^2} + \frac{1}{8}\right) \tag{7.57}$$

When $R/h >> 1$ the stress balance assumptions give converging results towards an equivalent mass m equal to:

$$m \approx m_1 \approx m_2 \approx \frac{\rho\pi R^4}{8h} \tag{7.58}$$

7.7 Case Study and Numeric Validation of the Simplified Kelvin-Voigt Model

The analytical approximations established in Sects. 7.4–7.6 result from the approximated computation of the Navier-Stokes equations. These equations have been reduced considering only the relevant terms in k, b or m-state, assuming some simplifications:

- The k-state characterises the stiffness of the meniscus. The liquid is at rest (velocity is zero), the pressure in the fluid is due to the free interface.
- The b-state characterises the friction of the meniscus. The velocity profile assumes the z component to be independent of r. The pressure in the fluid is due to the viscous term.
- The m-state characterises the inertia of the meniscus. The velocity profile assumes the z component to be independent of r. The pressure in the fluid is due to the accelerating term (just the second time-derivative term).

Practically, k, b and m can be found with Fig. 7.6, Eqs. (7.40) and (7.58), and henceforth the asymptotic gains defined in (6.8)–(6.18). At a specific pulse ω, the working state is defined by the largest gain G_k, G_b or G_m.

To go further these analytical approximations—and also to validate them digitally —, the computation of Navier-Stokes equations should be solved completely by numeric computation method like the finite elements method. We have applied this methodology to the case study of a cylindrical liquid bridge whose physical

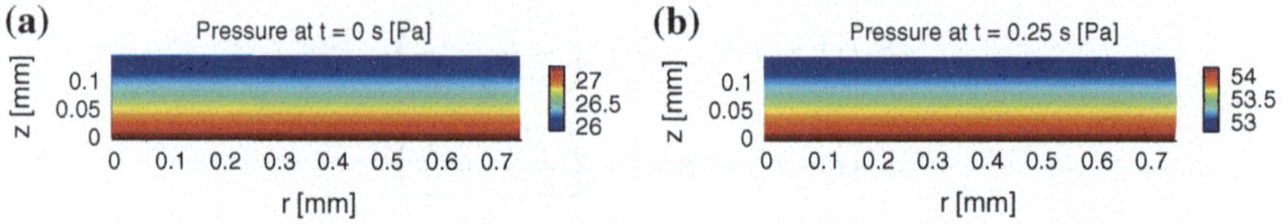

Fig. 7.12 Evaluation of pressure during k-state at $t = 0$ s and $t = 0.25$ s (*after half a period*), at low viscosity ($\mu = 1$ mPa s) and low frequency $f = 1$ Hz. Other parameters are $h = 0.15$ mm, $R = 0.75$ mm, $\theta = 90°$ and $\gamma = 20$ mN/m. *Note* The mesh represented is coarser than the mesh used during the numerical computation .**a** Initial pressure ($t = 0$ s). **b** Pressure at maximal velocity ($t = 0.25$ s)

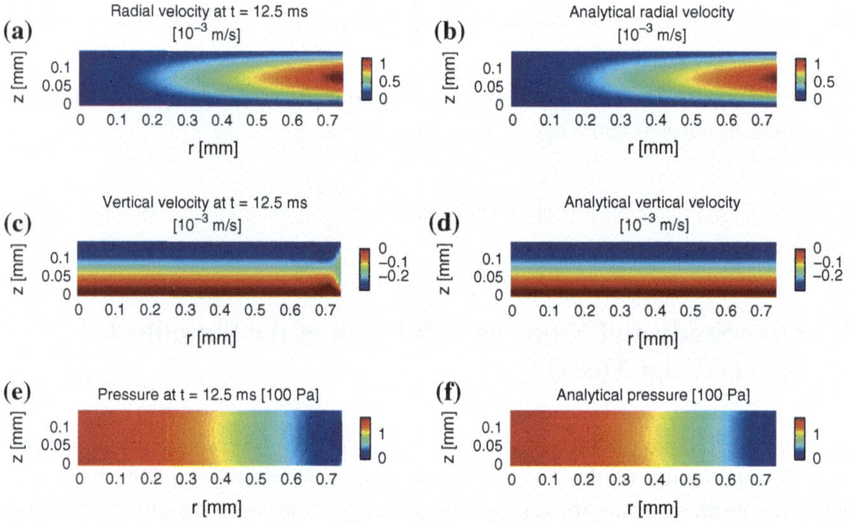

Fig. 7.13 Evaluation of velocity and pressure fields during b-state at $t = 12.5$ ms when the velocity of the *top* pad reaches a maximum (occurring at a phase shift of $\pi/4$), at high viscosity ($\mu = 1$ Pa s) and medium frequency $f = 100$ Hz. Other parameters are $h = 0.15$ mm, $R = 0.75$ mm, $\theta = 90°$ and $\gamma = 20$ mN/m. *Note* The mesh represented is not the one used for computation. **a** Radial velocity. **b** Analytical radial velocity. **c** Vertical velocity. **d** Analytical vertical velocity. **e** Pressure field. **f** Analytical pressure field.

properties ρ, μ and γ have been chosen to set the liquid bridge successively in k, b and m-states.

The comparison between the approximated method and the finite elements solving is provided in the following figures, each of them illustrating one of the three states: $k-$state in Fig. 7.12, $b-$state in Fig. 7.13 and $m-$state in Fig. 7.14.

- $k-$**state**: Fig. 7.12 shows the pressure field in the liquid bridge at initial time $t = 0$ s (Fig. 7.12a) and at maximal velocity $t = 0.25$ s (Fig. 7.12b). Particularly, at $r = 0$

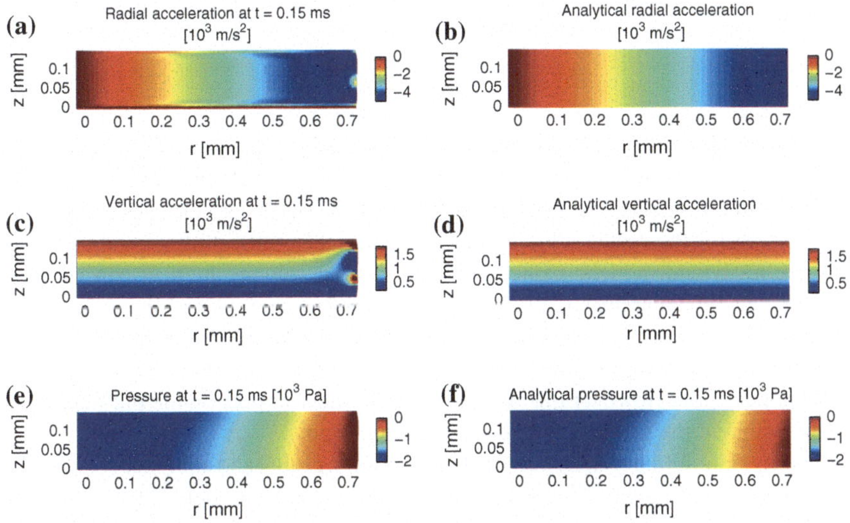

Fig. 7.14 Evaluation of velocity and pressure fields during m-state. Evaluation of velocity and pressure fields during m-state at $t = 0.15\,\text{ms}$ when the acceleration of the *top* pad reaches a maximum (occurring at a phase shift of π), at *low* viscosity ($\mu = 1\,\text{mPa s}$) and high frequency $f = 10\,\text{KHz}$. Other parameters are $h = 0.15\,\text{mm}$, $R = 0.75\,\text{mm}$, $\theta = 90°$ and $\gamma = 20\,\text{mN/m}$. *Note* The mesh represented is not the one used for computation. **a** Radial velocity. **b** Analytical radial velocity. **c** Vertical velocity. **d** Analytical vertical velocity. **e** Pressure field. **f** Analytical pressure field.

and $z = 0$ the pressure can be digitally measured[3] at respectively $p_{t=0\,\text{s}} = 27.40\,\text{Pa}$ and $p_{t=0.25\,\text{s}} = 54.23\,\text{Pa}$. The pressure difference $26.83\,\text{Pa} = 54.23\,\text{Pa} - 27.40\,\text{Pa}$ during this time range can be related to the Laplace pressure variation in the meniscus due to the displacement of the top plate. Referring to Table 7.1 to compute the curvature $2H$ and the contact angle derivative $\frac{d\theta}{dh}$ (for both circular or parabolic models) with $\theta = 90°$, $\Delta h = 1\,\mu\text{m}$, $\gamma = 20\,\text{mN/m}$), the pressure variation inside the meniscus is given by :

$$p(0.25\,\text{s}) - p(0\,\text{s}) = 1\,\mu\text{m}\gamma\left(\frac{3}{4Rh} - \frac{6R}{h^3}\right)$$
$$= 26.53\,\text{Pa}$$

which is a rather good estimate of the digital result provided here above. Similarly, the contact angle variation $\Delta\theta$ has also been estimated digitally at 5.86° (from the normal of the interface at the lower corner). According to Table 7.1, the contact angle variation is equal to:

[3] The reader may note that the gravity force was included during the numerical simulations, which explains the linear variation of the pressure along the vertical axis (due to hydrostatic effect). However it does not affect the results since gravity term is constant and we are interested in the variation of the pressure (and that of the force) during the displacement δh of the top plate).

$$\Delta\theta = -3\frac{R}{h^2}1\,\mu\text{m}$$
$$= 5.73^\circ$$

- $b-$**state**: Fig. 7.13 provides a comparison between digital and analytical results for the radial velocity field (Figs. 7.13a and b), for the vertical velocity field (Figs. 7.13c and d) and for the pressure field (Figs. 7.13e and f). These comparisons show good agreement for the radial velocity and pressure fields, indicating a good estimate by the analytical models. Concerning the vertical velocity field, it can be seen that the analytical approximation is quite fair on the major part of the domain. At the interface however, the estimation is poor. We could expect this result because when we have developed the damping coefficient, the interface stress equation was not verified. However this disagreement has low impact on the pressure field responsible of the force exerted by the liquid bridge.
- $m-$**state**: Fig. 7.14 provides a comparison between digital and analytical results for the radial velocity field (Figs. 7.14a&b), the vertical velocity field (Figs. 7.14c&d) and the pressure field (Figs. 7.14e&f). The radial velocity respects the no-slip condition (appearing clearly at the bottom interface but less at the top interface because of the interpolation technique for colour rendering) and increases to a value independent of z. This illustrates the viscous boundary layers. As for the b-state, the approximations are poor near the interface because the analytical assumptions do not balance perfectly the stresses at the free interface balance. Finally, the free interface present some oscillation at high frequencies, but the effect of Laplace pressure is weak with respect to inertial effect.

7.8 Gain Curves

The previous section showed that the approximations are relevant when a particular state occurs. To validate the approximations with respect to frequency, we compared graphically the gain curves computed with (6.8–6.18) and the analytical coefficient with the gain curves obtained from numerical simulations (see Figs. 7.15 and 7.16).

$$\log G_k(\omega) = \log k \tag{7.59}$$
$$\log G_b(\omega) = \log\omega + \log b \tag{7.60}$$
$$\log G_m(\omega) = 2\log\omega + \log m \tag{7.61}$$

The set of parameters used in these comparisons is summarised in Table 7.2.

The numerical experimental space (made up of all the combinations of the parameters) cover a wide range of Reynold numbers and capillary numbers:

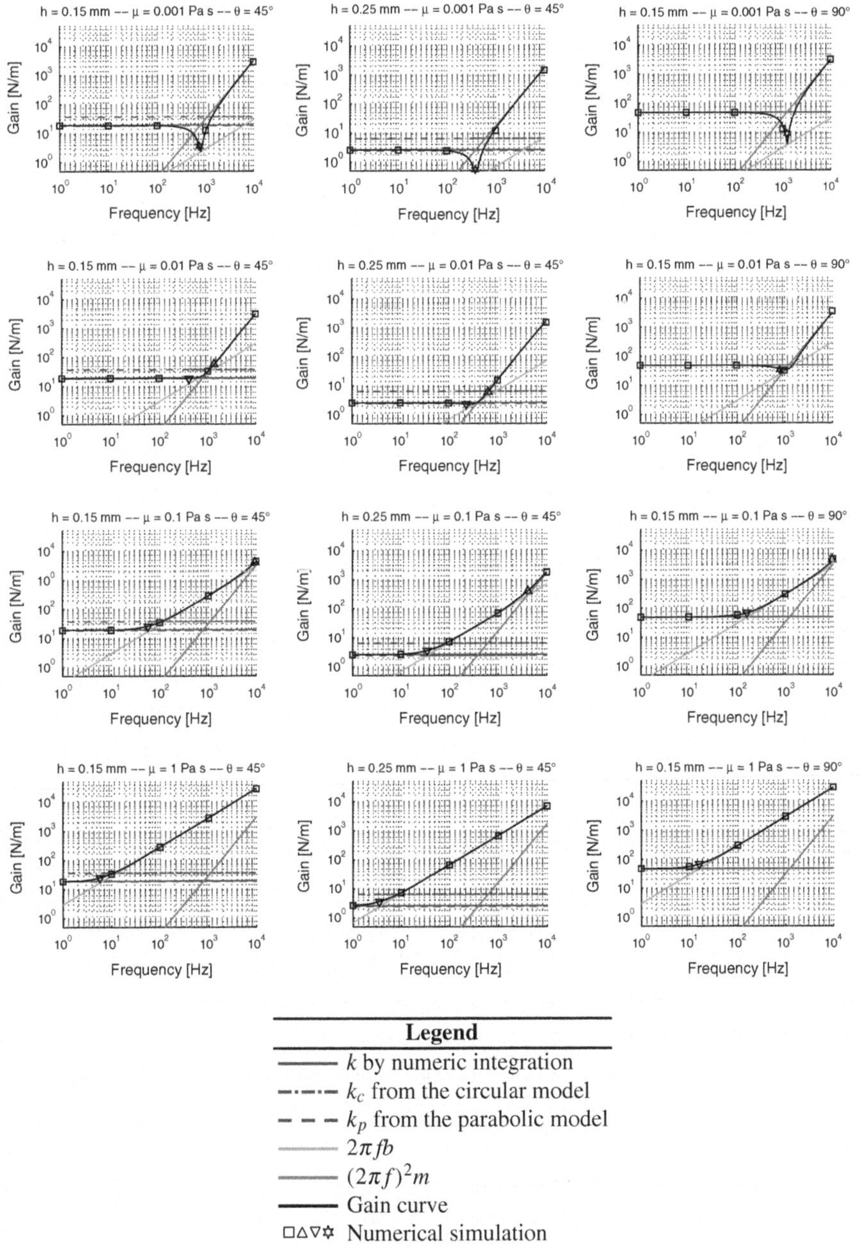

Fig. 7.15 Comparison of gain *curves* between numerical simulations and analytical approximations (part 1)

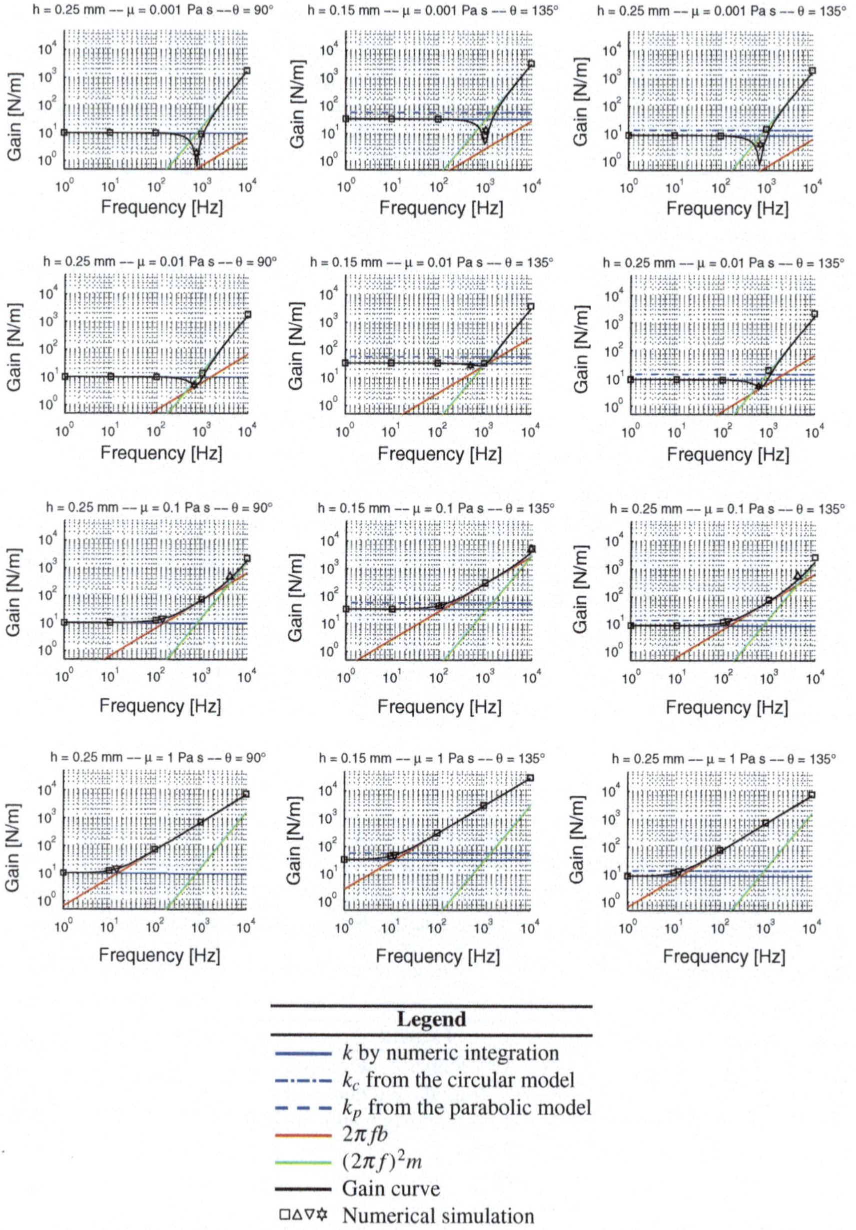

Fig. 7.16 Comparison of gain *curves* between numerical simulations and analytical approximations (part 2)

Table 7.2 Parameters of the digital simulations. This figure was published in [11].

Parameter	Symbol	Value
Radius	R	0.75 mm
Gap	h	0.15 mm, 0.25 mm
Contact angle	θ	45°, 90° 135°
Viscosity	μ	0.001–1 Pa s
Surface tension	γ	20 mN/m
Frequency	f	1 Hz to 10 KHz

$$\mathrm{Re} = \frac{\rho f h^2}{\mu} \quad 4\,10^{-5} \rightarrow 400 \tag{7.62}$$

$$\mathrm{Ca} = \frac{\mu f h}{\gamma} \quad 10^{-5} \rightarrow 100 \tag{7.63}$$

These values are shown in Fig. 7.17 for each experiment (because Reynolds and capillary numbers do not depend on the contact angle, only $\theta = 90°$ have been represented). The states vary according to the frequency. The m-state appears when the Reynolds number increases, when the frequency is bigger than f_{c^+} or f_m. When the capillary number increases, the b-state appears.

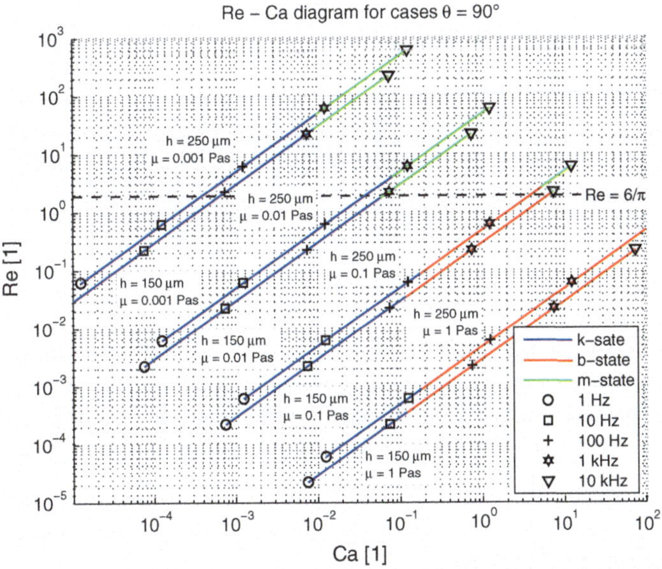

Fig. 7.17 Plot of Reynold and Capillary numbers defined in (7.62) and (7.63) for results showed in Figs. 7.15 and 7.16 when frequency increases (only for $\theta = 90°$, see Table 7.2 for parameters). For a *low* inertial system, the states change at frequencies f_{c^-}, f_{c^+}, for a *high* inertial system, the states change at f_m

In addition, we plot the cutting frequencies f_{c^-} (downward triangle) between states k and b, f_{c^+} (upward triangle) between states b and m, and the resonating frequency (hexagram):

$$f_{c^-} = \frac{\sqrt{b^2+4km}-b}{4\pi m} \tag{7.64}$$

$$f_{c^+} = \frac{\sqrt{b^2+4km}+b}{4\pi m} \tag{7.65}$$

$$f_r = \frac{\sqrt{km-b^2}}{2\pi m} \approx \frac{1}{2\pi}\sqrt{\frac{k}{m}} \tag{7.66}$$

At these resonance frequencies, the approximations are the less correct because two different oscillating states contribute equivalently to the gain. The assumption made for the different states are not compatible. For example at f_{c^+}, the coefficient b supposes the velocity profile to be driven only by viscous effect and the coefficient m assume only inertial effect.

An interesting results may be highlighted between states b and m. If we neglect the stiffness k in (7.65), f_{c^+} becomes:

$$f_{c^+} = \frac{b}{2\pi m} \tag{7.67}$$

Using the approximations (7.40) and (7.58), this reads:

$$f_{c^+} = \frac{1}{2\pi}\frac{3\pi\mu R^4}{2h^3}\frac{8h}{\rho\pi R^4} = \frac{6\mu}{\pi\rho h^2} \tag{7.68}$$

The Reynolds number (7.62) at f_{c^+} is:

$$\mathrm{Re}(f_{c^+}) = \frac{6}{\pi} \tag{7.69}$$

This means that the b-state exists only for Reynolds number $\mathrm{Re} < \frac{6}{\pi}$, if the system has a low inertia. This results has a physical origin: the Reynolds number is a ratio between inertial effects and viscous effects while the frequency f_{c^+} delimits a viscous state from an inertial state. They are somehow referring to the same phenomena.

7.9 Experimental Setup

This section presents the experimental setup designed to measure the transfer function of liquid bridges (Sect. 7.9.1), the experimental protocole (Sect. 7.9.2) and a brief description on the applied image processing (Sect. 7.9.3).

7.9.1 Experimental Bench

The particularity of the bench is the submillimetric size of the system. Size is below mm and force is below mN. The liquid bridges are confined between two circular solids constituting two parallel planes. As indicated in Fig. 7.18a, the bench is made of an actuator driven by an harmonic signal (to move the top of the liquid bridge), an imaging system to characterise the liquid meniscus, a force sensor on bottom of the liquid bridge and two circular pads able to confine the liquid bridge by pinning the triple lines.

The displacement was imposed thanks to a piezo-electric actuator [8, 9]. A commercial FEMTO- TOOLS force sensor was used (Fig. 7.18b). The principle of this sensor is a change of capacitance. This technique has two advantages: the sensor is very dynamic due to the absence of mechanical part and the stiffness is high (the value is not given by the constructor). The tip and the capacitor is mounted on a circuit board that integrates chips converting the change of capacitance into an output voltage. Each sensor is provided with its own unique characteristic (the sensitivity may vary from 9 to 1.1 mV/μN). A major inconvenience is the reduced range of measurable force. These sensors are quite cheap (190€) but the tips are extremely fragile. The maximum load of this sensor is 2 mN, with a resolution equal to 0.4 μN at 30 Hz and 2 μN at 1 KHz. Its stiffness is large enough to neglect the sensor deformation with respect to the gap amplitude imposed on the liquid bridge. The resonance frequency is about 6400 Hz, which is much more larger than with the first design. It will however limit our bandwidth.

This sensor was embedded in a design, whose principle is shown in Fig. 7.19: (1) the non contact displacement sensor LK-G10 points towards the top surface of a mechanical connector, linking the actuator imposing the harmonic displacement (2) to the top pad glued on the bottom of the connector. This harmonic displacement is measured by the signal u of the displacement sensor. The displacement of the top pad is therefore known with a constant offset u+constant. The bottom pad (3) is fixed to the tip of the force sensor (4). Since the force sensor is assumed to be stiff

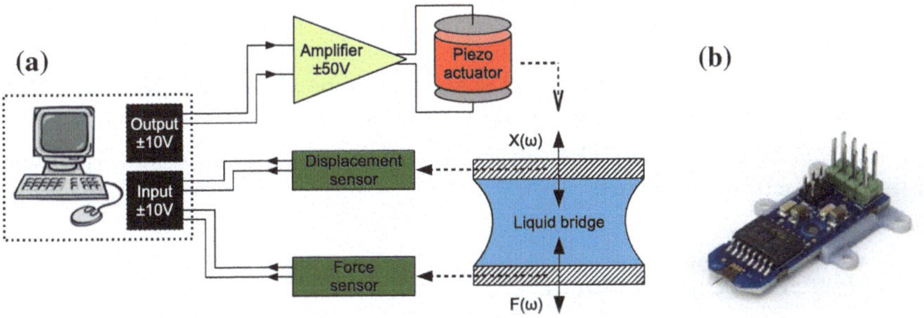

Fig. 7.18 **a** Principle of the experimental test bed; **b** Force sensor FT-S270. This figure was published in [11].

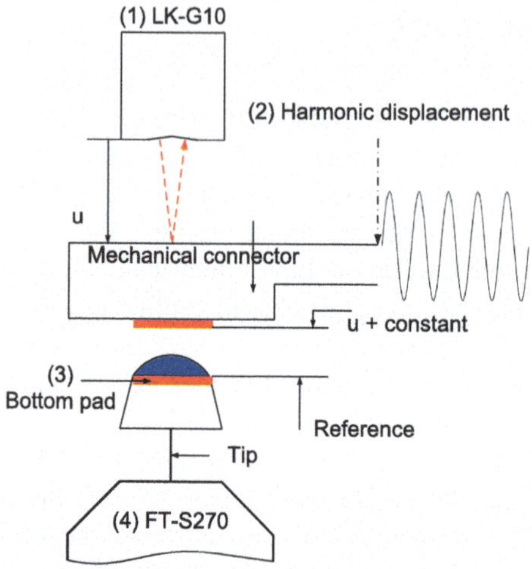

Fig. 7.19 Measurement principle

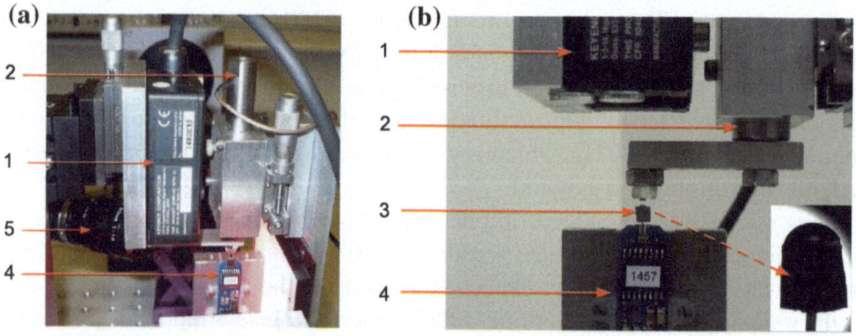

Fig. 7.20 Pictures of the test bed: (1) is the displacement sensor LK-G10, (2) is the piezo actuator imposing the harmonic displacement to the *top pad*, (3) is the *bottom pad*, (4) is the FT-S270 sensor and (5) is a lateral camera with optical axis perpendicular to the liquid bridge symmetry axis (vertical on these pictures). This figure was published in [11].

enough, the position of the bottom pad is assumed to be a reference. Note well that the gap between both pads is actually not known with a precision better than 25 μm the error was ±8 pixels, and 900 pixels represent 1.5mm. Figure 7.20 shows pictures of the manufactured test bench.

7.9.2 Experimental Protocol

Thanks to the data acquisition system, measurements were almost fully automated. For each experiment, the following steps were adopted:

1. The circular plates were cleaned with acetone and ethanol;
2. A small amount of liquid (around 1 μL) was dropped on the lower plate;
3. The upper plate was lowered until the contact with the liquid bridge;
4. The lower plate was accurately positioned and aligned thanks to the two cameras;
5. The gap was fixed. If the volume needed to be adjusted, the process restarted from step 2;
6. Two pictures were taken in order to compute the volume;
7. Dynamic parameters (frequency range, delays, amplitude of actuator) were defined;
8. A systematic acquisition program in LabVIEW[4] was used:
    a. An output was generated to the piezoelectric driver;
    b. According to the oil viscosity and the frequency, a delay was inserted to avoid any effect of transient response;
    c. Data were acquired and recorded in a text file;
9. Two pictures were taken to control the state of the meniscus.

The protocol was restarted from step 8 to successively acquire multiple experiments on the same system, from step 7 to get the effect of the amplitude of the actuator, from step 5 to change the gap and from step 1 to change the liquid.

7.9.3 Image Processing Towards Geometrical Parameters Acquisition

The pictures of the liquid bridges have been taken before and after each experimental sequence, with two cameras with intersecting optical axes, as shown in Fig. 7.21. In these images, the air-liquid interface here contains all necessary information to completely describe the geometry of the liquid bridge: their analysis provided the contact angle, the gap and the meniscus curvature.

7.10 Results

7.10.1 Experiments List

Experiments have been led with different liquids whose properties are summarized in Table 7.3.

An experiments list is also provided in Table 7.4, summarizing the geometrical data.

[4] http://www.ni.com/labview/

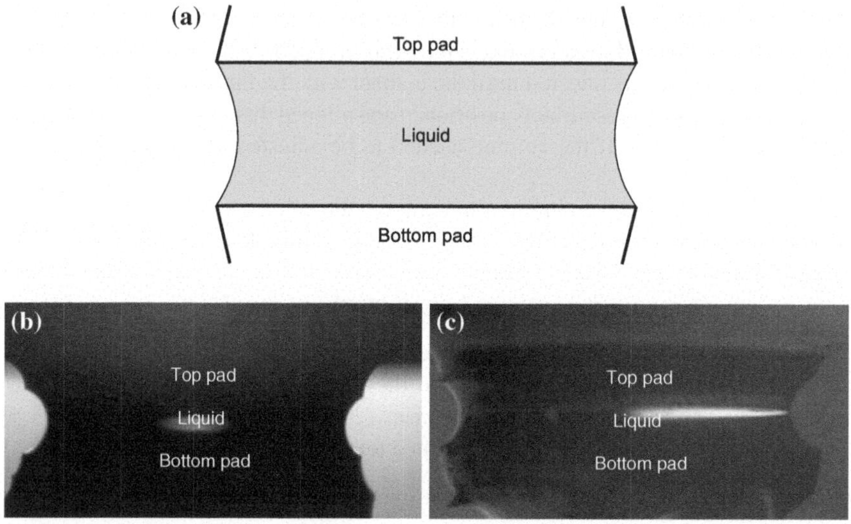

Fig. 7.21 Example of control pictures taken before and after each experiments. All rights reserved. **a** Sketch of the pinned liquid bridge. **b** Picture taken with the camera 1. **c** Picture taken with camera 2. This figure was published in [11]. Copyright © 2013 Elsevier Masson SAS.

Table 7.3 Main physical properties of liquid used.The value are given for a temperature of 25 °C.

Ref	Liquid	Density Kg/m^3	Dynamic viscosity Pas	Surface tension N/m
Oil1	R47V500	970	0.485	21.1 10^{-3}
Oil2	R47V5000	973	4.865	21.1 10^{-3}
Oil6	DC200FLUID100	960	0.096	20.9 10^{-3}
Oil7	DC200FLUID1000	971	0.971	21.2 10^{-3}
Oil8	DC200FLUID60000	976	58.56	21.5 10^{-3}

This figure was published in [11]. Copyright © 2013 Elsevier Masson SAS. All rights reserved

7.10.2 Gain and Phase Shift Curves

On each graph shown in Figs. 7.22 and 7.23, the experimental curve and numerical simulations have been superimposed. Four simulations were realised with each geometrical information, as well as a simulation with the average value of the parameters measured. The results are conclusive: indeed, all experimental data points are in between the numerical curves.

However, the variation of the geometric parameters gathered from image analysis produces an important dispersion of the stiffness. The dispersion may be explained for several reasons. First, the positioning error (mainly the tilt in both horizontal directions) gives slightly different profile of the liquid bridge for both cameras. Second, there is a small hysteresis inherent to the piezo actuator. Therefore, the gap and the edge angles may change accordingly. Finally, there is a small amount of

Table 7.4 Experimental results from both cameras

Experiment	Picture	Bottom radius (mm)	Top radius (mm)	Bottom angle (°)	Top angle (°)	Gap (mm)	Volume (mm^3)
1	1	0.717	0.696	15.9	23.6	0.284	0.354
1	2	0.735	0.721	15.3	20.1	0.298	0.386
1	3	0.713	0.703	18.6	22.6	0.262	0.335
1	4	0.738	0.732	13.9	16.2	0.281	0.374
1	Mean	0.726	0.713	15.9	20.6	0.282	0.362
2	1	0.747	0.747	43.2	43.2	0.218	0.345
2	2	0.77	0.746	36	46.7	0.242	0.391
2	3	0.736	0.722	18.6	24.7	0.231	0.325
2	4	0.747	0.729	20.2	28.1	0.235	0.342
2	Mean	0.75	0.736	29.5	35.7	0.231	0.351
3	1	0.756	0.746	28.8	32	0.323	0.473
3	2	0.745	0.748	34.7	33.6	0.317	0.463
3	3	0.755	0.738	21.8	27.3	0.319	0.449
3	4	0.767	0.755	21.4	25.2	0.322	0.468
3	Mean	0.756	0.747	26.7	29.5	0.32	0.463
4	1	0.749	0.751	45.8	45.2	0.284	0.446
4	2	0.755	0.752	46	47.4	0.28	0.449
4	3	0.755	0.746	24.5	27.9	0.28	0.412
4	4	0.758	0.735	22.4	31.1	0.273	0.400
4	Mean	0.754	0.746	34.7	37.9	0.279	0.427
5	1	0.75	0.751	50.7	50.1	0.255	0.412
5	2	0.76	0.752	46.5	50	0.246	0.404
5	3	0.751	0.75	41.6	42	0.251	0.396
5	4	0.766	0.751	36.8	43.3	0.243	0.393
5	Mean	0.757	0.751	43.9	46.3	0.249	0.401
7	1	0.743	0.747	27.4	24.9	0.158	0.243
7	2	0.752	0.748	24.2	27.1	0.16	0.255
7	3	0.744	0.744	40.8	40.8	0.155	0.250
7	4	0.758	0.752	43	47.2	0.163	0.271
7	Mean	0.749	0.747	33.9	35	0.159	0.255
8	1	0.747	0.741	41	42	0.554	0.743
8	2	0.754	0.751	40.1	40.6	0.562	0.764
8	3	0.747	0.747	40	40	0.569	0.756
8	4	0.754	0.746	39.1	40.5	0.575	0.768
8	Mean	0.75	0.746	40.1	40.8	0.565	0.758
9	1	0.748	0.744	65.5	66.3	0.479	0.758
9	2	0.765	0.754	65.5	67.8	0.485	0.800
9	3	0.745	0.744	52.1	52.3	0.482	0.714
9	4	0.752	0.754	50.7	50.3	0.485	0.727
9	Mean	0.752	0.749	58.5	59.2	0.483	0.75
10	1	0.743	0.75	25.4	22.4	0.26	0.381

(Continued)

Table 7.4 (Continued)

Experiment	Picture	Bottom radius (mm)	Top radius (mm)	Bottom angle (°)	Top angle (°)	Gap (mm)	Volume (mm^3)
10	2	0.766	0.76	22.9	25.3	0.27	0.412
10	3	0.748	0.745	29.8	30.8	0.266	0.396
10	4	0.748	0.754	31.6	29.5	0.269	0.406
10	Mean	0.751	0.752	27.4	27	0.266	0.399
11	1	0.734	0.751	30.5	24.6	0.293	0.422
11	2	0.746	0.747	27.2	27.1	0.297	0.430
11	3	0.739	0.752	32.9	28.1	0.295	0.432
11	4	0.755	0.75	27	28.5	0.3	0.443
11	Mean	0.743	0.75	29.4	27.1	0.296	0.432
12	1	0.753	0.757	11	9.83	0.334	0.435
12	2	0.746	0.759	16	12.4	0.343	0.451
12	3	0.746	0.761	15.9	11.7	0.335	0.445
12	4	0.744	0.764	17.7	12.2	0.345	0.458
12	Mean	0.747	0.76	15.2	11.5	0.339	0.447

Pictures 1 and 2 have been recorded before the experiment and picture 3 and 4 have been recorded after. This figure was published in [11].

liquid that is lost during the experiments, due to flooding outside the pad (in case the pinning was not perfect) or to evaporation.

These geometric errors are less visible on the *b*-state. Indeed, the equivalent damping depends on the volume that is relatively less sensitive to a variation of the free interface. We may see on experiments 11 and 12 that for high frequency, the curve goes under the linear asymptote. This is due to the sensor saturation. Consequently, the inertial state could not be observed experimentally because the level of force was too high.

The spring and anti-spring cases can be observed with the phase curves:

- Experiments 2, 4, 5, 7, 9 are spring cases because the phase starts at 0°.
- Experiments 1, 3, 8 are anti-springs because the phase starts at 180°.
- Experiments 10, 11, 12 are purely viscous because the phase starts around 90°. The k-state is observable at frequency lower than 0.1 Hz. Nevertheless, we can guess that experiments 10 is a spring case (the phase is below 90°) and experiments 12 is an anti-spring case (phase is higher than 90°).

7.11 Conclusions

In this chapter we characterised the behaviour of an axisymmetric liquid bridge under small vertical oscillations. The meniscus was modelled using a Kelvin-Voigt model, consisting of a spring, a damper and a mass in parallel. We proposed an abacus for

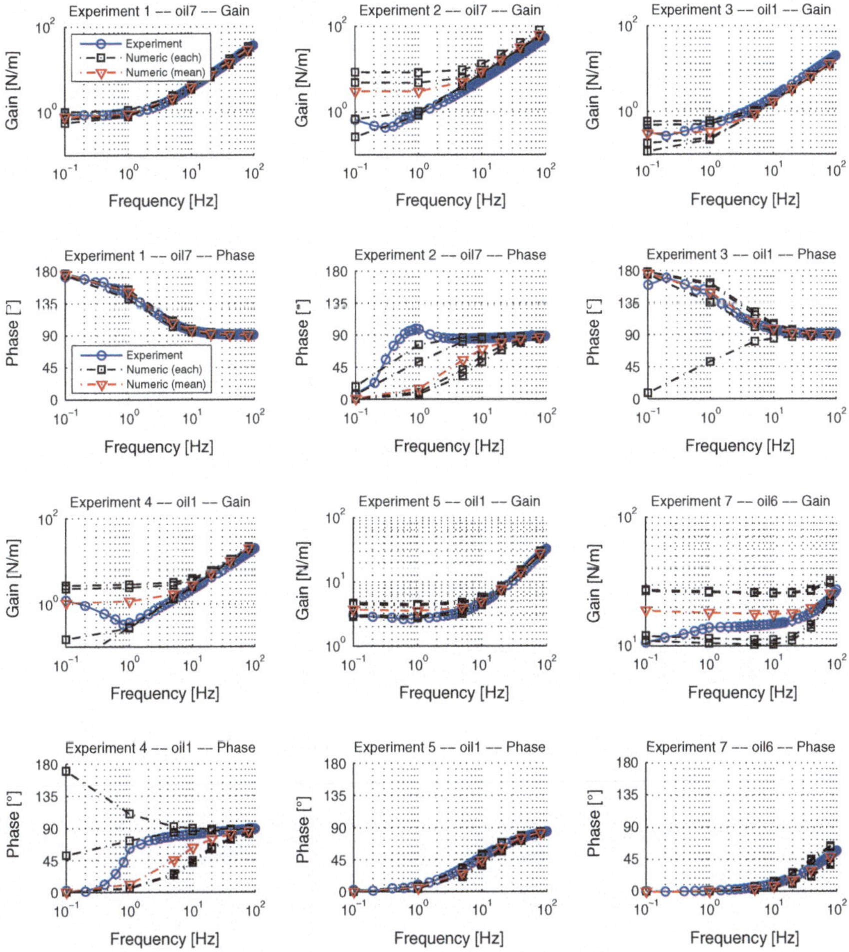

Fig. 7.22 Gain and phase *curves* for experiments 1–7. The *circle* represents experimental data, the *square* the individual parameters and the *triangle* the mean value of the geometric parameters. Experimental parameters are given in Table 7.4. This figure was published in [11]. Copyright © 2013 Elsevier Masson SAS. All rights reserved

the stiffness, and analytical expressions for the stiffness, the damping and the inertial coefficients.

The validation was provided through numerical simulations and experimental data. The numerical simulations acted as a buffer for results validation: they were first compared with analytical approximations (with a mirror symmetry with respect to the plane containing the neck of the meniscus), and then with experiments.

We showed that it is possible to characterise the meniscus by geometrical and physical parameters of liquids, and without downscaling the system to microscopic dimension. The proposed analytical laws were based on simplifications of the 2D

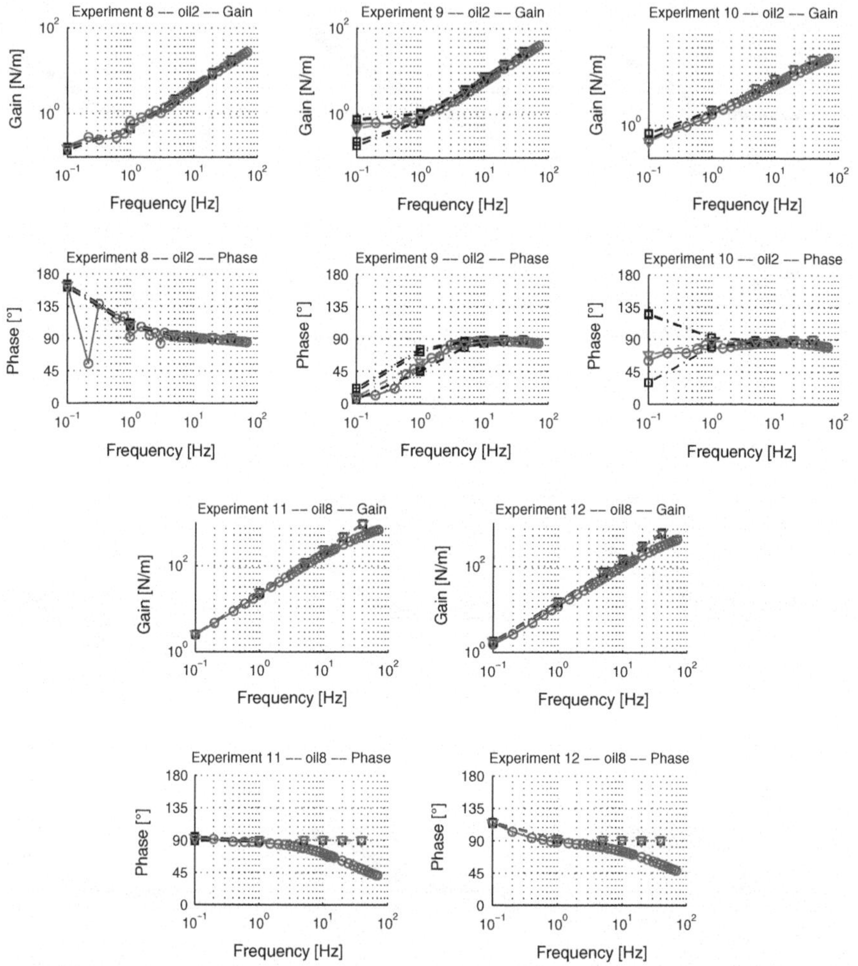

Fig. 7.23 Gain and phase *curves* for experiments 8–12. The *circle* represents experimental data, the *square* the individual parameters and the *triangle* the mean value of the geometric parameters. Experimental parameters are given in Table 7.4

axisymmetric Navier-Stokes equation. They can be used to quickly estimate the order of magnitude of the different parameters k, b, m. The first one involves the knowledge of the free liquid surface while the latter are based on the liquid volume. Consequently, the parameter k is more difficult to evaluate: although it is a good approximation, the experimental values of the edge angle and of the free surface curvature may be rather difficult to estimate.

The results showed excellent agreement between **analytical** and **numerical** models, validating the assumptions on the state of the fluid (static flow for spring state, viscous flow for the damping state and inertial flow for the inertial state).

The results showed good agreement between **experimental** data and **numerical** simulation, although measurements were difficult to achieve. The present experimental bench did not allow us to inspect the inertial state experimentally since the needed input frequencies are very high and forces generated are too large for the sensor hereby adopted.

The interest of this model lies in the fact that important information can be derived from these coefficients: the step response, the impulse response or any frequential response. For small displacement the differential Eq. (7.4) associated to the system can be easily solved and often gives an analytical solution. For specific input $F(t)$ a numeric integration can be quickly computed. The reader will remind that the coefficients change according to the gap of the meniscus. For large displacement of an interface, the coefficients vary. However, the variation of the coefficients can be included in the numeric integration.

Acknowledgments This work partially presented in [11] is funded by a grant of the F.R.I.A.—*Fonds pour la Formation à la Recherche dans l'Industrie et l'Agriculture.*

Appendix on Finite Element Method

In this appendix, we explain how we implement the free surface problem with COMSOL MULTIPHYSICS 3.5a. The software proposes built-in models for a large set of applications with predefined equations including Navier-Stokes and moving mesh. Moreover, the software allows to customise these equations to implement complex constraints, like surface tension in our case. Equations (Navier-Stokes and moving mesh) were formulated using the *vectorised weak formulation* [6].

Note that a general appendix on vectorial operators in provided in Appendix B at the end of this book.

Model

The geometry used for numeric simulations is depicted in Fig. 7.24. The following boundary conditions apply: the no-slip condition is imposed on the bottom boundary; a sinusoidal normal inflow moving with the mesh on the top boundary; the symmetry condition on the axis of symmetry boundary; the normal stress and the moving mesh on the free boundary. This is summarised in Table 7.5. The motion of the top boundary is downward and oscillate sinusoidally: $z_t(t) = h + A(\cos(2\pi f t) - 1)$. The amplitude A has been set to 0.5 μm, small enough to stay in linear state.

The force generated by the liquid bridge on the substrate is divided into the surface tension force and the viscous force [3]:

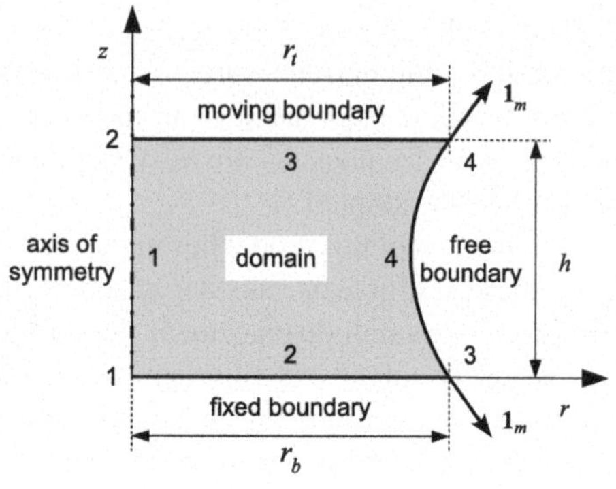

Fig. 7.24 Domain for the numerical model

$$\bar{F} = \left[2\pi R\gamma \sin\theta + \iint_{\Gamma} \left(-p + 2\mu \frac{\partial v}{\partial z} \right) \mathrm{dS} \right] \bar{1}_z \tag{7.70}$$

In the following, we will refer by *pressure force* the second term, and by the *viscous force* the third one.

Mathematical Background

The weak formulation of the partial differential equations system is required to implement the free surface condition as a boundary condition in a fluid flow problem. Basically, the pressure is given by the curvature $2H$ (Table 7.6):

$$\Delta p = 2H = \bar{\nabla} \cdot \bar{1}_n \tag{7.71}$$

Table 7.5 Boundary conditions for the numerical model. Appl.

Boundary	Appl.	Equation
Axial	All	$r = 0$
$\frac{A}{B}$	All	$\frac{\partial \cdot}{\partial r} = 0$
	NS	$u_r = 0$
	MM	$\mathrm{d}\dot{\bar{x}} \cdot \bar{1}_n = 0$
Bottom	NS	$\bar{u} = 0$
	MM	$\mathrm{d}\dot{\bar{x}} = 0$
Top	NS	$\bar{u} = -2\pi f A \cos 2\pi f t \bar{1}_z$
	MM	$\mathrm{d}\dot{\bar{x}} = u_z \bar{1}_z$
Free	NS	$-p + \tau = 2H\gamma \bar{1}_n$
	MM	$\mathrm{d}\dot{\bar{x}} = (\bar{u} \cdot \bar{1}_n)\bar{1}_n$

Stands for *Application Mode*, NS for *Navier-Stokes Application Mode* and MM for *Moving Mesh Application Mode* as in COMSOL terminology

Table 7.6 Simulation parameters used for the comparison of finite element model with analytical expressions and experimental data

Property	Value
Mesh	
Type	Triangular
Domain	$1.3\ 10^{-5}$m
Boundary 2	$0.5\ 10^{-5}$m
Boundary 3	$0.5\ 10^{-5}$m
Boundary 4	$0.4\ 10^{-5}$m
Vertex 3	$0.1\ 10^{-5}$m
Vertex 4	$0.1\ 10^{-6}$m
Growth rate	1.3
Tolerances	
Relative	0.2%
Absolute	0.05%
Time stepping	$1/100f$
Motion	
Number of cycles	3
Output points	25/cycle

Unfortunately, COMSOL is not able to compute the derivative of $\bar{1}_n$. In this section, we will present the mathematical development of the *conventional $\bar{u}$-p formulation and the stress-divergence form combined.* The reader will refer to [4, 6] for further reading. This conventional $\bar{u}$-p formulation exists in different forms, which is a scalar and a vector form. In literature, these forms are very common but the implementation of surface tension effect is rather rare. The implementation of surface tension in scalar form of our $\bar{u}$-p formulation has already been discussed in [13]. The vector form is drawn up here. The vector notation will be used to remain independent of any coordinate system.

The main idea is that if an equation is verified, we can multiply this equation by a *test-function* and integrate it over the whole domain. Then, we reduce the order of derivative by applying successive integral theorems. The test-function are managed by COMSOL. The reader will find further information in [6]. The starting point is the Navier-Stokes and continuity equations:

$$\rho\left(\frac{\partial\bar{u}}{\partial t} + \left(\bar{u}\cdot\bar{\nabla}\right)\bar{u}\right) = \rho\bar{g} - \bar{\nabla}p + \bar{\nabla}\cdot\tau \tag{7.72}$$

$$\bar{\nabla}\cdot\bar{u} = 0 \tag{7.73}$$

These equations are multiplied at both sides by *test functions* and integrated on the domain Ω. The test functions are $\bar{v}$ and q:

$$\iiint_{\Omega} \left[\rho \left(\frac{\partial \bar{u}}{\partial t} + (\bar{u} \cdot \bar{\nabla})\bar{u} \right) \cdot \bar{v} \right] \mathrm{dV} = \iiint_{\Omega} \left[-\bar{\nabla} p \cdot \bar{v} + \bar{v} \cdot \left(\bar{\nabla} \tau \right) + \rho \bar{v} \cdot \bar{g} \right] \mathrm{dV} \tag{7.74}$$

$$\iiint_{\Omega} q \bar{\nabla} \cdot \bar{u} \, \mathrm{dV} = 0 \tag{7.75}$$

We will start with the Navier-Stokes equation. The purpose is to reduce the order of the derivatives in the integral. Using some divergence identities:

$$\bar{v} \cdot \bar{\nabla} p = \bar{\nabla} \cdot (p \bar{v}) - p \bar{\nabla} \cdot \bar{v} \tag{7.76}$$

$$\bar{v} \cdot \left(\bar{\nabla} \cdot \tau \right) = \bar{\nabla} \cdot (\tau \cdot \bar{v}) - \tau : \bar{\nabla} \bar{v} \tag{7.77}$$

Reordering by isolating divergence terms, the integral now reads:

$$\iiint_{\Omega} \left[\rho \left(\frac{\partial \bar{u}}{\partial t} + (\bar{u} \cdot \bar{\nabla})\bar{u} - \bar{g} \right) \cdot \bar{v} - p \bar{\nabla} \cdot \bar{v} + \tau : \bar{\nabla} \bar{v} \right] \mathrm{dV} \tag{7.78}$$

$$= \iiint_{\Omega} \left[-\bar{\nabla} \cdot (p \bar{v}) + \bar{\nabla} \cdot (\tau \cdot \bar{v}) \right] \mathrm{dV} \tag{7.79}$$

Gauss' theorem is applied to the right-hand side term to give, with the commutative matrix product:

$$\begin{aligned} \iiint_{\Omega} \left[-\bar{\nabla} \cdot (p \bar{v}) + \bar{\nabla} \cdot (\tau \cdot \bar{v}) \right] \mathrm{dV} &= \iint_{\partial \Omega} \left[-(p \bar{v}) + (\tau \cdot \bar{v}) \right] \cdot \mathrm{d}\bar{\mathrm{S}} \\ &= \iint_{\partial \Omega} \left[-(p \bar{1}_n) + (\tau \cdot \bar{1}_n) \right] \cdot \bar{v} \, \mathrm{dS} \\ &= \iint_{\partial \Omega} \left[(-p\mathrm{I} + \tau) \cdot \bar{1}_n \right] \cdot \bar{v} \, \mathrm{dS} \end{aligned} \tag{7.80}$$

Hence, the weak form of the Navier-Stokes equation is finally:

$$\iiint_{\Omega} \left[\rho \left(\frac{\partial \bar{u}}{\partial t} + (\bar{u} \cdot \bar{\nabla})\bar{u} - \bar{g} \right) \cdot \bar{v} - p \bar{\nabla} \cdot \bar{v} + \tau : \bar{\nabla} \bar{v} \right] \mathrm{dV}$$

$$= \iint_{\partial \Omega} \left[(-p\mathrm{I} + \tau) \cdot \bar{1}_n \right] \cdot \bar{v} \, \mathrm{dS}$$

The latter term should be divided into as much integrals as boundaries. In the next section, we explicit this latter term according the boundary conditions. The continuity Eq. 7.75 remains as it is.

Implementing Boundary Conditions

On the boundaries, the Dirichlet conditions applied on a variable must be also applied on the test function associated ($\bar{u}$ with $\bar{v}$ and p with q). The boundaries are expressed in the integral $\mathscr{I}$ computed on the boundary concerned (7.81). This integral can be rewritten in a decomposition of normal and tangent components:

$$\mathscr{I} = \iint_{\partial\Omega} \left[-p + (\tau \cdot \bar{1}_n) \cdot \bar{1}_n \right] (\bar{v} \cdot \bar{1}_n) \, \mathrm{dS} + \iint_{\partial\Omega} \left[(\tau \cdot \bar{1}_n) \cdot \bar{1}_t \right] (\bar{v} \cdot \bar{1}_t) \, \mathrm{dS} \tag{7.81}$$

$$= \mathscr{I}_n + \mathscr{I}_t \tag{7.82}$$

- **Inlet/outlet velocity** This configuration include the no slip case, according to the value of $\bar{v}^*$. When both components are fixed, $\mathscr{I}$ is *removed* and we have:

$$\mathscr{I} = 0 \Rightarrow \begin{cases} \bar{u} = \bar{v}_\Omega + \bar{u}^* \\ \bar{v} = \bar{v}_\Omega + \bar{u}^* \end{cases} \tag{7.83}$$

  The reason is that by defining $\bar{u}$, stresses are automatically define. If we give also a value to the stress, the problem is over-constrained. To clearly understand, let's suppose that the stresses required to produce a particular velocity profile $\bar{v}_\Omega + \bar{u}^*$ at the boundary is $\bar{f}^*$. We could use:

$$\mathscr{I} = \iint_{\partial\Omega} \bar{f}^* \cdot \bar{v} \, \mathrm{dS} \tag{7.84}$$

$$\bar{v} = \bar{v}_\Omega + \bar{u}^* \tag{7.85}$$

  giving the solution $\bar{u} = \bar{v}_\Omega + \bar{u}^*$. A similar formulation is used when using Lagrange multipliers, see [2].

- **Viscous slipping** We have mixed conditions, on the normal velocity on one hand, and on the tangent component of the normal stress $(\tau \cdot \bar{1}_n) \cdot \bar{1}_t$. The same argument as previous boundary condition stands for the normal component of the normal stress. We have thus:

$$\mathscr{I}_t = \iint_{\partial\Omega} -\mu \frac{\bar{u} - \bar{v}_\Omega}{\beta} \cdot \bar{1}_t (\bar{v} \cdot \bar{1}_t) \, \mathrm{dS} \tag{7.86}$$

$$\mathscr{I}_n = 0 \Rightarrow \begin{cases} \bar{u} \cdot \bar{1}_n = \bar{v}_\Omega \cdot \bar{1}_n \\ \bar{v} \cdot \bar{1}_n = \bar{v}_\Omega \cdot \bar{1}_n \end{cases} \tag{7.87}$$

This configuration include the total slipping case, by considering $\beta \to \infty$ (and $\mathscr{I}_t = 0$).

- **Inlet/outlet pressure** The viscous stress is zero and the pressure is constrained. The equations are:

$$\mathscr{I}_n = \iint\limits_{\partial\Omega} -p\bar{v}\cdot\bar{1}_n \text{ dS} \tag{7.88}$$

$$\mathscr{I}_t = 0 \tag{7.89}$$

$$p = p^* \tag{7.90}$$

$$q = p^* \tag{7.91}$$

- **Surface force** The integrals read:

$$\mathscr{I}_n = \iint\limits_{\partial\Omega} f_n^*\bar{v}\cdot\bar{1}_n \text{ dS} \tag{7.92}$$

$$\mathscr{I}_t = \iint\limits_{\partial\Omega} f_t^*\bar{v}\cdot\bar{1}_t \text{ dS} \tag{7.93}$$

- **Surface tension interface** This is a particular case that needs some development because of the curvature, as mentioned previously. Actually, this boundary is the reason that requires the setting of the weak equations. Without loss of generality, we will consider that outer pressure is zero. Since the surface tension is constant, the tangential stresses are zero and therefore, we may write:

$$\mathscr{I} = \iint\limits_{\partial\Omega} 2H\gamma\bar{1}_n\cdot\bar{v} \text{ dS} \tag{7.94}$$

We will use the surface divergence theorem saying that:

$$\iint\limits_{\Gamma} \bar{\nabla}_s\cdot\bar{u} \text{ dS} = \int\limits_{\partial\Gamma} \bar{u}\cdot\bar{1}_m \text{ dl} + \iint\limits_{\Gamma} 2H\bar{u}\cdot\bar{1}_n \text{ dS} \tag{7.95}$$

The surface integral is

$$\mathscr{I} = \iint\limits_{\partial\Omega} \gamma\bar{\nabla}_s\cdot\bar{v} \text{ dS} - \int\limits_{\partial^2\Omega} \gamma\bar{v}\cdot\bar{1}_m \text{ dl} \tag{7.96}$$

where $\bar{\nabla}_s$ is the surface divergence operator and $\bar{1}_m$ is the binormal vector, such as $\bar{1}_t \times \bar{1}_n = \bar{1}_m$, in Fig. 7.25. This vector $\bar{1}_m$ gives the slope of the surface on the triple line. Therefore it contains the contact angle information. The surface

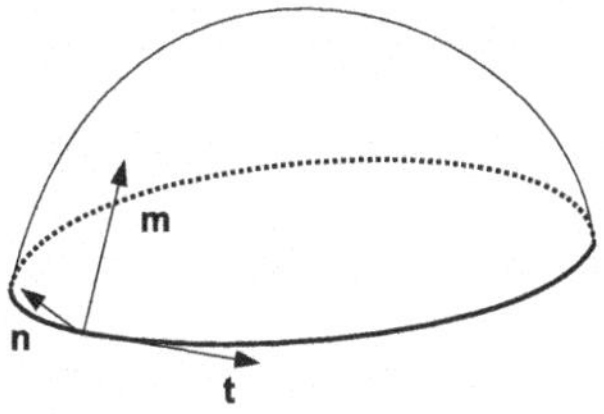

Fig. 7.25 Tangent, normal and binormal vectors. They are orthonormed and $\bar{1}_t \times \bar{1}_n = \bar{1}_m$

divergent operator may be computed after removing the derivative along the normal direction [13, 14]. Hence

$$\bar{\nabla}_s \cdot \bar{u} = \left(\mathsf{I} - \bar{1}_n \otimes \bar{1}_n\right) : \left(\bar{\nabla}\bar{u}\right) \tag{7.97}$$

References

1. D. Cheneler, M.C.L. Ward, M.J. Adams, Z. Zhang, Measurement of dynamic properties of small volumes of fluid using mems. Sens. Actuators B **130**(2), 701–706 (2008)
2. A.B. Comsol, Comsol Multiphysics V3.5a: User's Guide, 2008
3. P.-G. de Gennes, F. Brochart-Wyard, D. Quéré, Gouttes, bulles, perles et ondes. Belin (2002)
4. G. Dhatt, G. Touzot, E. Lefrançois, Méthode des éléments finis. Hermes Sciences
5. J. Engmann, C. Servais, A.S. Burbidge, Squeeze flow theory and applications to rheometry. A review. J. Non-Newtonian Fluid Mech. **132**(1–3), 1–27 (2005)
6. P.M. Gresho, R.L. Sani, Incompressible flow and the finite element method: Advection-diffusion, 2000
7. Piezomechaniks Gmbh, *Piezo-mechanical and electrostrictive stack and ring actuators: Product range and technical data* (Technical report, Piezomechaniks Gmbh, 2009)
8. Piezomechaniks Gmbh, *Piezo-mechanics: An introduction* (Technical report, Piezomechaniks Gmbh, 2009)
9. O. Pitois, P. Moucheront, X. Chateau, Liquid bridge between two moving spheres: An experimental study of viscosity effects. J. Colloid Interf. Sci. **231**(1), 26–31 (2000)
10. J.-B. Valsamis, A study of liquid bridges dynamics. PhD Thesis, Université libre de Bruxelles, 2010
11. J.-B. Valsamis, M. Mastrangeli, P. Lambert, Vertical excitation of axisymmetric liquid bridges. Eur. J. Mech. B/Fluids **38**, 47–57 (2013)
12. N. van Veen, Analytical derivation of the self-alignment motion of flip chip soldered components. J. Electr. Packag. **121**, 116–121 (1999)
13. M.A. Walkley, P.H. Gaskell, P.K. Jimack, M.A. Kelmanson, J.L. Summers, Finite element simulation of three-dimensional free-surface flow problems. J. Sci. Comput. **24**(2), 147–162 (2005)
14. C.E. Weatherburn, *Differential Geometry of Three Dimensions*. (Cambridge University Press, Cambridge, 1955)

Chapter 8
Lateral Capillary Forces (Dynamics)

Pierre Lambert

Abstract This chapter presents a study on the dynamics of lateral motion of a liquid meniscus confined by a circular pad and a circular chip moving parallely to the pad. This problem is a typical flip-chip case study, whose use is wide-spread in industrial assembly. The proposed model describing this dynamics is made of two coupled physics: the Navier-Stokes equation governing the liquid flow between the pad and the chip, and the Newton's law describing the motion of the chip. This coupled problem is solved with a spectral method based on Chebyshev polynomials, by assuming an analytical expression of the lateral stiffness of the meniscus in the cases of circular and square pads. The theoretical results are benchmarked with literature results and thoroughly experimentally validated. From these results, we propose a map giving the characteristic time of the chip dynamics according to only two non-dimensional parameters, constructed with the physical (density, surface tension, viscosity), geometrical (pad area, gap) or dynamical (chip mass) parameters of the problem. These results have been published in [8].

8.1 State of the Art and Definition of the Problem

As mentioned in previous chapters, most modeling works are quasi-static, i.e. based on surface energy minimization to find the equilibrium positions of the system, and on energy gradient to compute restoring forces [10, 14]. For small displacements, an analytical model to estimate the lateral restoring force developed by a meniscus between two square pads is sketched in [15]. Nevertheless, most of these studies does not address dynamical aspects such as characteristic damping time and resonance frequencies.

P. Lambert (✉)
BEAMS Department, Université libre de Bruxelles,
Avenue F.D. Roosevelt, 50, B-1050 Brussels, Belgium
e-mail: pierre.lambert@ulb.ac.be

P. Lambert (ed.), *Surface Tension in Microsystems*, Microtechnology and MEMS,
DOI: 10.1007/978-3-642-37552-1_8, © Springer-Verlag Berlin Heidelberg 2013

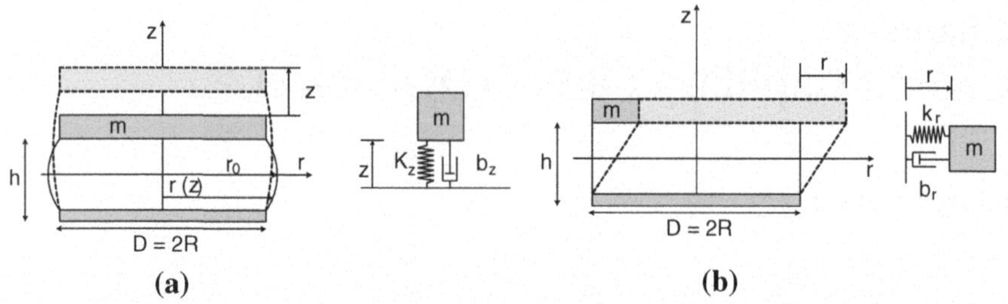

Fig. 8.1 Axial and radial models used by van Veen [16]. Reprinted from [8], with kind permission from Springer Science and Business Media. **a** Axial shift, **b** radial shift

In this direction, van Veen [16] derived analytical relations to model both the axial compressive motion and the lateral motion (Fig. 8.1). Kim et al. [6] proposed a study on dynamic modeling for resin self-alignment mechanism. Because the authors used a material with a low surface tension such as liquid resin, they claimed the alignment motion to be different from the oscillatory motion of the solder described by van Veen [16]. Recently, Lin et al. compared 2D numerical results with experiments [9]. To solve the motion equations, these authors used the CFD-ACE+ package making use of an iterative algorithm alternating one time step resolution of Navier-Stokes equations (and the continuity equation) with one time step resolution of the structural mechanics equations [9]. The authors observed a good match of their 2D simulation to experimental data in case of fluid meniscus aspect ratios larger than 3.

Lastly, Lu and Bailey [11] predicted the dynamical behavior of a chip in flip-chip alignment, whose dynamics is governed by the following Newton's law:

$$m\ddot{x} = -k(x)x - c\dot{x} \tag{8.1}$$

where $k(x)$ is the lateral stiffness of a meniscus linking a circular pad and a circular substrate, and the viscous force $-c\dot{x}$ is applied by the viscous stress on the chip. The lateral stiffness was computed with Surface Evolver and found to be almost constant, while the viscous force $c\dot{x} = \mu S \frac{\partial u}{\partial y}|_{y=h}$ was computed from the liquid flow between the chip and the pad, governed by the Navier-Stokes equation (μ is the dynamic viscosity of the liquid, S is the area of the chip in contact with liquid, u is the velocity along x-axis, y is the coordinate perpendicular to the chip):

$$\rho \frac{\partial u}{\partial t} - \mu \frac{\partial^2 u}{\partial y^2} = 0 \tag{8.2}$$

The coupled problem defined by (8.1) and (8.2) is thus described by two unknowns: firstly the position of the chip $x(t)$, and secondly the velocity profile inside the meniscus $u(y, t)$.

The first equation (8.1) is associated with the following initial conditions (δ is the initial elongation):

$$x(t=0)=\delta \tag{8.3}$$

$$\dot{x}(t=0)=0 \tag{8.4}$$

while the initial and boundary conditions for the second equation (8.2) are:

$$u(y=0,t)=0 \text{ (no slip condition)} \tag{8.5}$$

$$u(y=h,t)=\dot{x}(t) \text{ (coupling term)} \tag{8.6}$$

$$u(y,t=0)=0 \text{ (zero initial velocity)} \tag{8.7}$$

To decouple the equations, Lu and Bailey solved this problem by assuming a linear velocity profile $u(y,t)=(y/h)\dot{x}$. They also considered the application of a constant force along the top surface of the meniscus to compute the velocity profile. Their developments led to several response times: they showed that considering the chip dynamics only, the pseudo-period of the oscillating motion was of the order of 3.77 ms (using the underlying assumption of a linear velocity profile). On the other hand, they also showed that the characteristic time required for the velocity profile to become linear was of the same order of magnitude (5.3 ms). Therefore, they concluded that both physics (chip dynamics and fluid dynamics inside the meniscus) were strongly coupled and should be solved accordingly. They consequently proposed an iterative scheme including the alternate use of Surface Evolver to compute the meniscus stiffness and of a CFD package to compute the liquid flow.

To solve the same problem more efficiently, we hereby propose to use an analytical model for the lateral stiffness and a resolution based on Chebyshev polynomials, which transform the coupled problem of (8.1, 8.2) into the form of a system of constant-coefficients ordinary differential equations (ODEs).

We finally mention the need to study the other degrees-of-freedom (dof): beside the lateral motion, it is necessary to study the axial and tilt motions dynamically. Concerning the first one and additionally to the study of van Veen, partial information can be found in [3] (Stefan equation to estimate the axial viscous force), [12, 13] (the latter one studied the compression flow modeling). As far as the tilt motion is concerned, we refer to the work of [5], where use is made of water and also glycerin ($\gamma = 0.0635$ Nm^{-1} and $\mu = 0.900$ Pas) to study the oscillation of a tilted circular pad on a droplet for the self-alignment process. With their model, these authors estimated the damping ratio and the oscillating frequencies.

8.2 Lateral Stiffness of Liquid Meniscus

General Approach

The equilibrium shape of a meniscus is ruled by the so-called Laplace equation [7] which in the axially symmetric case is written as a non-linear second-order differential equation. Except in the case of a cylindrical meniscus or in the case of the catenoid, this problem does not have analytical solutions. Nevertheless, analytical approximations can be found in the case of circular and square pads. These approximations can be benchmarked using numerical solutions obtained with Surface Evolver [2]. It will be interesting to note that in both cases, the stiffness is proportional to half-perimeter of the wetted chip and to the surface tension, and inversely proportional to the gap.[1] Another way to interpret this result is to note that the lateral stiffness is proportional to the ratio of the component size over the gap of the meniscus.

Circular Pads

To estimate analytically the lateral stiffness of a meniscus confined between two circular pads, we compute firstly the lateral area of a tilted cylinder of height h, radius R and shift s (also called elongation x in Fig. 8.2): this means that the cylinder axis is not perpendicular to both circles but inclined with an angle ϕ given by $\tan\phi = s/h$.

The vectorial equation of a point P of this cylinder is given by:

$$\begin{aligned} \bar{OP} &= z\bar{1}_z + s(z)\bar{1}_x + R\bar{1}_r \\ &= \left(\frac{z}{h}s + R\cos\theta\right)\bar{1}_x + (R\sin\theta)\bar{1}_y + z\bar{1}_z \end{aligned} \tag{8.8}$$

The lateral area is consequently equal to:

$$S = R\int_0^h dz \int_0^{2\pi} \sqrt{1+\cos^2\theta\frac{s^2}{h^2}}\,d\theta \tag{8.9}$$

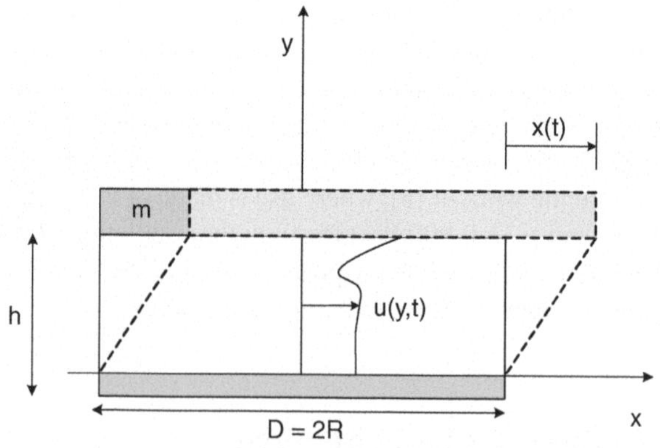

Fig. 8.2 Problem modeled in this work: a moving top pad is linked to a fixed bottom pad through a liquid meniscus whose velocity profile is assumed to be described by the horizontal velocity component u only. Reprinted from [8], with kind permission from Springer Science and Business Media

[1] This conclusion has been very recently extended to whatever polygonal shapes [1].

and the total surface energy of the system is here equal to:

$$E = \gamma S = \gamma Rh \int_{0}^{2\pi} \sqrt{1 + \cos^2\theta \frac{s^2}{h^2}}\, d\theta \tag{8.10}$$

where γ is the surface tension.

The lateral restoring force is consequently equal to:

$$F = -\frac{\partial E}{\partial s} = -\frac{\gamma Rs}{h} \underbrace{\int_{0}^{2\pi} \frac{\cos^2\theta}{\sqrt{1 + \cos^2\theta \frac{s^2}{h^2}}}\, d\theta}_{I} \tag{8.11}$$

The integral I in the latter equation can be numerically computed (see the result in Fig. 8.3) or expressed in terms of elliptic integrals of the first (EllipticK) and second (EllipticE) kinds:

$$I = 4\frac{h^2}{s^2}\left[\text{EllipticE}\left(\sqrt{-\frac{s^2}{h^2}}\right) - \text{EllipticK}\left(\sqrt{-\frac{s^2}{h^2}}\right)\right] \tag{8.12}$$

Nevertheless, it is interesting to calculate it in our domain of interest (i.e. for small s/h ratio, $\sqrt{1 + \cos^2\theta \frac{s^2}{h^2}} \approx 1$ in (8.11)), which gives the following analytical relationship:

$$F = -\frac{\partial E}{\partial s} = -\pi R\gamma \frac{s}{h} \tag{8.13}$$

which corresponds to a constant stiffness k_c given by:

$$k_c = \frac{\pi R\gamma}{h} \tag{8.14}$$

The existence of an analytical expression of the stiffness allows the numerical simulation presented in Sect. 8.3, where the advantages of such a resolution will be discussed.

This formulation has been benchmarked using Surface Evolver, in the case of $r = 50\,\mu\text{m}$, $h = 70\,\mu\text{m}$, $\gamma = 0.325\,\text{Nm}^{-1}$ and a volume of liquid given by $V = \pi r^2 h$. The comparison is plotted in Fig. 8.3. The discrepancy between the numerical and Surface Evolver results comes from the fact that the cylindrical geometry of (8.8) is not physically exact, since such a shape does not have a constant curvature.

Square Pads

Tsai [15] already presented a model to compute the lateral stiffness of a square pad shifted by a distance s along one of its edges. This model is based on the assumption

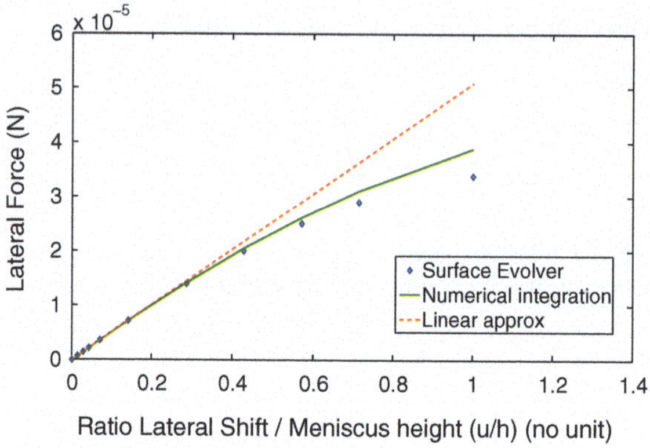

Fig. 8.3 Lateral restoring force and related stiffness of a cylindrical meniscus: comparison between the Surface Evolver benchmark, the analytical approximation of (8.13) and the numerical evaluation of (8.11) ($R = 50\ \mu$m, $h = 70\ \mu$m and $\gamma = 0.325$ Nm^{-1}). Note that the '-' sign of the force has been omitted here. Reprinted from [8], with kind permission from Springer Science and Business Media

of a prismatic meniscus whose volume is equal to the area of the square pad multiplied by the gap. In this case, a shift of the component keeps the volume constant at constant gap, but the lateral area Σ of the liquid-vapor interface is increasing, given by:

$$\Sigma = 2ch + 2c\sqrt{h^2 + s^2} \tag{8.15}$$

Consequently, the surface energy E in this problem is given by:

$$E = \gamma\Sigma + C \tag{8.16}$$

where C is an arbitrary constant. The derivation with respect to the shift s leads to the restoring force:

$$F = -\frac{2\gamma cs}{\sqrt{s^2 + h^2}} \approx -\frac{2cs\gamma}{h} \tag{8.17}$$

The approximation is consistent with the assumption that $s < h$ and consequently $s^2 \ll h^2$. To fix the ideas, let us consider $s < h/10$. Note that the force direction is opposite to the shift. The force derivative leads to the stiffness (Fig. 8.4):

$$k_s = \frac{2ch^2\gamma}{(s^2 + h^2)^{\frac{3}{2}}} \approx \frac{2c\gamma}{h} \tag{8.18}$$

Note well that $2c$ in this formula is half perimeter of the pad, which was also the case in the circular pad ($\pi R\gamma/h$). The comparison between (8.18) and Surface Evolver is shown in Fig. 8.5, where again there is a small difference between the model and Surface Evolver because the shape of Fig. 8.4 does not have a constant curvature.

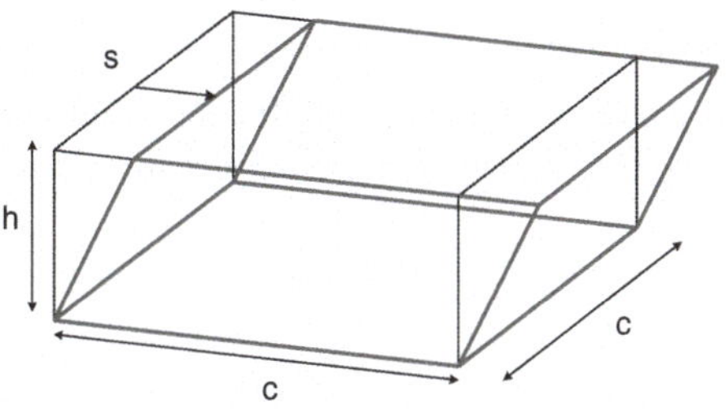

Fig. 8.4 Geometrical model used to compute the lateral stiffness of a square pad. Reprinted from [8], with kind permission from Springer Science and Business Media

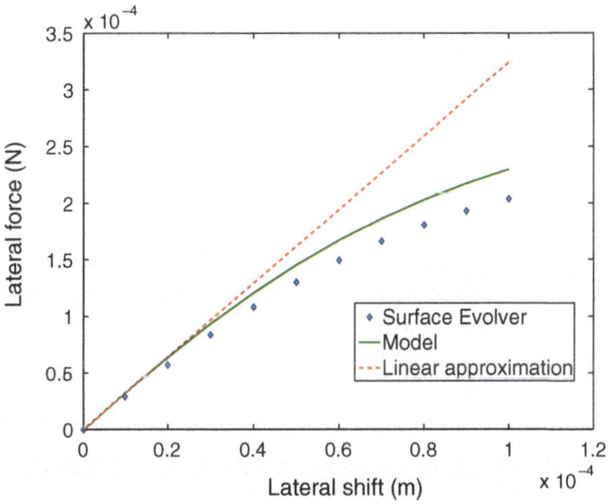

Fig. 8.5 Lateral capillary force in case of a square pad with edge $c = 500\,\mu\text{m}$, a gap $h = 100\,\mu\text{m}$, a surface tension $325\,\text{m}\,\text{Nm}^{-1}$. Reprinted from [8], with kind permission from Springer Science and Business Media

8.3 Coupled Problem Resolution with a Pseudo-Spectral Method Based on Chebyshev Polynomials

The problem described by (8.1, 8.2) is firstly normalized using characteristic length δ and time τ:

$$x = \delta\tilde{x} \tag{8.19}$$

$$t = \tau\tilde{t} \tag{8.20}$$

$$u = \frac{\delta}{\tau}\tilde{u} \tag{8.21}$$

where δ is the initial position of the chip and $\tau = \sqrt{\frac{m}{k}}$ is a characteristic time of the chip dynamics. The y coordinate ranging from 0 to h is replaced by $\tilde{y}$ ranging from -1 to 1 ($y = (1+\tilde{y})\frac{h}{2}$). This leads to two non dimensional equations:

$$\ddot{\tilde{x}} = -\tilde{x} - \underbrace{\frac{\rho S h}{2m}}_{\tilde{m}^{-1}} \underbrace{\frac{4\nu\tau}{h^2}}_{\alpha} \left.\frac{\partial \tilde{u}}{\partial \tilde{y}}\right|_{\tilde{y}=+1} \tag{8.22}$$

$$\frac{\partial \tilde{u}}{\partial \tilde{t}} = \frac{4\nu\tau}{h^2}\frac{\partial^2 \tilde{u}}{\partial \tilde{y}^2} \tag{8.23}$$

with the following boundary and initial conditions:

$$\tilde{u}(\tilde{y} = -1, \tilde{t}) = 0 \tag{8.24}$$

$$\tilde{u}(\tilde{y} = +1, \tilde{t}) = \dot{\tilde{x}} \tag{8.25}$$

$$\tilde{u}(\tilde{y}, \tilde{t} = 0) = 0 \tag{8.26}$$

$$\tilde{x}(\tilde{t} = 0) = 1 \tag{8.27}$$

$$\dot{\tilde{x}}(\tilde{t} = 0) = 0 \tag{8.28}$$

It can be seen from both non dimensional equations that the problem only depends on two non dimensional parameters: (i) $\alpha = 4\nu\tau/h^2$ the diffusion coefficient and (ii) $\tilde{m} = 2m/(\rho S h)$ the mass ratio.

In the following the symbol $\tilde{}$ has been dropped everywhere and the superscript $^{(i)}$ indicates the ith derivative with respect to y. With these conventions, the unknown velocity field u can be written as a series of Chebyshev polynomials $T_k(y)$:

$$u(y,t) = \sum_{k=0}^{N} a_k(t) T_k(y) \tag{8.29}$$

After some developments, a new (linear) system of equations can be written as:

$$M\dot{X} = AX \tag{8.30}$$

where the unknown vector X with $N+2$ unknowns contains the unknown Chebyshev coefficients a_i and the chip position x:

$$X = (a_0\ a_1 \ldots\ a_i \ldots\ a_N\ x)^T \tag{8.31}$$

The corresponding initial condition are:

$$X(t=0) = X_0 = (\underbrace{0 .. 0}_{1\times N+1}\ \ 1)^T \tag{8.32}$$

M and A matrices will be detailed.

This constant coefficient system of ODEs (8.30) is easily solved analytically:

$$X_i = \sum_{j=1}^{N+2} r_{ij} w_j e^{\lambda_j t} \tag{8.33}$$

where λ_j is the jth eigen value, r_{ij} the ith component of the jth eigenvector associated with λ_j and w_j the jth component of the vector $\mathbf{w}$ given by $\mathbf{w} = V^{-1}X_0$, and V is the matrix whose columns are the eigenvectors r_j.

8.4 Results and Discussion

Benchmark

We have benchmarked our method with the theoretical results of Lu and Bailey [11], as indicated in Fig. 8.6 by superimposing the results of our work on the results of Lu and Bailey (Fig. 9 of their paper). The difference is about 4 %. The main advantage of the proposed resolution method is to reduce the computing time drastically while still keeping the physics coupled (224 couples ($\tilde{m}$,α) can be computed in 5–10 minutes), by comparison with the method of Lu and Bailey using Surface Evolver and a CFD package for the flow problem. This advantage permits the parametric study presented in the next section.

Numerical Results

For any couple of parameters (α, $\tilde{m}$), (8.33) gives $x(t) = X_{N+2}(t)$ the position of the plate as a combination of different modes whose amplitude decreases with time (real parts of λ_j are negative). Consequently, the decrease of x with time is governed

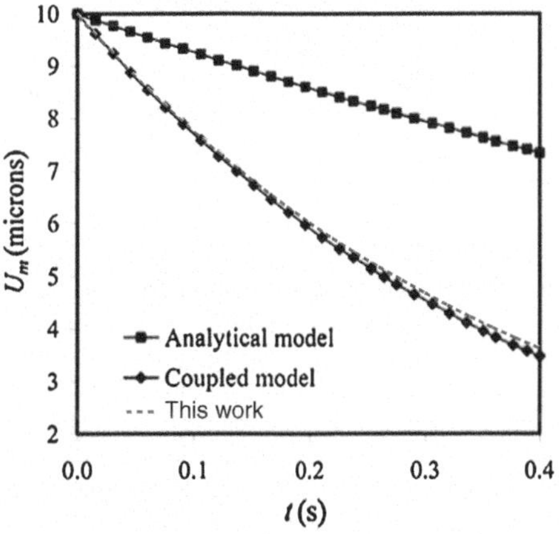

Fig. 8.6 Benchmark of our solution by comparison with Fig. 9 from Lu and Bailey. For $\alpha = 0.18$, $\tilde{m} = 99$, $\delta = 10$ µm, $N = 11$, we found a non dimensional damping time $\tau_d = 0.3958$ where Lu and Bailey found $\tau_d = 0.3784$ (4 % difference). Reprinted from [8], with kind permission from Springer Science and Business Media

by the slowest mode, i.e. by the mode with the eigen value λ whose (negative) real part has the smallest absolute value. τ_c is then equal to $-1/\lambda$. The combinatory space defined by $\alpha = 40 - 4000$ (logspace with 16 points) and $\tilde{m} = 400 - 1000$ (logspace with 14 points) has been explored, outputing a grid of 16×14 points. The τ_c at these points have been divided by their respective characteristic time $\tau = \sqrt{m/k}$ and interpolated with Matlab using the contour function in order to display the iso-lines of Fig. 8.7.

Experimental Results

Thanks to a dedicated set up described in Chap. 3, a large set of experiments has been performed, as indicated in Table 8.1, whose column "liquid ID" points to the liquid properties given in Table 8.2. The good signal-to-noise ratio given by the ratio of the mean over the standard deviation of characteristic time τ_c is reported. Missing numbers in the table are experiments that were discarded because of smaller signal-to-noise ratio, that would have led to a large scattering of results ; they correspond to less viscous liquids, i.e. water ($\mu = 1$ mPas) and Dow Corning DC200FLUID10 oil ($\mu = 9.3$ mPas), leading to longer damping times up to 32 s: since $u(t)$ has not

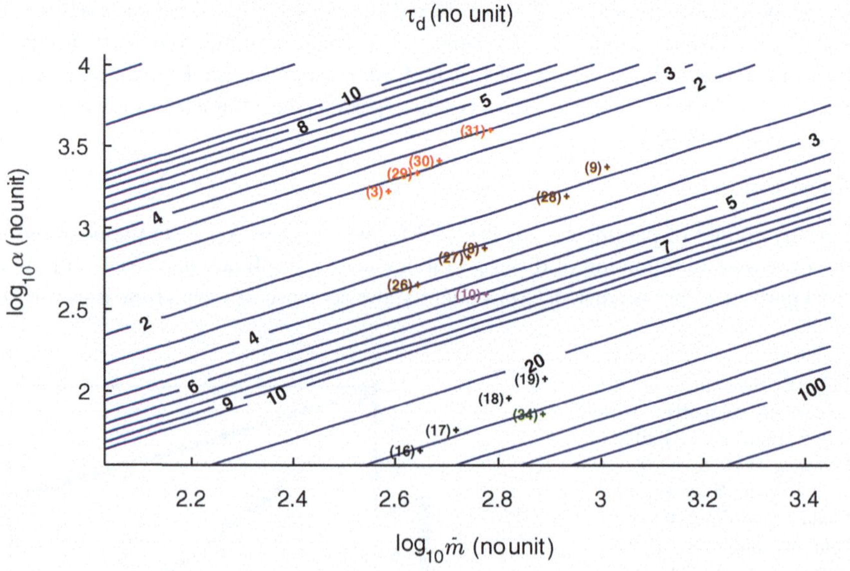

Fig. 8.7 Non-dimensional map of damping times, for experiments 3,29-31 (Oil 2); 10 (Oil 1); 16–19 (Oil 6); 8–9, 26–28 (Oil 7); 34 (Dymax 628 VLV). The family set of parallel lines represents the iso-values of $\tau_{d(sim)}$ obtained by simulation. Each experiment number is indicated between brackets, the '+' mark to its right showing the exact location of the experimental couple $(\tilde{m}, \alpha)$. The numerical comparison between simulation and experiment can be found in Table 8.1, this map of iso-values of τ_d can be used to estimate the damping time. For example, experiment 30 ($\alpha = 2610$, $\tilde{m} = 484$) lies between the iso-line 2 and the iso-line 3 (the measured value is 2.3). Reprinted from [8], with kind permission from Springer Science and Business Media

Table 8.1 Experiments are stacked by liquid ID (a table will be provided)

Exp. no	Liquid ID	α (no unit)	$\tilde{m}$ (no unit)	Times τ_{c-mean} (s)	τ_{c-std} (s)	$\frac{\tau_{c-mean}}{\tau_{c-std}}$ (no unit)	τ (s)	$\tau_d = \frac{\tau_{c-mean}}{\tau}$ (no unit)
34	Dymax	72.6	770	1.749	0.081	21.6	0.097	18.0
16	6	43.2	445	4.396	0.094	46.9	0.108	40.6
17	6	57.8	522	3.523	0.110	32.1	0.105	33.5
18	6	90.8	661	2.306	0.138	16.7	0.103	22.4
19	6	119	777	2.147	0.153	14.0	0.098	21.9
10	1	395	597	0.686	0.070	9.8	0.107	6.3
8	7	761	593	0.272	0.012	23.1	0.101	2.6
9	7	2390	1030	0.101	0.013	7.7	0.110	0.9
26	7	451	440	0.475	0.010	48.0	0.113	4.2
27	7	669	552	0.366	0.007	50.9	0.106	3.4
28	7	1580	855	0.236	0.009	26.8	0.104	2.2
3	2	1700	386	0.254	0.029	8.6	0.110	2.3
29	2	2190	439	0.193	0.010	20.2	0.109	1.7
30	2	2610	484	0.249	0.008	29.8	0.107	2.3
31	2	4030	609	0.305	0.024	12.6	0.104	2.9

We indicate for each experiment the gap. The characteristic time τ_c is the characteristic time obtained from the dynamic response of the shuttle $u(t)$. As indicated in the text, these data have been averaged and the standard deviation has been used to compute a signal-to-noise ratio, which is shown to be always larger than 7.7. The characteristic time $\tau = \sqrt{m/k_d}$ is recalled — where m is the equivalent mass of the shuttle and k_d is the sum of the shuttle stiffness $k = 1\,\mathrm{N\,m^{-1}}$ and the meniscus stiffness measured in that experiment (not shown here). The last column gives τ_d, the non dimensional ratio of τ_c over τ, which is plotted in Fig. 8.7

Table 8.2 Liquid properties

Liquid ID	ρ ($\mathrm{Kg\,m^{-3}}$)	μ (Pas)	γ ($\mathrm{N\,m^{-1}}$)	Supplier
1	970	0.485	0.0211	Rhodorsil R47V500
7	971	0.971	0.0212	Dow Corning DC200FLUID1000
2	973	4.865	0.0211	Rhodorsil R47V5000
Dymax	1050	0.055	0.025	ρ and μ: Dymax 628-VLV data γ: own measurement

been registered on a sufficiently large period of time, the amplitude decrease was not large enough to ensure reliable fit, hence a larger standard deviation. For the other experiments, we can see in Fig. 8.7 the fair concordance between simulation and experimental results. Thanks to both non dimensional parameters α and $\tilde{m}$, the map of Fig. 8.7 contains all information concerning the dynamical lateral behavior of a liquid meniscus.

Over- and Underdamping

Theoretically, a mechanical system governed by (8.1) is underdamped if $c^2 - 4\,km < 0$. Since c is related to the dimensional characteristic time τ_c by the relation $c =$

$2m/\tau_c$, the former condition for overdamping becomes $\tau_c > \sqrt{\frac{m}{k}}$ which can also expressed using the non-dimensional damping time τ_d as $\tau_d > 1$. Experimentally, the last column of Table 8.1 indicates a border between both behaviors between 2 and 3 instead of 1.

We have shown that solving the coupled problem requires a numerical approach. Nevertheless, assuming a linear velocity profile in the meniscus, (8.1) can be rewritten into:

$$m\ddot{x} + \frac{\mu S}{h}\dot{x} + kx = 0 \tag{8.34}$$

$$\ddot{x} + 2\underbrace{\frac{\alpha}{4\tilde{m}}}_{\equiv\lambda}\frac{1}{\tau}\dot{x} + \frac{1}{\tau^2}x = 0 \tag{8.35}$$

This equation admits two roots:

$$r_{1,2} = \frac{1}{\tau}\left(-\lambda \pm \sqrt{\lambda^2 - 1}\right) \tag{8.36}$$

- Overdamping: $\lambda > 1$
  Therefore, $x(t)$ is given by:

$$x(t) = A\exp\left(-\frac{t}{\tau}(\lambda - \sqrt{\lambda^2 - 1})\right) + \underbrace{B\exp\left(-\frac{t}{\tau}(\lambda + \sqrt{\lambda^2 - 1})\right)}_{\text{more quickly damped}} \tag{8.37}$$

  the characteristic damping time of which τ_c being equal to $\tau/(\lambda - \sqrt{\lambda^2 - 1})$. Consequently, τ_d is given by:

$$\tau_d = \frac{\tau_c}{\tau} = \frac{1}{\lambda - \sqrt{\lambda^2 - 1}} \tag{8.38}$$

  and

$$1 \leq \tau_d \leq \infty \tag{8.39}$$

  Since $\lambda \equiv \alpha/(4\tilde{m})$, we can express τ_d as a function of α and $\tilde{m}$:

$$\log\alpha = \log\tilde{m} + \log\left(\frac{2(1 + \tau_d^2)}{\tau_d}\right) \tag{8.40}$$

- Underdamping: $\lambda < 1$
  In case of underdamping, the general solution $x(t)$ is given by:

$$x(t) = A \exp\left(-\frac{\lambda}{\tau}t\right)\cos(\omega t + \phi) \tag{8.41}$$

The characteristic damping time τ_c is here equal to:

$$\tau_c = \frac{\tau}{\lambda} \tag{8.42}$$

leading to:

$$\tau_d = \frac{\tau_c}{\tau} = \frac{1}{\lambda} \geq 1 \tag{8.43}$$

Once again, we can express τ_d as a function of α and $\tilde{m}$:

$$\log\alpha = \log\tilde{m} + \log\left(\frac{4}{\tau_d}\right) \tag{8.44}$$

We see that in both under- and overdamping situations, $\tau_d \geq 1$. Moreover, if we remember that a typical diffusion time is equal to $h^2/(4\nu)$ [4], the hereabove mentioned equations can be expected to be valid when the characteristic time τ is larger than this typical diffusion time, i.e. when $\alpha \gg 1$. All these results are shown in Fig. 8.8:

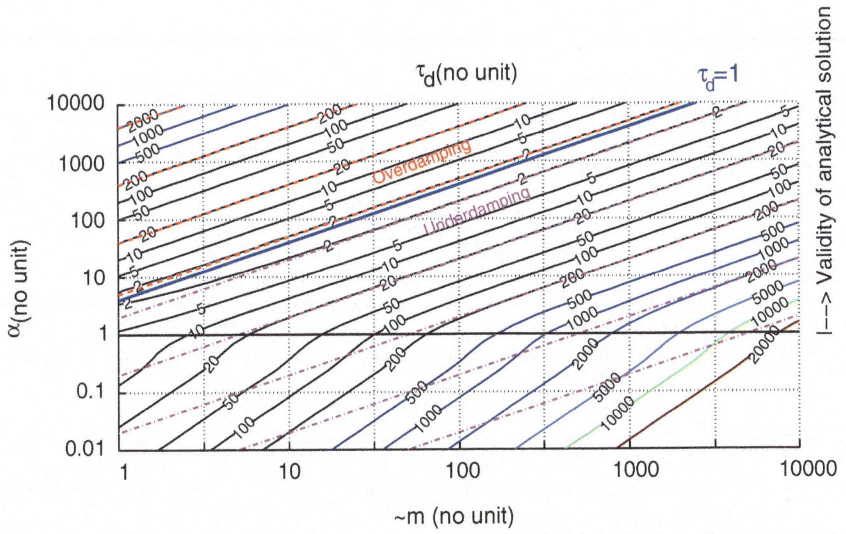

Fig. 8.8 Map of the non-dimensional damping time τ_d as a function of the normalized mass $\tilde{m}$ and the normalized viscosity α. τ_d is always larger than 1, but overdamping is located above the straight line $\tau_d = 1$, while underdamping is located below this line. The domain above $\alpha = 1$ is the validity domain of the analytical expressions of the damping time (8.40) and (8.44)

- overdamping domain is located above the line $\tau_d = 1$;
- underdamping domain is located below the line $\tau_d = 1$;
- the analytical expression is valid when $\alpha \gg 1$.

8.5 Conclusions

The coupled problem of liquid and chip dynamics could be efficiently solved with a pseudo-spectral method (Chebyshev polynomials). Beside the fair agreement observed during the thorough experimental validation, some discrepancies were however observed for low viscosity configurations. It could nevertheless be shown that for various experiments covering the experimental space $\alpha = 40 - 4000$ and $\tilde{m} = 400 - 1000$, the numerical proposed resolution and the subsequent modeling have been proven reliable. A discussion was finally led to provide criteria for under- and overdamping behaviours.

Acknowledgments This work is conducted with financial support from the project Hybrid Ultra Precision Manufacturing Process Based on Positional and Self assembly for Complex Micro-Products (HYDROMEL NMP2-CT-2006-026622) funded by the European Commission. A special thanks to B. Tartini. Thanks to Bruno Tartini for manufacturing the mechanical components of the experimental set up, and thanks to my students for their help in these developments, especially Jeanne Boute and Carsten Engel.

References

1. J. Berthier, S. Mermoz, K. Brakke, L. Sanchez, C. Frétigny, L. Di Cioccio, Capillary self-alignment of polygonal chips: a generalization for the shift-restoring force. Microfluid. Nanofluid. **14**(5), 845–858 (2012)
2. K. Brakke, The surface evolver. Exp. Math. **1**(2), 141–165 (1992)
3. J. Engmann, C. Servais, A.S. Burbidge, Squeeze flow theory and applications to rheometry: a review. J. Nonnewton. Fluid. Mech. **132**(1–3), 1–27 (2005)
4. E. Guyon, J.-P. Hulin, L. Petit, Hydrodynamique physique, CNRS Editions (2012)
5. M. Kaneda, M. Yamamoto, K. Nakaso, T. Yamamoto, J. Fukai, Oscillation of a tilted circular pad on a droplet for the self-alignment process. Precis. Eng. **31**, 177–184 (2007)
6. J.M. Kim, Y.E. Shin, K. Fujimoto, Dynamic modeling for resin self-alignment mechanism. Microelectron. Reliab. **44**, 983–992 (2004)
7. P. Lambert, *Capillary Forces in Microassembly: Modeling, Simulation, Experiments, and Case Study. Microtechnology and MEMS* (Springer, New York, 2007)
8. P. Lambert, M. Mastrangeli, J.-B. Valsamis, G. Degrez, *Spectral Analysis and Experimental Study of Lateral Capillary Dynamics (for Flip-Chip Applications)*. Microfluid. Nanofluid **9**, 797–807 (2010). Published online
9. C. Lin, F. Tsen, H. C. Kan, C.-C. Chieng, *Numerical Studies on Micropart Self-Alignment using Surface Tension Forces*, Microfluid. Nanofluid. **6**(1), 63–75 (2008)
10. W. Lin, S.K. Patra, Y.C. Lee, Design of solder joints for self-aligned optoelectronic assemblies. IEEE Trans. Adv. Packag. **18**(3), 543–551 (1995)
11. H. Lu, C. Bailey, Dynamic analysis of flip-chip self alignment. IEEE Trans. Adv. Packag. **28**(3), 475–480 (2005)

12. M.-H. Meurisse, M. Querry, Squeeze effects in a flat liquid bridge between parallel solid surfaces. J. Tribol. **128**(3), 575–584 (2006)
13. N.W. Pascarella, D.F. Baldwin, Compression flow modeling of underfill encapsulants for low cost flip chip assembly, in *Proceedings. of 1998 International Symposium of Underfill Encapsulants for Low Cost Flip Chip Assembly* (1998), p. 33
14. K.F. Bohringer, S.-H. Liang, X. Xiong, Towards optimal designs for self-alignment in surface tension driven self-assembly in *Proceedings the 17th IEEE International Conference on MEMS*, IEEE (2004) pp. 9–12
15. C.G. Tsai, C.M. Hsieh, J.A. Yeh, Self-alignment of microchips using surface tension and solid edge. Sens. Actuators A **139**, 343–349 (2007)
16. N. van Veen, Analytical derivation of the self-alignment motion of flip chip soldered components. J. Electron. Packag. **121**, 116–121 (1999)

Chapter 9
Conclusions

Pierre Lambert

This chapter ends up the Part III of this book devoted to dynamic modelling of capillary forces. It has been shown in Chaps. 7, 8 that a meniscus with an infinite number of degrees-of-freedom could be modelled by second-order systems with only one degree-of-freedom for the axial translation and one degree-of-freedom for the lateral shift. The coupling of both degrees-of-freedom was not investigated. As a result, this part provides analytical estimates to compute the coefficients of the second-order systems, and, henceforth, to quickly assess the dynamics behaviour. The meniscus dynamics can for instance been applied to self-centering in hybrid assembly.

This also closes the theoretical part of this book and introduces Part IV devoted to case studies, i.e. a capillary micromotor (Chap. 10), a capillary based sealing (Chap.11), surface tension-driven self assemblies (Chap. 12), surface tension-driven actuation (Chap. 13) and liquid nanodispensing (Chap. 14).

P. Lambert (✉)
BEAMS Department, Université libre de Bruxelles, 50, av. F.D. Roosevelt,
1050 Brussels, Belgium
e-mail: pierre.lambert@ulb.ac.be

P. Lambert (ed.), *Surface Tension in Microsystems*, Microtechnology and MEMS,
DOI: 10.1007/978-3-642-37552-1_9,

Part IV
Case Studies

Chapter 10
Capillary Micro Motor

Atsushi Takei

Abstract A micro structure supported on a droplet is subjected to capillary force and aligned depending on its shape as we explained in Chap. 4. If the droplet's boundary condition at the bottom and the micro structure are non-circular, a capillary torque is exerted on the structures. The direction of the torque is determined by the boundary condition and the position of the structure. By changing the boundary condition continuously, the rotational motion of the plate was achieved. The boundary condition of the droplet was controlled by electrowetting. We patterned electrodes in annular shape on the plate supporting the droplet. By changing the voltage-applied electrodes, the boundary condition is changed and the plate is rotated. With this method, the droplet and the plate work as a capillary motor. In this chapter, we describe the relationship between the characteristics of the capillary motor and its rotational motion. We sandwiched 3.0 μL water droplet between 2 mm plate and achieved the rotational motion of 720 rpm at the maximum (Content of this chapter, including all pictures and tables, originally published in [21] © 2010 Royal Society of Chemistry (available at: http://pubs.rsc.org/en/Content/ArticleLanding/2010/LC/c001211d). Presented hereby in revised form with permission of The Royal Society of Chemistry).

10.1 Introduction

Fluid has been applied to optical systems in lab-on-a-chip, using its transparency and high refractive index. For example, by controlling the stream of the fluid, optofluidic lenses have been achieved in microchannels [1, 2]. Other optical functions have recently been attained by combining microstructures and fluid. Tunable prisms have been achieved by controlling the angle of a micro-plate on a droplet or at the interface

A. Takei (✉)
Institute of Industrial Science, The University of Tokyo,
4-6-1 Komaba, Meguro-ku, Tokyo 153-8505, Japan
e-mail: atsushi.takei@espci.fr

P. Lambert (ed.), *Surface Tension in Microsystems*, Microtechnology and MEMS,
DOI: 10.1007/978-3-642-37552-1_10, © Springer-Verlag Berlin Heidelberg 2013

of fluid [3, 4]. While tilting motion has been accomplished, it is still hard to attain rotational motion. Although a silicon micro-motor driven by electrostatic force has previously been reported [5], it required the rotor and stator to be precisely fabricated and transparent material was not suitable for fabricating it. Therefore, it is hard to apply micro-motors to optical or optofluidic systems. Since optical devices such as diffractive gratings or polarizers activate or switch their functions in rotational positions, another method of rotating microstructures is required. Rotational, horizontal, or vertical movements in micro-structures can be seen in self-alignment using capillary force. Capillary force becomes dominant on a small scale (see Fig. 1.1) and has been used for fabricating micro-structures. As reported in Chap. 12, self-alignment of Integrated Circuit (IC) chips has previously been reported [6–8]. A droplet was placed on a hydrophilic domain patterned on a plate, and then an IC chip was placed on the droplet. The chip moved to a particular position, since the potential energy of the droplet was proportinal to its own surface area and force was applied to the chip to make its surface take a minimum value. We explained the relationship between capillary torque and the shape of the liquid in relation to alignment in the rotational direction in Chap. 4. A plate placed on a droplet maintains its horizontal position. If the droplet is sandwiched between two non-circular plates, the droplets exert torque on the plates. Self-alignment using capillary force has generally dealt with the static position of microstructures, since boundary conditions such as those of the hydrophilic domain do not change dynamically. Electrowetting, on the other hand, is a method of actuation to control the boundary conditions of liquid (see Sect. 1.7) and has been used in optical systems [9] or drop-delivery systems [10, 11]. If a microstructure is floating on a droplet and the bottom edge of the droplet is continuously changed by electrowetting, the floating plate can be rotated. In comparison with other methods (i.e, those involving thermocapillaries [12], structured surfaces [13], and electro-osmosis [14]), a droplet actuated by electrowetting has high velocity. Moreover, the inertial moment becomes minute on a small scale. Therefore, quick motion can be expected. This chapter presents a motor driven by electrowetting. Our motor is composed of two plates and a droplet sandwiched between them. The droplet is placed on one of the flat plates and the other non-circular plate is floated on the droplet. Electrodes are patterned with an annular shape on the base plate and the bottom edge of the droplet is surrounded by the electrodes. By applying voltage to part of the electrodes, the bottom edge of the droplet is deformed and the stable rotational position of the floating plate is changed. Electrowetting has generally been used to manipulate droplets vertically or horizontally [15–19]. By combining a droplet with microstructures, rotational motion was achieved. Lee and Kim [20] found a droplet circulates in a circular channel, on the other hand the droplet of our motor is rotated without sliding. We carried out experiments to establish the relationship between rotational motion and three parameters, i.e., the shape of the plates, the volume of the sandwiched droplet, and the weight of the floating plate. Our motor did not require any stator or rotor and could be made solely of transparent materials. This makes it advantageous for optical or optofluidic systems.

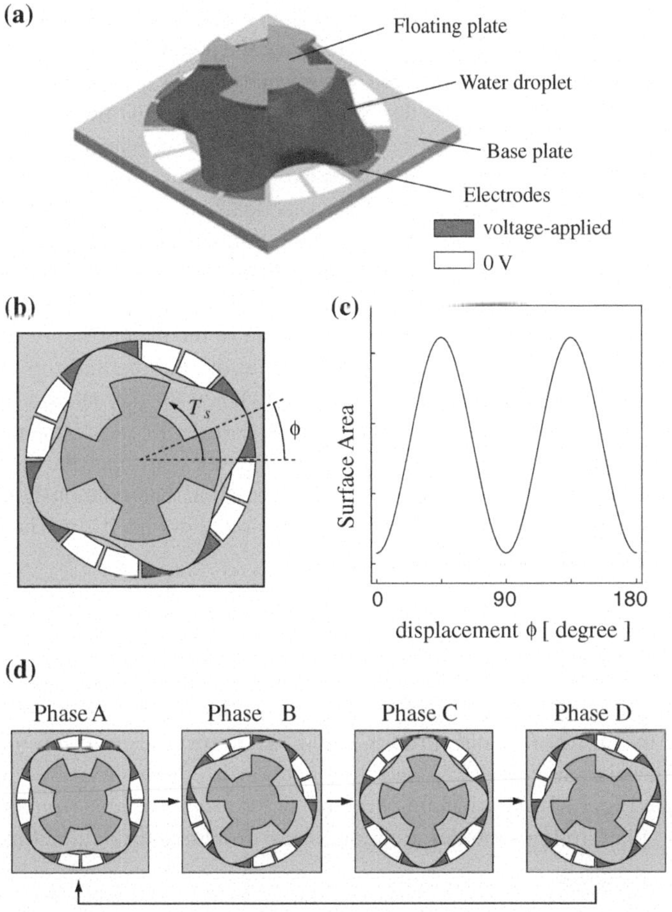

Fig. 10.1 **a** Schematic of capillary motor. Plate with four cogs is floated on droplet and droplet is supported by other plate. On base plate, electrodes are patterned and bottom edge of droplet can be controlled by electrowetting. **b** Schematic of top view of capillary motor. **c** Surface area of droplet as a function of rotational displacement of floating plate. Surface area takes minimum value at $\phi = 0$. **d** Rotational position of floating plate can be controlled by changing the voltage-applied electrodes. [21]—Reproduced by permission of the Royal Society of Chemistry

10.2 Principle and Method

10.2.1 Principal

Figure 10.1 depicts the capillary motor schematically. A droplet is placed on a plate and a non-circular plate is floated on the droplet. Electrodes are patterned in an annular shape on the base plate, and the edge of the droplet is controlled by choosing the voltage-applied electrodes. The rotational position of the floating plate is determined by the boundary conditions of the base plate, since the potential energy of the droplet is proportional to its surface area and the floating plate moves to minimize the surface. In other words, the droplet exerts the restoring moment on the floating plate, once the floating plate is displaced from the stable position. To estimate the stable position and the direction of the restoring moment, we need to take into consideration the conditions in Fig. 10.1b. The plate with four cogs floats on the droplet and the droplet is pulled in four directions. The ϕ angle is the rotational displacement between the cog symmetry line and that of the maximum displaced point of the droplet. We previously reported the relationship between the change in the surface area of a sandwiched droplet and the rotational displacement of the plate (Chap. 4). From the presented theory, when the boundary conditions at the bottom and top planes are repeated every 90°, the surface area of the sandwiched droplet takes a maximum value at $\phi = 45° + 90° \times n$ and a minimum at $\phi = 0° + 90° \times n$. This means that the stable position of the floating plate is at $\phi = 0° + 90° \times n$, since the the potential energy of the droplet is proportional to its surface area. The relationship between rotational displacement ϕ and the surface area is plotted in Fig. 10.1c. The restoring moment applied to the floating plate is given by

$$T = -\gamma \frac{\partial A}{\partial \phi} \tag{10.1}$$

where A is the surface area and γ is the surface free energy of the sandwiched liquid (i.e. its surface tension). From Chap. 4, we know that restoring force takes a maximum value at $\phi = 22.5°$ and the direction is counterclockwise under the conditions in Fig. 10.1b. By changing the edge of the droplet as seen from Fig. 10.1d, the floating plate follows the deformation of the droplet and rotational moment is achieved.

10.2.2 Fabrication

The capillary motor is composed of a floating plate, a base plate and liquid sandwiched between them. Electrodes are patterned on the base plate and the boundary conditions of the droplet can be controlled by applying voltage. The base plate is made of Indium Tin Oxide (ITO) on a glass substrate and its surface is coated with a hydrophobic and insulating layer of Cytop (an amorphous fluorocarbon polymer, Asahi Glass Co., Ltd.). The thickness of the Cytop layer is 0.65 μm and its relative permittivity is $\varepsilon = 2.1$. The Cytop layer was spin-coated at 3000 rpm. If a flat electrode is situated under a flat cytop layer that is 0.65 μm thick and voltage is applied between a droplet on the surface and the electrode, the contact angles of the droplet at 0 and 60 V correspond to 110° and 80°. Sixteen electrodes were patterned in a circular shape and one electrode connected to the ground was patterned as shown in Fig. 10.2b. The angle of the electrodes is 22.5°, since, from theoretical estimates,

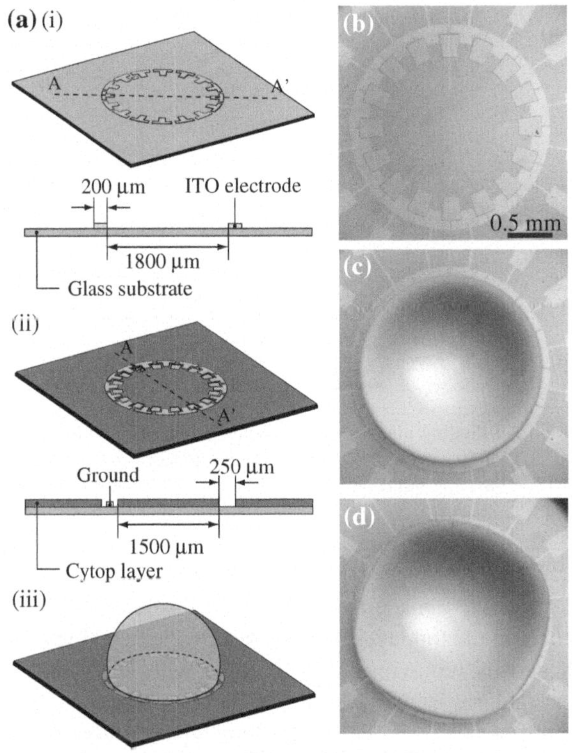

Fig. 10.2 **a** Schematic of fabrication process for base plate. Hydrophobic layer is etched in annular shape and edge of droplet matches exposed area of glass. **b** Photograph of base plate and **c** that of base plate with droplet. **d** By applying voltage to part of electrodes, edge of droplet moves toward the voltage-applied electrodes. [21]—Reproduced by permission of the Royal Society of Chemistry

maximum torque is generated at rotational displacement $\phi = 22.5°$. The width of the electrodes is 200 μm. The hydrophobic layer is etched in an annular shape and the glass layer is exposed. Once a droplet is placed on the base plate, its edge stays in this glass-exposed area as shown in Fig. 10.2c. The distance between the outer edges of the glass-exposed area and the annular electrodes is 100 μm. From the experimental results obtained by Sen and Kim [17], the magnitude of the contact-line oscillaton of a 4 μL water droplet actuated at lower than 100 Hz is larger than 100 μm. Thus, the edge of the droplet is supposed to reach the outer edge of the electrodes by applying voltage. The ground-connected electrode is also exposed and the droplet is electrically connected to the ground. With no voltage applied, the droplet stays spherical. However, the boundary conditions of the droplet are deformed by applying voltage as shown Fig. 10.2d. We applied 60 V under the conditions in Fig. 10.2d. As seen in the figure, 60 V was adequate to make the edge of the droplet move to the outer edge of the electrode. By changing the voltage-applied electrodes, the edge of the droplet is tunable.

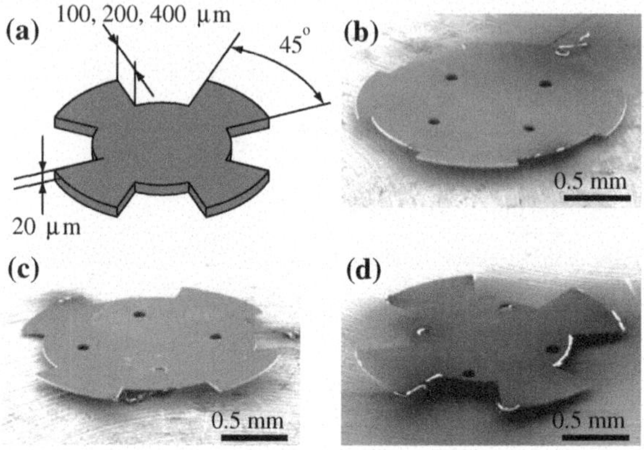

Fig. 10.3 **a** Schematic of floating plate. Outer diameter, thickness, and the cog angle of floating plate correspond to 2, 20 or 50 μm, and 45°. SEM photographs of floating plates of **b** type 1 (L = 100 μm), **c** type 2 (L = 200 μm) and **d** type3 (L = 400 μm). [21]—Reproduced by permission of the Royal Society of Chemistry

The floating plates were made from silicon on insulator (SOI) wafer. First, the plates were formed on the top layer by deep reactive ion etching (DRIE). Then the plates were released by removing the back layer with DRIE and the glass layer with hydrofluoric acid, respectively. Figure 10.3a has a schematic of the plates. The floating plate has four cogs. We fabricated four types of floating plates. The cog length, L, and the thickness, t, of the plates we fabricated were: L = 100 μm and t = 20 μm (type 1), L = 200 μm and t = 20 μm (type 2), L = 400 μm and t = 20 μm (type 3) and L = 400 μm and t = 50 μm (type 4). The cog angle was 45° and their outer diameter was 2 mm. The scanning electronic microscopy (SEM) images of three of the fabricated plates (types 1–3) are shown in Figs. 10.3b–d. Four holes were patterned on the plates to clarify the rotational movement for computational image tracking.

A droplet of deionized water was placed on the base plate and then the floating plate was placed on the droplet. The rotational position of the floating plate was determined by the voltage-applied electrodes. Figure 10.4 shows the four different states in the rotational positions. The angle shift in each state is 22.5°.

10.3 Experiments

10.3.1 Step Motion

We observed rotational motion during its change in state with a high-speed camera (Photron FASTCAM 1024PCI). Figure 10.5a shows sequential images during rotation. The floating plate remained in a stable position determined by the voltage-applied

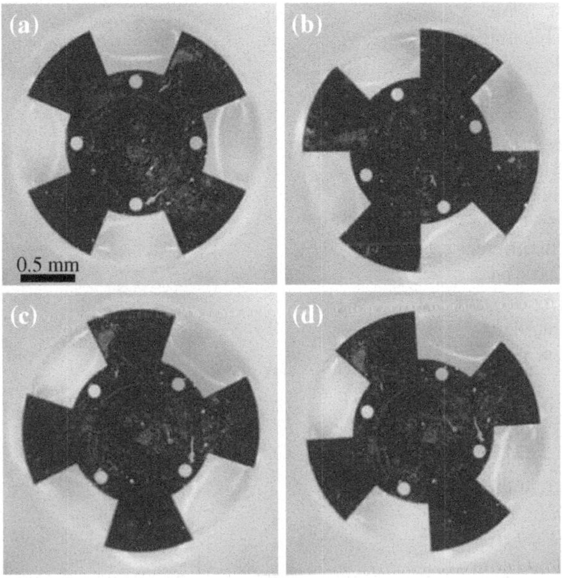

Fig. 10.4 Shift in rotational position of floating plate. Voltage-applied electrodes of all phases are described in Fig. 10.1d. **a** Phase A, **b** Phase B, **c** Phase C, **d** Phase D. [21]—Reproduced by permission of the Royal Society of Chemistry

electrode (Fig. 10.5a–i). Once the voltage-applied electrodes were changed, the shape of the liquid changed and then the floating plate started to move (Fig. 10.5a–ii). After overshoot (Fig. 10.5a–iii), the floating plate stayed in the new stable position (Fig. 10.5a–iv). We observed a relationship between the step response and the characteristics of the capillary motor by changing the floating plates and volume of the liquid. The amplitude of the applied voltage was DC 60 V and its polarity was positive in the experimentsduring this session. The voltage was supplied by a piezo actuator (MESS-TEK M-2680, ±300 V and ±50 mA at maximum). We set the time to 0 at the moment velocity in the floating plate was observed. The theoretical shift of the rotational position was 22.5°.

Figure 10.5b shows the step response of the liquid motor with different volumes of liquid. We used a type-3 floating plate in this experiment, shown in Fig. 10.3d, and placed droplets of liquid of 2.5, 3.0 and 3.5 μL on the base plate. The times to reach the maximum displacement corresponded to 21, 33, and 48 ms. The rising speed was increased while the volume of the liquid was reduced. The step responses with different cog lengths are shown in Fig. 10.5c. In this experiment, types 1–3 floating plates were used and the volume of the liquid was 3 μL. With 100, 200 and 400 μm cog lengths, the times to reach the maximum displacement corresponded to 77, 46, and 33 ms. The rising speed increased when the cog length was longer. Finally, we observed rotational motion in the capillary motor with the type-4 floating plate (50-μm thick). Figure 10.5d shows the step response. The step response of the

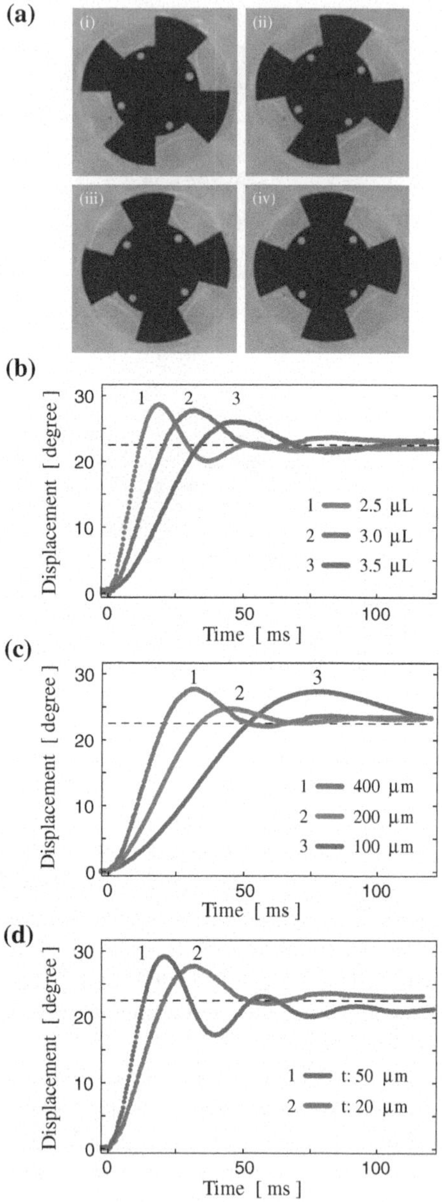

Fig. 10.5 **a** Step response of rotational motion. Floating plate stays in stable position (*i*). By changing the voltage-applied electrodes, floating plate starts to rotate (*ii*). After overshoot (*iii*), floating plate returns to new stable position (*iv*). **b** Step response of capillary motor with different droplet volumes (*1* 2.5 μL, *2* 3.0 μL, and *3* 3.5 μL). **c** Step response of capillary motor with different cog lengths (*1* 100 μm, *2* 200 μm, and *3* 400 μm). Volume of liquid was 3.0 μL. **d** Step response of capillary motor with plates of different thickness (*1* 20 μm and *2* 50 μm). Volume of liquid was 3.0 μL. [21]—Reproduced by permission of the Royal Society of Chemistry

capillary motor with the type-3 floating plate (20-μm thick) has also been plotted in Fig. 10.5d for comparison. The inertial moments of the types-3 and -4 floating plates were 0.7×10^{-13} Kg m^2 for the former and 1.7×10^{-13} Kg m^2 for the latter. Even though the inertial moment of the floating plate was increased, the rising speed increased. With the types-3 and -4 floating plates, the times to reach the maximum displacement were 33 ms for the former and 23 ms for the latter. The amplitude of the overshoot observed in the experiments was less than 5°.

The relationship between the cog length, the volume of the liquid, and the rising speed can be explained by taking into consideration the torque exerted on the floating plate. The magnitude of the torque was the derivative value of the surface area with respect to rotational displacement ϕ. The change in the surface area was generated by the deformation caused by the non-circular boundary conditions of the two plates. The longer cog length increased the deformation, and enlarged the change in the surface area. However, larger droplets increased the gap between the two plates and the deformation caused by the two plates decreased affecting each other. Thus, larger droplets decreased torque. Fast rotational motion could be obtained with a capillary motor made of small droplets with a plate having long cogs. The properties required for quick rotation are the same as those for large torque as explained in Chap. 4. The increase in the rising speed of the capillary motor with the 50-μm-thick floating plate can be explained by taking into consideration the gap between the two plates. The heavier plate compressed the droplet due to gravity and the height of the droplet was reduced. Then, the deformation caused by the two plates affected each other and the torque that was generated was increased. According to the inertial moment of the capillary motor, not only the floating plate, but also the droplet was rotated during rotational motion. The height of the droplet was around 1 mm, while the thickness of the plate was 20 or 50 μm. This means that the droplet was 20 times thicker than the floating plate, and that the water in the droplet was dominant in determining the inertial moment of the capillary motor. In the experiment detailed in Fig. 10.5d, both cases have the same volume of liquid, while the floating plate that is 50-μm-thick provides a large amount of torque. Therefore, faster rotation is achieved with the floating plate that is 50-μm-thick. In the experiments, ±2° error from the theoretically stable position was observed. Defects or nonuniformity in the hydrophobic layer were the reason for the error.

10.3.2 Continuous Rotation

We verified that the liquid motor could be rotated continuously by changing the voltage-applied electrodes. In this experiment, we used a type-3 floating plate and placed a 3 μL droplet of liquid on the base plate. The amplitude of the applied voltage was 60 V. The voltage-applied electrodes is shifted by 22.5° in each step. From phases A to D, the voltage-applied electrodes were changed four times and the floating plate was shifted by 90°. Defining the phase shifts from A to D as one cycle, we applied phase shifts of 8, 24, 48 and 52 cycles/s and observed rotation with a high-speed camera. Figure 10.6a shows a photograph of the rotating liquid motor with the phase shift of 8 cycles/s. Figure 10.6b shows the rotational displacement as a function of time. At low frequency (8 cycles/s), vibration was observed during rotation. The vibration was reduced when applying the high-frequency phase shift. In our experiment, the floating plate was rotated at 720 rpm at maximum with the

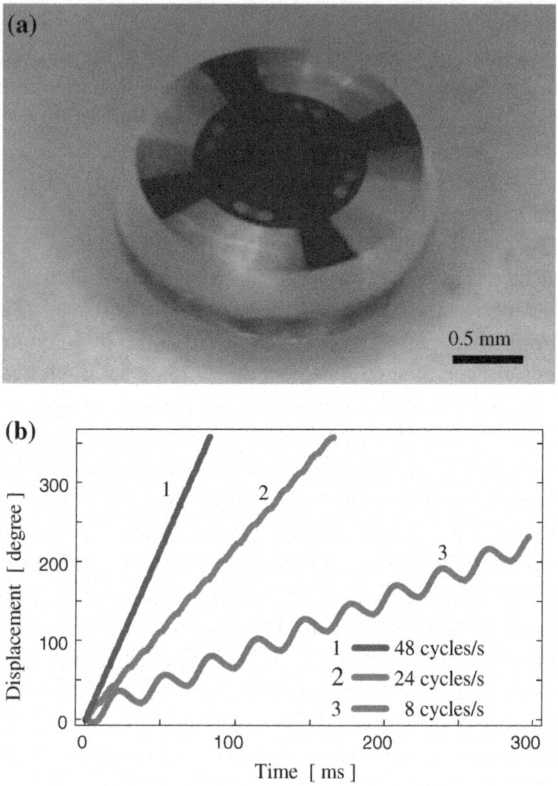

Fig. 10.6 **a** Photograph of rotating capillary motor at 120 rpm. **b** Rotational displacement as a function of time. Floating plate of type-3 was used and 3.0 μL droplet was placed on the base plate. Phase shifts of 8, 24, and 48 cycles/s were applied. [21]—Reproduced by permission of the Royal Society of Chemistry

phase shift of 48 cycles/s. With phase shifts higher than 48 cycles/s, the floating plates could not follow the change in liquid deformation and remained in the same position.

10.4 Conclusion

We achieved a rotating capillary motor by using electrowetting. The diameter of the droplet was 2 mm. A relationship between rotational motion, the floating plate, and the volume of liquid was experimentally observed. The rotating speed was increased, while a longer cog plate and smaller droplets were used. With a 2.5 μL droplet, the time to reach maximum displacement was less than 20 ms. The overshoots in the experiments were within 5°. The maximum rotational speed was 720 rpm with a 3 μL droplet and a type-3 floating plate. With frequencies higher than 48 cycles/s, the floating plate could not follow the change in the bottom edge of the droplet and could not be rotated. Various optical components can be equipped on the floating plate, and the capillary motor can be applied to optical or optofluidic systems.

References

1. C. Song, N.T. Nguyen, S.H. Tan, A.K. Asundi, Lab Chip **9**, 1178 (2009)
2. X. Mao, S.C.S. Lin, M.I. Lapsley, J. Shi, B.K. Juluri, T.J. Huang, Lab Chip **9**, 2050 (2009)
3. L. Hou, N.R. Smith, J. Heikenfeld, Appl. Phys. Lett. **90**, 25114 (2007)
4. A. Takei, E. Iwase, K. Matsumoto, I. Shimoyama, J. Microelectromech. Syst. **90**, 1537 (2007)
5. Y.C. Tai, R.S. Muller, Sens. Actuators, A **21**, 180 (1990)
6. X. Xiong, Y. Hanein, J. Fang, W. Wang, D. Schwarz, K. Böhringer, J. Micromech. Microeng. **16**, 721 (2006)
7. J. Fang, K. Böhringer, J. Microelectromech. Syst. **16**, 721 (2006)
8. U. Srinivasan, D. Liepmann, R.T. Howe, J. Microelectromech. Syst. **10**, 17 (2001)
9. B. Berge, J. Peseux, Eur. Phys. J. E **3**, 159 (2000)
10. S.-K. Fan, H. Yang, T.-T. Wang, W. Hsu, Lab Chip **7**, 1330 (2007)
11. S.K. Cho, H. Moon, C.-J. Kim, J. Microelectromech. Syst. **1**, 70 (2003)
12. M.A. Burns, Phys. Rev. Lett., Proc. Natl. Acad. Sci. U. S. A. **93**, 5556 (1996)
13. O. Sandre, L. Gorre-Talini, A. Ajdari, J. Prost, P. Silberzen, Phys. Rev. E Stat. Phys. Plasmas Fluids Relat. Interdiscip. **60**, 2964 (1999)
14. R.B.M. Schasfoort, S. Schlautmann, J. Hendrikse, A. van den Berg, Science **286**, 942 (1999)
15. H.-W. Lu, F. Bottausci, J.D. Fowler, A.L. Bertozzi, C. Meinhart, C.-J. Kim, Lab Chip **8**, 456 (2008)
16. P. Paik, V.K. Pamula, R.B. Fair, Lab Chip **3**, 253 (2003)
17. P. Sen, C.J. Kim, Langmuir **25**, 4302 (2009)
18. S.H. Ko, H. Lee, K.H. Kang, Langmuir **24**, 1094 (2008)
19. J.M. Oh, S.H. Ko, K.H. Kang, Langmuir **24**, 8379 (2008)
20. J. Lee, C.J. Kim, J. Microelectromech. Syst. **9**, 469 (2000)
21. A. Takei, K. Matsumoto, I. Shimoyama, Lab Chip **10**, 1781 (2010)

Chapter 11
Capillary Based Sealing

Michael De Volder

Abstract This chapter provides a practical illustration of how the favourable scaling of surface tension can provide new engineering solutions at the microscale. The problem studied here is the fabrication of dynamic seals for fluidic microactuators . To date, fluidic microactuators are uncommon in microsystems because they are difficult to seal. However, they offer great potential for applications requiring high actuation force and strokes. The sealing problems is originating from the fact that standard seals such as rubber O-rings are unsuitable for microdevices, since they generate either high leakage or friction. Because of the advantageous scaling of surface tension at microscale, the authors were inspired to replace the classic rubber "O-rings" by liquid seal rings that are maintained by surface tension. This technology provides an elegant solution for this sealing problem since it eliminates friction and leakage problems. In this chapter, we will first provide a strong physical analysis of this problem based on capillary laws. This exercise shows how the physical phenomena described in Chap. 1 such as contact angle hysteresis, interfacial energy, and Young's law are needed to understand this problem. Next we continue by pointing out how the selection of the sealing liquid strongly influences the performance of these capillary based devices. It is important to note that at the microscale the amount of material required to fabricate certain devices can be very low. Therefore, certain products that are prohibitively expensive in normal applications can still offer solutions in microsystems. We will present for instance devices based on liquid gallium and low melting point eutectic alloys. Further, we will describe how these capillary seals are fabricated and implemented in devices. We will discuss in detail all the fabrication aspects that are relevant for these systems, including coatings that are needed to optimally profit from the surface tension of the seals. Finally, measurements performed on prototypes are discussed, and we will provide insight in differences between analytical predictions and results obtained with experimental prototypes. We believe that this chapter not only provides an elegant example of how microsystems

M. De Volder(✉)
imec and KULeuven, 3001Heverlee, Belgium
e-mail: Michael.DeVolder@mech.kuleuven.be

P. Lambert (ed.), *Surface Tension in Microsystems*, Microtechnology and MEMS,
DOI: 10.1007/978-3-642-37552-1_11,

can profit from surface tension, but also provides a good example of the design of surface tension based systems, which is particularly helpful for engineers that are developing new surface tension based microdevices.

11.1 Introduction

Over the last decades, the development of microfluidic systems gained a lot of interest from the scientific community, and has led to the development of some of the most innovative microsystems [1, 2]. While capillary forces play an important role in many of these devices, we will here focus on one specific problem, namely microseals. Whenever two media containing a flowing liquid need to be separated from each other, seals are used. In most cases, rubber contact seals are used such as for instance o-rings, or press-fits. However, when a seal is needed between two components that are sliding with respect to each other, such static seals are inadequate as they hamper the movement of the two bodies and generate high friction. A typical example of applications of such dynamic seals can be found in piston-cylinder actuators which are depicted in Fig. 11.1 [3]. This figure shows a typical hydraulic double acting actuator, which is able to generate both pulling and pushing forces depending whether the pressure is applied to the pressure supply 1 or 2 respectively. These actuators are commonly used in classic mechanical devices including cranes for roadworks. Recently, these actuators have gained the interest of the microsystems community to achieve high forces and power densities at the microscale [3, 4]. However, one of the main challenges in miniaturization of these devices lies in the development of low friction microseals. Classic seal technologies such as rubber O-rings are unsuitable at microscale since small shape or size variations of the rubber O-ring, the seal house or the piston result in either high friction or leakage. Interestingly, surface tension offers an elegant solution to avoid these strict production requirements [5–7]. Seals based on surface tension have been described in patents for precision flow measurement systems, the oldest one goes back to 1960 [8].The implementation of surface tension seals is comparable to that of rubber O-rings, but the seal ring is made of a liquid instead of rubber. These liquid O-ring are locked in a hydrophobic cavity

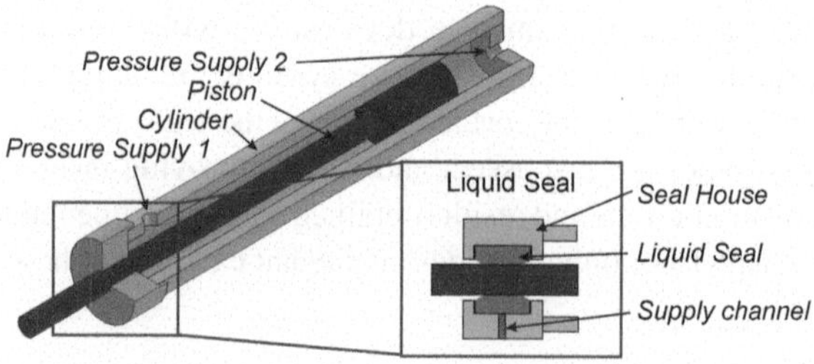

Fig. 11.1 Outline of the surface tension seal discussed in this chapter

and they resists the pressure drop over the seal relying on surface tension, as will be described in details further on. Unlike rubber seals, the liquid nature of surface tension seals allows them to adjust their shape to production errors on the piston or the seal house, and thus reduce seal leakage. In addition, because surface tension seals are non-contact seals (i.e. do not require contact between the seal and piston), they are entirely leak free, and show no seal wear.

Despite a couple of promising older patents on surface tension seals [8–11], nowadays, most of these seals have fallen in disuse since mainly toxic fluids such as mercury have a sufficient surface tension to seal high pressures. However, two important factors make surface tension interesting for sealing microdevices. First of all, scale laws predict that surface tension is one of the strongest forces at microscale (see Chap. 1, Fig. 1.1), and therefore, it is possible to use non-toxic liquids with a lower surface tension than that of mercury at microscale. Second, the amount of material required for making liquid microseals is small, which makes it possible to use more expensive, exotic seal materials such as gallium (Ga) or galinstan (eutectic alloys mainly consisting of gallium, indium, and tin). Figure 11.2 shows a cross-section of a surface tension seal. The seal house comprises a cavity which is filled with the sealing fluid (here Ga) that separates the actuating pressure from the environmental pressure. The seal fluid is applied and kept in the cavity with a pressurized supply channel (see Fig. 11.2).

Before developing a physical model of the surface tension seals, it is interesting to gain some intuitive insight in their operation. The operation of the seals is sketched in Fig. 11.2, which is based on a mathematical model that will be described later on. This figure shows a cross-section of the surface tension seal for different pressure drops over the seal. The seal is first forced in a non-wetting cavity by pressurizing the seal fluid through the Ga supply channel (Fig. 11.2a, b). If the pressure on the right hand side of the seals is increased, the curvature of the seal surface at the right hand side will decrease as far as the contact angle hysteresis allows (Fig. 11.2c). The curvature of the seal on the left hand side that is in contact with the atmospheric pressure, is determined solely by the pressure applied on the Ga supply channel and thus stays constant. If the pressure drop over the seal increases further, the contact line of the seal and the piston will move backwards as far as allowed by the contact angle between the seal liquid and the seal cavity as illustrated in Fig. 11.2d. Above the maximum seal pressure (p_{max}), the seal is swept away as illustrated in Fig. 11.2e. Beyond this point, seal leakage can occur.

11.2 Operating Principle and Physics

In what follows, we derive the maximum seal pressure of surface tension seals based on the following assumptions:

- **The gravity force can be neglected:** Given the small volume of typical surface tension seals, gravity typically plays a minor role. The relative importance of

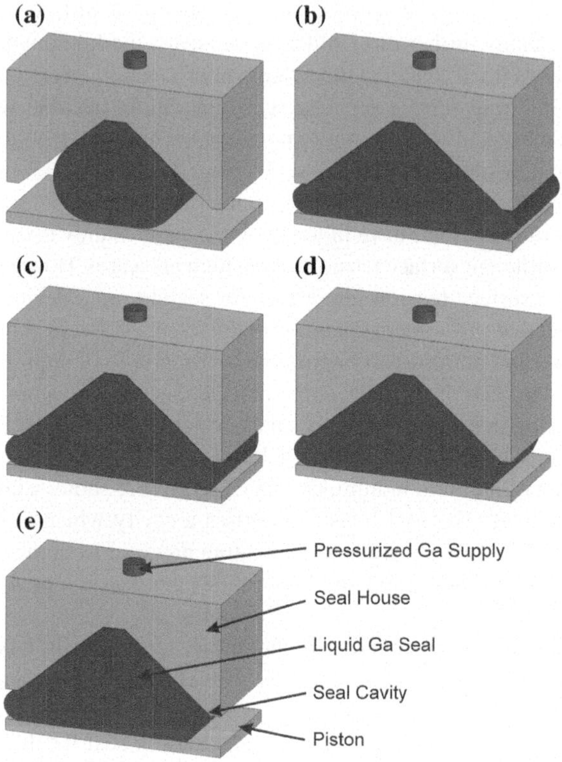

Fig. 11.2 Evolution of the front of the surface tension seal when the pressure applied on the seal increases from the environmental pressure p_{low} to the pressure at which the seal fails p_{max}

gravitation with respect to surface tension is expressed by the Bond number Bo (see Sect. 1.3 or [12]):

$$\begin{aligned} \mathrm{Bo} &\equiv \frac{\rho g \ell}{\gamma/\ell} = \frac{\text{hydrostatic pressure}}{\text{capillary pressure}} \\ &= \frac{6100\,\mathrm{Kg\,m^{-3}} \cdot 9.81\,\mathrm{m\,s^{-2}} \cdot (10 \times 10^{-6}\,\mathrm{m})^2}{0.7\,\mathrm{N\,m^{-1}}} = 8.5 \times 10^{-6} \end{aligned} \tag{11.1}$$

with ρ the fluid density, γ the surface tension and ℓ a characteristic length (here the clearance between the piston and the cylinder). Bo is calculated for a Ga seal based on data from [13] and typical actuator geometries (see further). For the microseals discussed here (see experimental work in Sect.11.4), the Bond number is much smaller than one, which means that the influence of gravity can be neglected.

- **Dynamic effects can be neglected:** As introduced in Sect. 1.3, the Weber number We is a dimensionless parameter that describes the relative importance of fluid inertia to surface tension:

$$\begin{aligned}\text{We} &\equiv \frac{\rho v^2 \ell}{\gamma} = \frac{\text{Flow inertia}}{\text{Surface tension}} \\ &= \frac{6100\,\text{Kg}\,\text{m}^{-3} \cdot (0.01\,\text{m}\,\text{s}^{-1})^2 \cdot (100 \times 10^{-6}\,\text{m})}{0.7\,\text{N}\,\text{m}^{-1}} = 8.7 \times 10^{-5} \end{aligned} \tag{11.2}$$

with v the velocity of the fluid (here approximated by the piston velocity) and ℓ a characteristic length. In this case $We \ll 1$ and therefore the inertia of the fluid induced by the movement of the piston can be neglected. Similarly, the capillary number Ca (see Sect. 1.3) is defined to relate viscous drag to surface tension [12].

$$\begin{aligned}\text{Ca} &\equiv \frac{\mu v}{\gamma} = \frac{\text{Viscous force}}{\text{Surface tension}} \\ &= \frac{(2.04 \times 10^{-3}\text{Kg}\,\text{m}^{-1}\,\text{s}^{-1})\,0.01\,\text{m}\,\text{s}^{-1}}{0.7\,\text{N}\,\text{m}^{-1}} = 2.9 \times 10^{-5}\end{aligned} \tag{11.3}$$

with μ the dynamic viscosity. Ca also suggests drag induced by the movement of the piston can be neglected.

As a consequence, little pressure gradients can be fairly assumed to be due to for instance gravity, and therefore, the liquid seals retain a circular cross section during operation. This greatly simplifies the calculation of the maximum pressure drop that these seals can withstand. Based on these assumptions, an energy method (Sect. 2.2.1) proposed by Man et al. [14] and implemented in order to determine the maximum seal pressure based on interfacial energy:

$$\gamma = \frac{\partial W}{\partial A} \tag{11.4}$$

Equation (11.4) describes the interfacial energy U stored by the surface tension γ in a surface A. This energy is a result of the unbalance between the molecules in the bulk liquid material and those at the interface with another medium. Young's Equation (1.2) relates the surface energies in a solid-liquid-gas interface to the contact angle θ:

$$\gamma_{SV} = \gamma_{\text{SL}} + \gamma \cos\theta \tag{11.5}$$

As recalled in Chap. 2 by (2.1), the total interfacial energy W stored in a system where a gas, a liquid and a solid are in contact with each other is described by (11.6).

$$W = W_{\text{SL}} + W_{\text{SV}} + W_{\text{LV}} = \gamma_{\text{SL}} S_{\text{SL}} + \gamma_{\text{SV}} S_{\text{SV}} + \gamma \Sigma \tag{11.6}$$

with S_{SL}, S_{SV} and Σ the solid-liquid, solid-vapor and liquid-air interface areas and γ_{SL}, γ_{SV} and γ the corresponding surface energies per unit area. By substituting (11.5) in (11.6), we deduce (11.7).

$$W = (S_{\text{SL}} + S_{\text{SV}})\gamma_{SV} - S_{\text{SL}}\gamma \cos\theta + \Sigma\gamma \tag{11.7}$$

By equating W as a function of the volume V, the pressure drop Δp over the liquid seal surface can be determined by the following equation [14]:

$$\Delta p = -\frac{\partial W}{\partial V} \tag{11.8}$$

Taking into account that the area $S_{\mathrm{SL}} + S_{SV}$ is a constant in our system, (11.7) and (11.8) collapse into:

$$\Delta p = \cos\theta \frac{\partial S_{\mathrm{SL}}}{\partial V}\gamma - \frac{\partial \Sigma}{\partial V}\gamma \tag{11.9}$$

The latter equation allows studying the seal pressure based on the surface tension γ and the contact angle θ. Both parameters need however careful consideration. First of all, γ tends to decrease by adsorption, i.e. absorbsion of impurities at the interface (See Chap. 1). Second, the contact angle θ varies between the so-called receding contact angle θ_r and the advancing contact angle θ_a, and this is influenced by surface topography, as described earlier on in the book [15, 16].

To simplify calculations, the gap h between the piston and the cylinder is unfolded and modelled as two parallel plates with a length w equal to the circumference of the piston. The designations used during the development of the physical model are shown in Fig. 11.3. Next we write the interfacial energy W as a function of the volume of the seal V (see Eq. (11.8)) for the different operating regimes of the seal:

1. **First Regime:** As the seal gets pressurized (i.e. the outer pressure increases), the contact angle will first change from θ_a to θ_r, as illustrated in Fig. 11.2b, c. In this case x_i is constant and equal to L.

$$S_{\mathrm{SL}} = (1 + \frac{1}{\sin\beta})Lw \tag{11.10}$$

$$\Sigma = \left(\theta_i - \frac{\pi}{2}\right)\frac{h}{\sin\left(\theta_i - \frac{\pi}{2}\right)}w \tag{11.11}$$

$$\begin{aligned} V = {} & \frac{L^2}{2\tan\beta}w + Lhw \\ & + \left(\left(\frac{h}{2\sin\left(\theta_i - \frac{\pi}{2}\right)}\right)^2\left(\theta_i - \frac{\pi}{2}\right) - \frac{h^2}{4\tan\left(\theta_i - \frac{\pi}{2}\right)}\right)w \end{aligned} \tag{11.12}$$

and

$$\Delta p = \cos\theta \frac{\partial S_{\mathrm{SL}}}{\partial V}\gamma - \frac{\partial \Sigma}{\partial V}\gamma \tag{11.13}$$

with $\theta_i \in [\theta_r,\ \ \theta_a]$

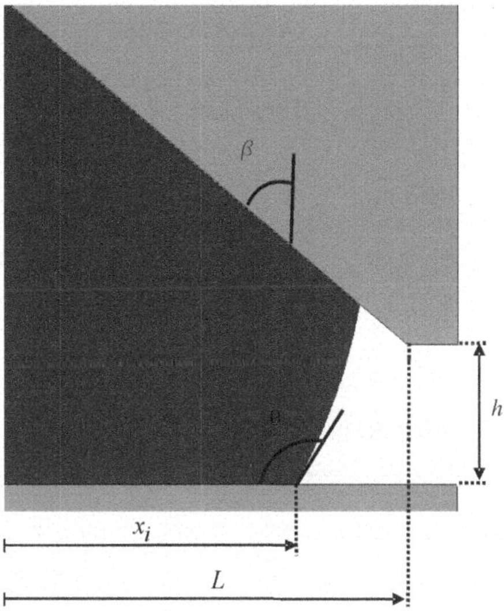

Fig. 11.3 Variables used during the development of the physical model. Reprinted from [5], with kind permission from Springer Science and Business Media

2. **Second Regime:** In this regime, the contact line between the seal liquid and the piston starts receding, as shown in Fig. 11.2c, d.

$$S_{SL} = \left(x_i + \frac{L}{\sin\beta} \right) w \tag{11.14}$$

$$\Sigma = \left(\tan^{-1}\left(\frac{h}{L - x_i} \right) + \theta_r - \pi \right) \left(\frac{h}{\sin\left(\tan^{-1}\left(\frac{h}{L-x_i}\right)\right) \sin\left(\tan^{-1}\left(\frac{h}{L-x_i}\right) + \theta_r - \pi\right)} \right) w \tag{11.15}$$

$$V = \frac{(L^2 w)}{2\tan\beta} + \frac{(L + x_i)hw}{2} + \left(\tan^{-1}\left(\frac{h}{L - x_i} \right) + \theta_r - \pi \right) \left(\frac{h}{2\sin\left(\tan^{-1}\left(\frac{h}{L-x_i}\right)\right) \sin\left(\tan^{-1}\left(\frac{h}{L-x_i}\right) + \theta_r - \pi\right)} \right)^2 w$$

$$-\frac{h\cos\left(\tan^{-1}\left(\frac{h}{L-x_i}+\theta_r-\pi\right)\right)\frac{h}{2\sin\left(\tan^{-1}\left(\frac{h}{L-x_i}\right)\right)}}{2\sin\left(\tan^{-1}\left(\frac{h}{L-x_i}\right)\right)\sin\left(\tan^{-1}\left(\frac{h}{L-x_i}\right)+\theta_r-\pi\right)}w \tag{11.16}$$

and

$$\Delta p = \cos\theta\frac{\partial S_{\mathrm{SL}}}{\partial V}\gamma - \frac{\partial \Sigma}{\partial V}\gamma \tag{11.17}$$

with $x_i \in \left[\left(L - \frac{h}{\tan\left(\frac{\pi}{4}+\frac{\beta}{2}\right)}\right), L\right]$

3. **Third Regime:** In the last regime, the seal will recede on both the piston and the seal house flange, as shown in Figs. 11.2d, e.

$$S_{\mathrm{SL}} = \left(2x_i + \left(\frac{1}{\sin\beta} - 1\right)(L + h\tan\beta)\right)w \tag{11.18}$$

$$\Sigma = \left(\sin\left(\frac{\pi}{4} - \frac{\beta}{2}\right)\frac{(L + h\tan\beta - x_i)}{\sin\left(\frac{\beta}{2}+\theta_r-\frac{3\pi}{4}\right)}\right)2w\left(\frac{\beta}{2}+\theta_r-\frac{3\pi}{4}\right) \tag{11.19}$$

$$\begin{aligned}
V &= (L + h\tan\beta)^2\frac{w}{2\tan\beta} \\
&\quad - (L + h\tan\beta - x_i)^2\sin\left(\frac{\pi}{4}-\frac{\beta}{2}\right)\cos\left(\frac{\pi}{4}-\frac{\beta}{2}\right)w \\
&\quad + \left(\frac{\beta}{2}+\theta_r-\frac{3\pi}{4}\right)\left(\frac{\sin\left(\frac{\pi}{4}-\frac{\beta}{2}\right)(L + h\tan\beta - x_i)}{\sin\left(\frac{\beta}{2}+\theta_r-\frac{3\pi}{4}\right)}\right)^2 w \\
&\quad - \frac{\left(\sin\left(\frac{\pi}{4}-\frac{\beta}{2}\right)(L + h\tan\beta) - x_i\right)^2}{\tan\left(\frac{\beta}{2}+\theta_r-\frac{3\pi}{4}\right)}w
\end{aligned} \tag{11.20}$$

and

$$\Delta p = \cos\theta\frac{\partial S_{\mathrm{SL}}}{\partial V}\gamma - \frac{\partial \Sigma}{\partial V}\gamma \tag{11.21}$$

with $x_i \in \left[0, \left(L - \frac{h}{\tan\left(\frac{\pi}{4}+\frac{\beta}{2}\right)}\right)\right]$

By substituting the above equations in (11.9), the (maximal) seal pressure can be calculated. The resulting total energy W and the pressure drop Δp are shown in Fig. 11.4 as a function of the volume of the seal (volume 1 corresponds to the situation depicted in Fig. 11.2b). These calculations are repeated for different values of the sidewall angle β.

Based on Fig. 11.4, the seal operation can be optimized as follows. First, a pressure of 90 kPa is applied to the pressurized Ga channel, to fill the seal cavity as much as possible without leakage (relative volume 1 in Fig. 11.4, situation shown in Fig. 11.2b). When the pressure in the cylinder increases (see Fig. 11.2c.), the seal volume is reduced, resulting in a pressure drop as shown in Fig. 11.4 right, regime 1. The un-balance caused by the external pressure is restored by the change in contact angle. As the pressure applied on the seal further increases, the pressure drop over the left hand side of the seal decreases until it reaches a situation in which the pressure drop over one side of the seal falls down to approximately zero. In that case, a pressure of 90 kPa can be sealed (no pressure drop over the right hand side of the seal fluid surface). A remarkable phenomenon illustrated in Fig. 11.4 is that the pressure drop over the surface can become negative for $0° \leq \beta \leq 20°$. Although, this may seem surprising at first sight, it is logic, that for the combination of certain side wall angles and receding contact angles, the curvature of the seal becomes concave, as illustrated in Fig. 11.5c and d. In that case the total seal pressure can be increased with almost 20 KPa as illustrated in Fig. 11.4 for $\beta = 0°$. This results in a maximum seal pressure of about 110 KPa for the dimensions used in Fig. 11.4.

A simplified approach to the above problem is possible by using the Laplace equation (1.1) that describes the pressure drop over a curved liquid surface. This equation relates the curvature of a fluid surface R_1 and R_2 to a pressure drop Δp.

$$\Delta p = \gamma \left(\frac{1}{\rho_1} + \frac{1}{\rho_2} \right) \tag{11.22}$$

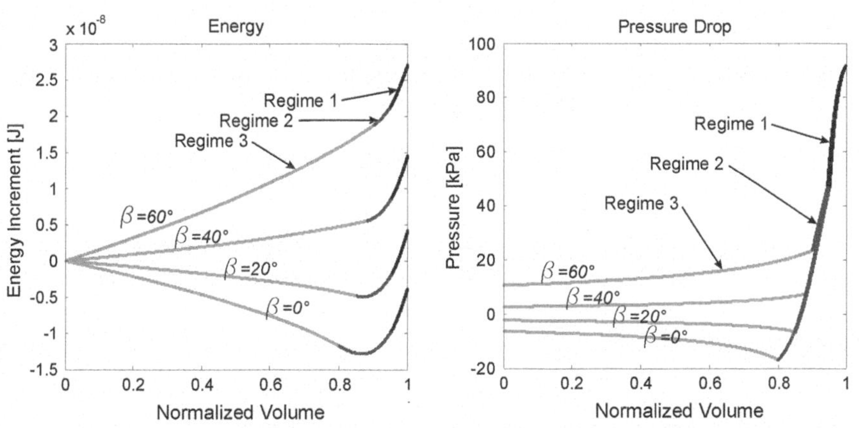

Fig. 11.4 *Left* Energy stored in the seal as a function of the normalized seal volume. *Right* Pressure drop over the seal as a function of the normalized seal volume. ($L = 40\,\mu$m, $h = 15\,\mu$m, $w = 2\pi \times 250\,\mu$m, $\gamma = 0.7\,\mathrm{N\,m^{-1}}$, $\theta_a = 170°$, $\theta_r = 120°$). Reprinted from [5], with kind permission from Springer Science and Business Media

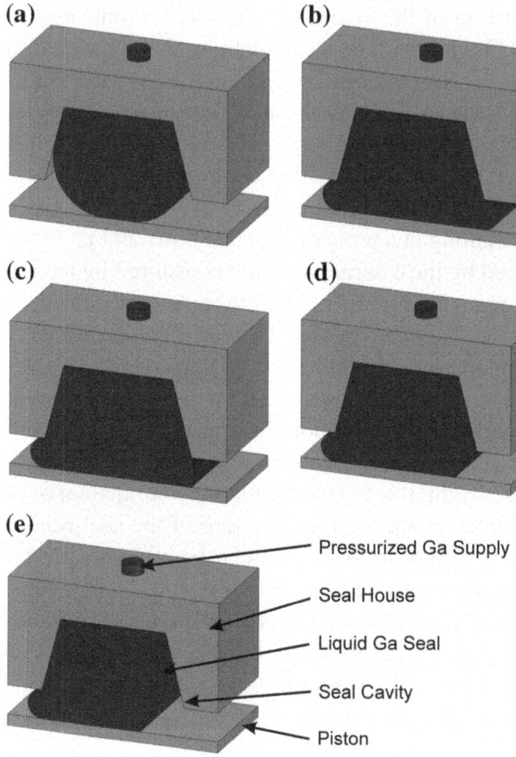

Fig. 11.5 Evolution of the front of the surface tension seal when the pressure applied on the seal increases from the environmental pressure to the pressure at which the seal fails

For a circular seal shape, $\rho_2 = \infty$ and ρ_1 can be expressed as a function of h and θ_a. Therefore, the pressure drop Δp is given by:

$$\Delta p = \frac{2\gamma \cos \theta_a}{h} \tag{11.23}$$

For the numerical values given in Fig. 11.4, the latter equation leads to a seal pressure equal to 92 KPa, in accordance to the energy formulation described previously. Following conclusions can be drawn from the above models for the design of surface tension seals:

- The sidewall angle β should be as small as possible. In the experimental setups discussed further on in this chapter, β is always equal to zero.
- A second implication, is that the gap between the piston and the cylinder should be minimized (see (11.23)). This however results in tight tolerances on the seal house and the piston, making the fabrication more challenging for high pressure seals.

Table 11.1 Properties of the investigated seal fluids

Seal fluid	Surface tension ($N\,m^{-1}$)	Melt temperature (°C)	Toxicity
H_2O	0.07	0	+
Hg	0.5	−40	−
Ga	0.7	30	0
Eut. Ga-In	0.6	15	0
Eut. Ga-In-Sn	<0.6	10	0

- Finally, (11.23) predicts a linear increase of the seal pressure (Δp) with the surface tension (γ). Liquid metals are therefore interesting, as they typically have a high surface tension.

While mercury toxicity prevents it to be used in many applications, other metallic liquid such as Ga and Ga alloys are promising sealing fluids as illustrated in Table 11.1. Water is an interesting alternative for Ga in applications where strict material regulations are imposed, such as in the food industry or for minimally invasive surgery. However, the surface tension of water is an order of magnitude lower than that of Ga, and the evaporation rate is much higher. The high price of Ga is not prohibitive for most seal applications, as the required amount of material is usually small (in our prototypes typically less than 1 mg per seal). Yet another interesting property of Ga, is that it shows good lubrication properties. For instance, Przybyszewski et al., showed that a Ga layer decreased the friction coefficient between metallic layers by half [17].

While Ga readily stays undercooled at room conditions [18–20], alloys with a melting point below 20 °C are preferable for some applications. For instance mixing approximately 20 weight percent of In with Ga, reduces the melting temperature to 15 °C. A disadvantage of these alloys is that the surface tension of the Ga-alloy decreases steeply when adding In [5, 6, 13, 20]. One of the practical problems that still needs further investigation is stiction of Ga to the piston. For this we are currently investigating Teflon and other coatings.

11.3 Fabrication Process and Assembly

In what follows, we will go through exemplary fabrication processes of surface tension seals. We emphasize the fabrication of miniature piston-cylinder actuators using surface tension seals, as these devices are particularly difficult to seal using standard rubber seals [21].

To increase the seal pressure, the clearance between the seal house and the piston should be as small as possible. The seal house was fabricated by micro-electro-discharge-machining (μEDM) as shown in Fig. 11.6a. First, the seal orifice is produced using μEDM-drilling, next the seal cavity is made using a WEDG-formed

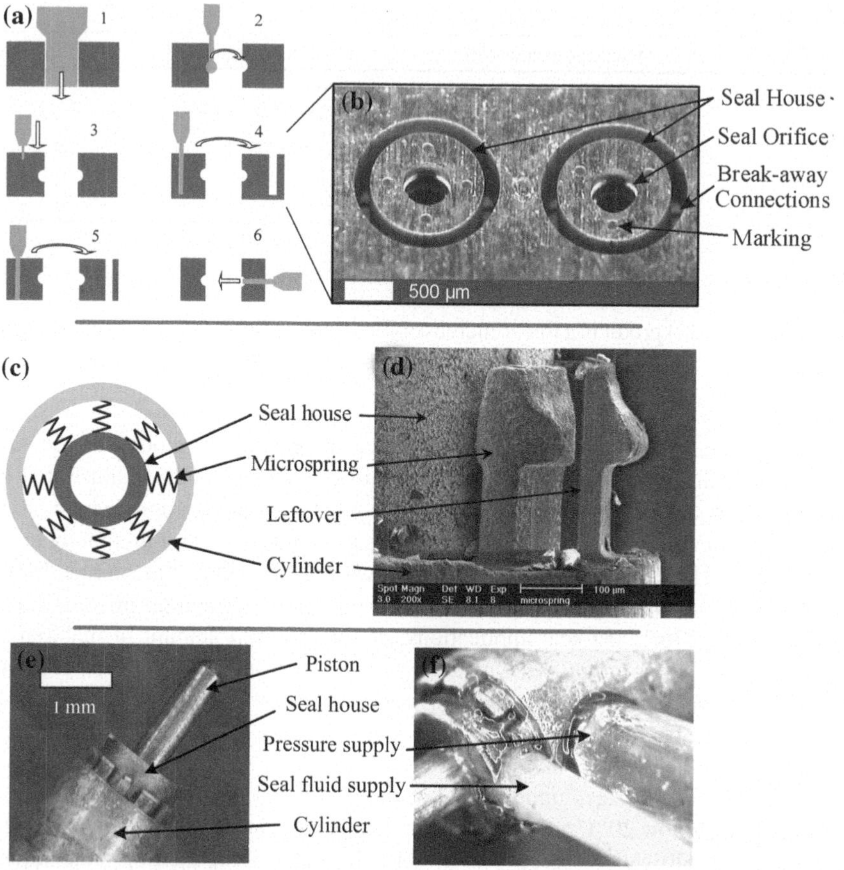

Fig. 11.6 **a** Production flow-chart for surface tension seals. **b** Seals during manufacturing. **c–d** illustration of a microspringbed to align the seal on the cylinder. **e–f** Assembled piston-cylinder actuators with a surface tension seal

electrode, then the seal houses are released, and finally, a channel is eroded in the seal wall that allows conveying the seal fluid to the seal cavity. Fig. 11.6b shows stainless steel seals during and after the fabrication. The seal cavity is 20 µm deep.

Good alignment between the seal and the cylinder is essential in order to avoid friction. In our test setups, the seal house and the cylinder have been aligned with spring-beds using on-the-machine assembly (see Fig. 11.6) [6]. Alternatively, this could be achieved using ultra precision machining of tight fittings. More details are provided in *WO2005/106296* [22]. Finally, plastic tubing is glued on the actuator to provide both pressurized fluid and seal liquid (see Fig. 11.6f).

11.4 Experimental Section

In what follows, experiments performed to verify the above theory will be discussed. First, we used an inclined drop method to measure the advancing and receding contact angle of Ga on a steel plate with a roughness of 5 μm Ra. The in this particular experiment, advancing contact angles up to 163° and receding angles of 117° were measured.

Next, the operation of surface tension seals is tested on a glass cylinder, providing direct visual information on the deformation of the seal under the applied pressure. As depicted in Fig. 11.7a, the piston of this setup comprises internal channels to convey Ga from a reservoir to an annular seal cavity. The piston is made by turning on a precision lathe, and the seal cavities are fabricated by wire-μEDM. The diameter of the piston is 1.6 mm and the clearance between the piston and the cylinder is about 20 μm. As the seal pressure is steadily increased, it reaches a point where seal failure occurs as illustrated in Fig. 11.7c. In a test setup with a total clearance of approximately 22 μm, the maximum seal pressure was 90 KPa, and in the case of a total clearance of approximately 44 μm, the seal pressure was 50 KPa. This corresponds to the theoretical expectations provided by (11.23), taking into account the uncertainties on the seal gap and the surface tension. During these experiments it was also ascertained that the seals are leak-tight in normal operation conditions.

Next a more complex seal design was tested on the actuators described in previous section. The average seal pressure for eutectic Ga-In seals is shown in Table 11.2. The variation in seal pressure is caused by variations in eccentricity of the piston as well as because it is difficult to determine the exact pressure at which the seal fails. The average seal pressure using Ga-In is 74 kPa, while using pure Ga, the average seal pressure was 80 kPa. This is in accordance with the higher surface tension of Ga.

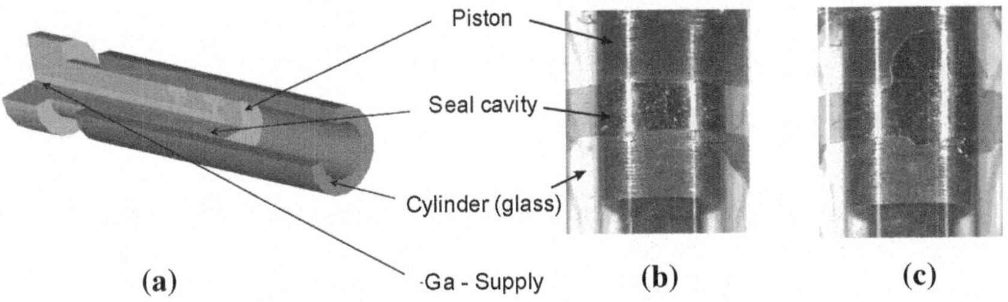

Fig. 11.7 **a** Schematics of the test setup. **b** Photographs of Ga-seals through a glass cylinder under normal operation conditions, and **c** after seal failure. Adapted from [6]

Table 11.2 Maximum seal pressure in KPa

Measurement	1	2	3	4	5	Average
Seal pressure	66	70	80	78	75	74

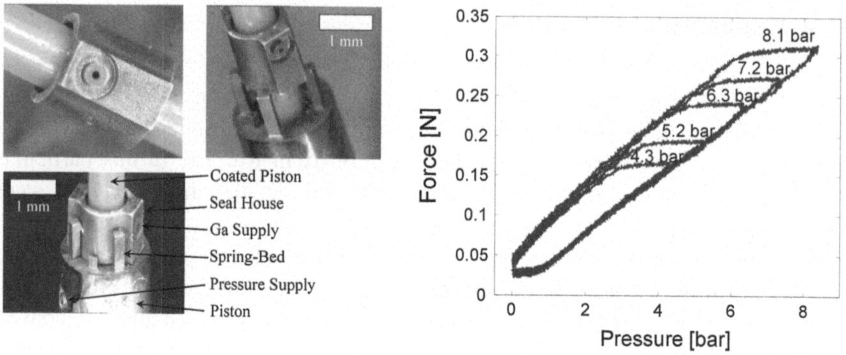

Fig. 11.8 *Left* Microactuator with an hybrid surface tension seal. *Right* Output force of an actuator with a hybrid surface tension seal as a function of the applied pressure. Reprinted from [5], with kind permission from Springer Science and Business Media

The seal pressure can be increased by reducing the seal gap, or by using a slightly more complex seal design, we will refer to as a "hybrid surface tension seal". In the latter approach, clearance seals and surface tension seals are used in series. The first seal is a simple restriction seal or a small gap between the piston and the cylinder. As the pressurized fluid is first forced through this small clearance, part of the pressure load is dissipated before it is reaching the surface tension seal. In this hybrid seal design, the clearance enables to seal higher pressures, while the surface tension seal provides leak tight operation [5, 7]. Using such a seal configuration, we achieved seals operating leak tight up to more than 800 kPa using a Ga-In. At those pressures, a microactuator with a piston diameter of 1 mm developed a force of 0.3 N as shown in Fig. 11.8. Interestingly, this is higher than what is achieved by electrostatic or electromagnetic actuators of the same dimensions.

11.5 Conclusions and Outlook

Because surface tension is one of the dominant forces at microscale, it enables engineering new microscale devices. Here we presented a low friction seal technology based on confining a seal liquid ring in a circular cavity. We showed how to dimension, design and fabricate these seals. The approach discussed in this chapter is using microfabrication methods including micromilling and micro electrodischarge machining, and resulted in millimeter sized devices. It is however possible to fabricate even smaller devices using capillary seals by silicon processing. Sniegowski et al. developed the smallest existing piston-cylinder actuator using a two-level poly-silicon process. The piston height of 2 μm and a width of 6 μm [3]. These devices, as well as those presented in this chapter achieve higher force outputs than electromagnetic or electrostatic devices of the same dimensions. This is obviously only possible because these actuators are sealed by a low friction seal.

On a larger scale, surface tension seals are interesting as well, for instance in precision actuators and flow meters. In principle these seals operate well for a wide variety of piston diameters, as long as the gap between the seal and the piston is small enough. Gaps in the range of 10 μm or below are needed to seal about 100 KPa. The seals discussed in this chapter are using high surface tension seal fluids such as gallium or gallium alloys as seal liquid. Using high precision manufacturing to further reduce the seal gap, it should be possible to replace gallium by water as seal liquid. The latter would enable application of these actuators in for instance surgical tools.

References

1. M.T. Guo, A. Rotem, J.A. Heyman, D.A. Weitz, Droplet microfluidics for high-throughput biological assays. Lab Chip **12**(12), 2146 (2012)
2. T. Thorsen, S.J. Maerkl, S.R. Quake, Microfluidic large-scale integration. Science **298**, 580–584 (2002)
3. M. De Volder, D. Reynaerts, Pneumatic and hydraulic microactuators: a review. J. Micromech. Microeng. **20**, 43001 (2010)
4. M. De Volder, J. Coosemans, R. Puers, D. Reynaerts, Characterization and control of a pneumatic microactuator with an integrated inductive position sensor. Sens. Actuators, A **152**, 234–240 (2009)
5. M. De Volder, D. Reynaerts, A hybrid surface tension seal for pneumatic and hydraulic microactuators. Microsyst. Technol. **15**(5), 739–744 (May 2009)
6. M. De Volder, J. Peirs, D. Reynaerts, J. Coosemans, R. Puers, O. Smal, B. Raucent, A novel hydraulic microactuator sealed by surface tension. Sens. Actuators A **123–124**, 547–554 (2005)
7. M. De Volder, D. Reynaerts, Development of a hybrid ferrofluid seal technology for miniature pneumatic and hydraulic actuators. Sens. Actuators A **152**, 234–240 (2009)
8. G.K. Porter, Pneumatic float with mercury ring. US Patent 2,927,829, 1960
9. G.R. Jackson, D.C. Frick, Method and apparatus for measuring gas flow. US Patent 4,479,377, 1984
10. W. Hoffman, P. Wapner, Microdevices based on surface tension and wettability that function as sensors, antuators, and other devices. US Patent 6,152,181, 2000
11. G.K. Porter, Flow rate calibration. US Patent 3,125,879, 1964
12. K.A. Polzin, E.Y. Choueiri, A passive propellant feeding mechanism for micropropulsion using capillarity. J. Phys. D: Appl. Phys. **36**, 3156–3167 (2003)
13. J.N. Koster, Directional solidification and melting of eutectic GaIn. Cryst. Res. Technol. **34**, 1129–1140 (1999)
14. P.F. Man, C.H. Mastrangelo, M.A. Burns, D.T. Burke, *Microfabricated capillarity-driven stop valve and sample injector* (Proc. MEMS, IEEE Micro Electro Mechanical Systems Conference, 1998), pp. 45–50
15. C.W. Extrand, Y. Kumagai, Liquid drops on an inclined plane: the relation between contact angles, drop shape, and retentive force. J. Colloid Interface Sci. **170**, 515–521 (1995)
16. F.M. Fowkes, *Contact Angle, Wettability and Adhesion*. Advances in Chemistry Series, vol. 43 (American Chemical Society, Washington, DC, 1964)
17. J.S. Przybyszewski, *Tungsten as a slippering material for use with gallium lubrication in ultrhigh vacuum*. Nasa Technical Note TN D-6184 (1971)

18. R.J. Good, W.G. Givens, C.S. Tucek, The contact angle at the gallium mercury interface on glass, ed. by F.M. Fowkes, *Contact Angle, Wettability and Adhesion*. Advances in Chemistry Series, vol. 43 (American Chemical Society, Washington, DC, 1964), pp. 211–221
19. J.W. Mellor, *A Comprehensive Treatise on Inorganic and Theoretical Chemistry* (Longmans, Green and Co., London, 1924)
20. T.G.T. van Venrooij, M. Sluyters-Rehbach, J.H. Sluyters, Indium adsorbtion at the surface of liquid In + Ga alloy electrodes in contact with an aqueous I M NaClO4 solutions. J. Electroanal. Chem. **462**, 111–126 (1999)
21. M. De Volder, F. Ceyssens, D. Reynaerts, R. Puers, A PDMS lipseal for hydraulic and pneumatic microactuators. J. Micromech. Microeng. **17**, 1232–1237 (2007)
22. Caterpillar. *Hydraulic System Support*. Caterpillar catalogue PEDP5027-02, 2004

Chapter 12
Surface Tension-Driven Self-Assembly

Massimo Mastrangeli

Abstract This chapter reviews the capillary self-assembly (SA) of heterogeneous systems composed by milli- to micrometer-sized building blocks. The representative, though not exhaustive, examples of microsystems presented belong to four main classes: two-dimensional assemblies at fluid-fluid interfaces, three-dimensional assemblies, two-dimensional die-to-substrate assembly, and deployment of three-dimensional architectures and polyhedra. Technology (processing), material issues (surface treatment, compatibility) and performance of the SA processes are critically assessed and supported by models where available. Surface-tension-driven SA is rapidly gaining relevance for precision manufacturing of microsystems. Nonetheless, it is far from exhausting the potentialities of SA for system engineering and fabrication, leaving ample room for extensive and innovative research investments.

12.1 Introduction

The predominance of surface phenomena at (sub-)millimetric length scales[1] has motivated their active exploitation for the assembly of miniaturized functional systems since the end of the 1990s. Capillarity, when combined with geometrical

[1] The *capillary length* (see (1) in the Preface) of most common liquids is in the order of millimeters.

M. Mastrangeli (✉)
Distributed Intelligent Systems and Algorithms Laboratory (DISAL), School of Architecture, Civil and Environmental Engineering, Ecole Polytechnique Fédérale de Lausanne, 1015 Lausanne, Switzerland
e-mail: mastrangelim@gmail.com

M. Mastrangeli
Microsystems Laboratory (LMIS1), School of Engineering, Ecole Polytechnique Fédérale de Lausanne, 1015 Lausanne, Switzerland

P. Lambert (ed.), *Surface Tension in Microsystems*, Microtechnology and MEMS,
DOI: 10.1007/978-3-642-37552-1_12, © Springer-Verlag Berlin Heidelberg 2013

and material constraints, can be engineered to instill a thermodynamic drive toward predefined configurations. Geometrical arrangements of building blocks—being them devices, or constituent panels—can be designed to be predictable and to correspond to minima of the free energy of the system. The desired system configuration is normally the one *globally* minimizing the system free energy. Once set up, the assembly process progresses with the addiction of kinetic energy, needed to move the system away from local minima of the associated energy landscape. Under these conditions, the process can run essentially unsupervised: hence the designation of *self*-assembly (SA).

Numerous physical interactions—including e.g. inertial, electric, magnetic and chemical forces[2]—can enable SA. Among them, capillarity affords an unmatched degree of accuracy in component registration. This superior placement precision is the joint result of (i) the lubrication provided by liquid joints (i.e. menisci), which avoid dry friction between contacting surfaces, (ii) self-alignment arising from a combination of capillarity and geometrical shape-matching, and (iii) the favorable downscaling of surface tension effects with size (see Fig. 1.1). High precision together with relative simplicity of implementation justifies the paramount technological relevance currently being gained by capillary SA and self-alignment in precision engineering applications.[3] Furthermore, the dominance of capillary forces at smaller and smaller scales well suits the higher importance of SA techniques for smaller and smaller components: as direct handling faces hard issues [1], non-contact handling, particularly fluid-mediated, is a viable option to perform microassembly tasks.

Fluidic approaches to SA extensively build upon the modulation of the chemical properties of the patterned surfaces of the system components. Control over the chemical affinities, wetting properties and energies of the interfaces is required to tailor the selectivity, directionality, spatial range and magnitude of interactions among the components. Moreover, on one hand for Bo $<<$ 1 the minimization of a fluid system's free energy coincides with the minimization of its surface energy, thereby including Laplace (i.e. pressure/curvature) and interfacial energy components (see Chap. 2). On the other, fluid interfaces, especially if highly-energetic, are mechanically metastable as they are susceptible to minimize their surface. Therefore, they are suitable to drive the system toward the desired assembled state.

Fluid/fluid interfaces hosted the earliest experiments in millimetric capillary SA, which showed that flotation forces could predictably drive the crystallization of geometrical patterns of pre-conditioned components (Sect. 12.2). A rather general assembly technique, combining both capillary and hydrophobic interactions with geometrical shape-matching, was developed almost at the same time for (sub-) millimetric units (Sect. 12.3). The self-placement and self-alignment made possible by the minimization of molten-solder/air interfaces is established since a few decades in the field of bonding and soldering [34]. Capillary die-to-substrate SA (Sect. 12.4) assolved a two-fold purpose: an extension of flip-chip assembly

[2] For an extensive review on self-assembly of microsystems, see [59].

[3] See Chap. 3 for examples.

techniques—whereby up to the entire surface of a die (and of the corresponding binding site on the substrate) is exploited for self-alignment, as compared to solder bumps only—as well as a specialization of capillary and hydrophobic SA to populate planar templates with devices. In die-to-substrate assembly, die feeding can be provided either by robotic feeders (*hybrid microassembly* [78]) or by fluidic mass transport (*fluidic* SA [82]). With reference to the spatial and temporal availability of components in proximity of the target sites during the assembly process, the latter approach can be either deterministic, as in railed microfluidic SA [22], or stochastic. Stochastic fluidic SA has the advantage of massive parallelism, and also enables fully three-dimensional assembly of functional devices, as shown in Sect. 12.3. Recently, the surface tension of water was used for the accurate self-alignment of patterned wafers [57, 91], as well. That of relatively-low melting point materials can be used to actuate out-of-plane three-dimensional MEMS architectures and even nanoscopic polyhedra, as described in Sect. 12.5.

The far-reaching technological potentialities of surface-tension-related effects for microsystem assembly are being demonstrated by an ever growing number of works. Their adoption in industrial assembly lines is being seriously considered. Still, a complete characterization of process parameters, design rules, scaling properties and limitations remains to be worked out, leaving ample room for systematic and innovative research.

12.2 SA at Fluid-Fluid Interfaces

A prerequisite for capillary SA processes to take place at a fluid/fluid interface is that the components to be assembled be able to float at the interface itself. This can be obtained by exploiting surface tension forces and Archimedean buoyancy, as related to volume, density and shape of the components. A thickness-to-width ratio much smaller than 1 has been normally used for such components, leaving the choice of materials rather free.[4]

A fluid/fluid interface is a privileged space for (sub-)millimetric capillary SA. Physically, when floating objects deform a fluid interface, they can experience capillary *flotation* forces [48]. The shape of the fluid menisci along the walls of the objects depends on the wetting character of their surfaces. Walls wet (not wet) by the fluid form positive (negative) menisci. The force then depends on the distance between the objects, the wetting angle of the menisci and the size of the objects. Importantly, the force can be either attracting or repelling depending on the types of facing menisci: menisci of the same (opposite) type produce repelling (attracting) interactions. Given this, flotation forces can be made selective and anisotropic.

Theoretically, the confinement of the assembly units within a bounded two-dimensional (sub)space like the interface between two immiscible fluid phases

[4] If the thickness-to-width ratio becomes too small, the component may be deformed by capillary forces: this is the domain of ultra-thin chips and elastocapillarity.

reduces the dimensionality of the search space where the SA process takes place. The encounters between components can thereby be more frequent and effective.

The Whitesides Group at Harvard University and Shimoyama's at Tokyo University pioneered this field of research.

In the original study by Shimoyama and colleagues [40], a set of identical, flat parts were floated and magnetically stirred at a water/air interface. Such parts featured a round edge as well as two right corners and an acute tip. Fractions of the parts also protruded from the reference plane of the interface. They found that the fractions of the parts posed at the same relative height experienced attractive forces, pulling the parts into direct contact. The force was otherwise repulsive. The interaction could also be spatially modulated. They demonstrated—both experimentally and by the numerical solution of the associated Laplace equation [40]—that the force had higher magnitude around the sharp edges of the parts, in analogy to electrostatic fields in proximity of tips. The parts were consequently designed to assemble into a finite number of two-dimensional configurations. A discrete-time model, inspired by chemical kinetics [39], was proposed to capture the dynamics and steady-state distributions of the population of assemblies. Such reasonably successful model constitutes one of the earliest instances of the use of the chemical reaction networks (CRN) formalism to describe artificial assembling systems [60].

The Whitesides Group's approach consisted in thin flat components of polygonal shapes floating at the interface between water and perfluorodecalin. The components were chemically conditioned to obtain on their sidewalls complementary wetting characteristics, such as hydrophobic and hydrophilic. The shape, weight and different wetting properties of the components determined the spatial orientation of the components as well as their vertical position relative to the fluidic interface (Fig. 12.1a). The components consequently experienced lateral capillary interactions, whose sign depended on the shapes of facing liquid menisci, as explained above, and whose magnitude depended on their relative height [11]. Such *capillary bond* drove the coordination of two-dimensional, crystal-like arrays, which could also be designed to host voids [9]. Systematic studies of the complete sets of coplanar aggregates and extended structures assemblable from sets of components with identical shapes but different chemical pre-conditioning were performed. The properties of the capillary bond were studied both experimentally and numerically [8–10, 37, 92]. A closed-form analytical description of the lateral capillary force was proposed, which could not be derived from an underlying potential field. As an additional analogy to crystal growth, the possibility of introducing templates—in the form of e.g. drops of water suspended in an apolar fluid or vice versa, or coated objects [20]—as nucleation sites and catalysts to promote assembly was demonstrated. Moreover, self-assembled aggregates reconfigurable through environmental changes (for instance, liquid density change) were shown [56] (Fig. 12.1b).

Klavins developed a *waterbug* model to describe and control tile-based, two-dimensional self-assembling systems [44]. The basic element of the model is a waterbug with four feet, whose attractive or repulsive interactions could be controlled by their wettability. He derived a Lagrangian description of the dynamics of the system based on the physics of flotation forces. He also proposed strategies

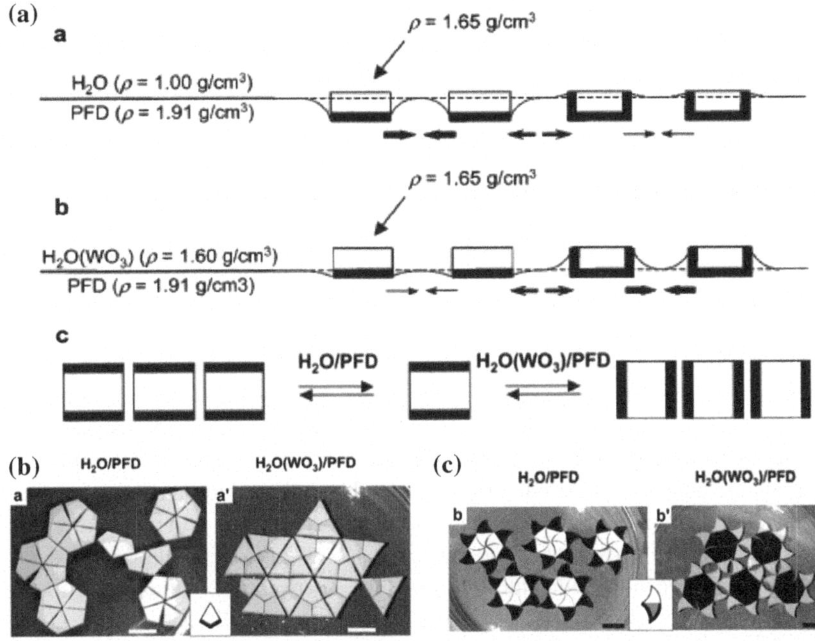

Fig. 12.1 SA at liquid-liquid interface mediated by flotation forces. **a** The shape of menisci depends on the wettability of the surfaces of the parts by adjacent liquids. Thin (thick) lines are hydrophilic (hydrophobic). Like (unlike) menisci attract (repel), and they can be modified by environmental parameters, such as fluid density in this case. **b**, **c** Aggregates self-assembled at a fluid/fluid interface reconfigurable by change in fluid density (Reprinted with permission from [56]. © (2002) American Chemical Society)

to avoid defects and to introduce pre-designed structures that would terminate the assembly.

A two-dimensional fluidic system featuring lateral capillary forces between sets of planar, millimetric tiles was also studied by Rothemund [71]. The tiles modulated their mutual lateral interactions by means of coded, hydrophobic or hydrophilic patches. Rothemund's was a partially-successful attempt to experimentally demonstrate the computational potentialities of tile-based self-assembling systems.[5]

Grzybowski and colleagues exploited the water/perfluorodecalin interface to float sets of parts that dynamically interacted under the competing influence of attractive magnetic forces and repulsive hydrodynamic ones, both induced by a spinning permanent magnet. Such prototypical setup and several variations were used to investigate the self-organization of dissipative structures, whose thermodynamics is still largely unknown [28].

[5] This work preluded to Rothemund's later breakthroughs with the SA of DNA-based tiles [72, 73].

Recently, adsorption and preferential orientation of micrometric dies at a liquid/liquid interface was exploited to boost the ordered transfer of the dies onto templated substrates (see Sect. 12.4.1.4).

12.3 Three-Dimensional SA

The Whitesides Group pioneered a general methodology for SA using capillary and hydrophobic effects. To this end, they reproduced at mesoscale two key features of supramolecular chemistry, namely bond *selectivity* and *directionality*.

In the method, the parts and the target templates are endowed with complementary geometrical features. The parts are then conditioned to allow the selective deposition of binder fluids on their superficial patterns. For instance, hydrophobic surfaces are coated with hydrophobic lubricants or adhesives in a polar hosting liquid, and metal pads are coated with low-melting-point solders. This combination of shape-matching and chemical pre-conditioning let the parts recognize and mate correctly. When immersed and stirred into fluid carriers (i.e. solvents, heated in the presence of solder), localized fluid/fluid interfaces of high energy are obtained to work as receptors. As the parts approach closely, their dynamics is locally constrained. Upon contact between matching geometries, the original fluid/fluid interface is replaced by a solid/fluid one of lower energy. As a consequence, capillary forces and torques lead or restore the parts into mutual alignment.

The role of the intermediate thin fluid film is crucial, because: (i) its interface provides the excess energy to drive the assembly, (ii) it spatially extends the capture cross-section of the binding sites, (iii) it provides lubrication to reduce friction between the parts as they approach their optimal spatial configuration, and (iv) it can mechanically bind (after e.g. polymerization, or solidification) the final assembly.

Using molten solders for the receptors, the bonds also form actual electrical interconnections. However, oxidation of the surface of the solder needs to be avoided. Oxidation hinders the wetting of metal patterns by the solder material, thus preventing self-alignment and bonding. Superficial oxide films are normally removed by diluting small amounts of acids (e.g. acetic, hydrochloric, sulphuric) in the fluid carriers to decrease their pH.

Such comprehensive technique was applied to parts ranging in size from tens of microns to millimetres [13, 23, 35, 88]. Notable examples of solder-directed SA include the fabrication of arrays of light emitting diodes (LEDs) on flexible substrates [41], of working three-dimensional electrical networks [36], and of sets of electronic devices into asymmetric structures in a way reminding of lock-in [89] and protein folding [6]. Electrical interconnection of self-assembled electronic devices, and the fabrication of functional parts specifically designed for three-dimensional SA were highlighted as important open issues [5]. The self-folding of planar patterned structures into closed polyhedra was envisioned as a partial solution to the latter problem (see Sect. 12.5).

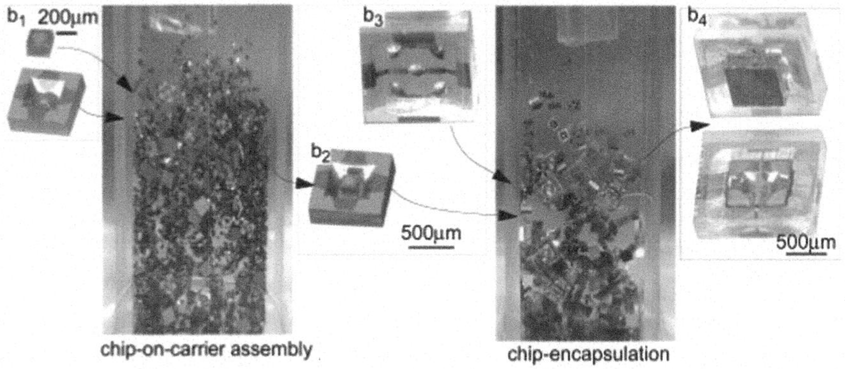

Fig. 12.2 Three-dimensional fluidic SA of a three-component sensor microsystem aided by capillarity and shape matching. The sequential, non-interfering assembly steps include: LED (for optical communication) onto silicon carrier, and encapsulation and electrical connection by a Pyrex cap. Assembly steps performed in an acidic ethylene glycol solution at 100 °C under external stirring (Reprinted with permission from [99], © (2005) John Wiley and Sons)

Jacobs and coworkers at University of Minnesota developed a process in which, in analogy with the biochemical formation of heterodimers, several functionally-different segments are sequentially assembled in a pre-designed three-dimensional configuration [98]. As an example, semiconductor devices were first assembled into glass carriers, and afterwards these aggregates were in turn assembled onto encapsulation units to form the disposable electric systems (Fig. 12.2). The solder bumps were hidden from unintended segments inside recesses. This, combined with shape matching, avoided *frustration*, i.e. the accumulation and binding of multiple blocks onto the same site. Stirring was provided by a turbulent flow induced in the carrier fluid by a piston pump. This agitation method is claimed to possess better energy transfer capacities, as compared to externally-induced acceleration (i.e. tumbling or shaking) and in presence of parts with differing densities and dimensions. In fact, fluidic drag forces scale with the square of the parts' linear dimensions—unlike the cube which is the case for inertia. Assembly yield as high as 97 % was demonstrated. A time-continuous analytical model in agreement with experimental results was also introduced to estimate assembly times and yield rates [99]. Such ODE model was later generalized to account also for finite disassembly rates [59]. Its predictions were reproduced, and its scope extended to include spatial parameters, by a numerical agent-based model [63].

A method for self-assembling modular, heterogeneous, three-dimensional systems was presented by Cannon et al. [14]. Arbitrary devices are thereby encapsulated into polymethylmethacrylate (PMMA) cubes, which are geometrically-homogeneous and easy-to-assemble building blocks. The blocks are cross-wrapped with strips of copper tape. The strips are subsequently coated with low-melting-point solder. The prepared blocks undergo molten solder-driven SA in a heated, diluted solution of potassium bromide. After assembly and cooling, the solder establishes mechanical bonding and electrical interconnections among the modules, and if required PMMA can be finally selectively removed by chloroform.

12.4 Die-to-Substrate SA

12.4.1 Stochastic SA

The Whitesides Group's approach to mesoscale SA, combining capillary and hydrophobic effects as discussed in the previous section, partly inspired the die-to-substrate SA technique developed by Srinivasan, Howe and coworkers at UC Berkeley [84]. This method (Fig. 12.3) combined together several concepts useful for microsystem integration: (i) *contactless handling* of microdevices, which avoids altogether handling issues at sub-millimetric scales [65] and complies with the ongoing downscaling of device dimensions; (ii) *massively parallel assembly* of devices, consistent with their batch fabrication mode; (iii) very-high *positioning accuracy* thanks to capillarity, as already discussed; (iv) *selectivity of interactions*, as obtained by chemical surface pre-conditioning and geometrical shape-matching between devices and binding sites; and (v) *high assembly rates* thanks to the adoption of a two-dimensional template in the form of the target substrate.

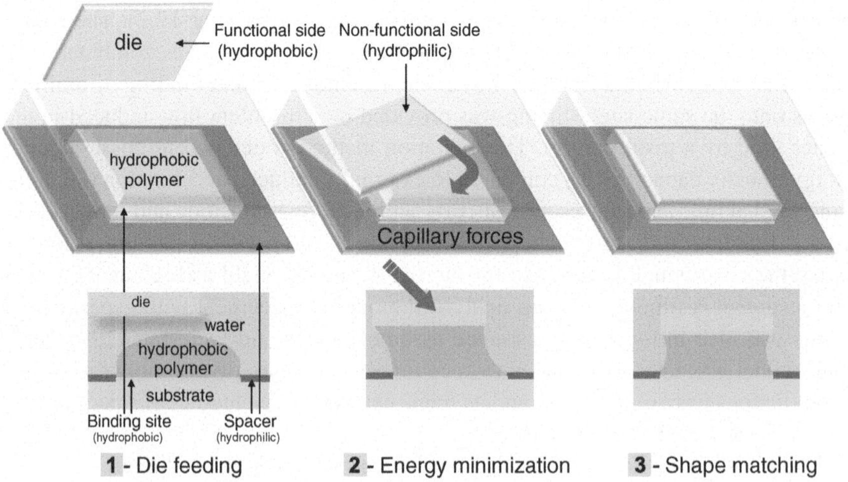

Fig. 12.3 Principle of capillary die-to-substrate self-alignment, as originally proposed by Srinivasan et al. [84]. When combined with stochastic or deterministic (robotic) die feeding, it yields *die-to-substrate SA* or *hybrid microhandling*, respectively (Adapted from [58])

In the original proposal, a hydrophilic substrate (i.e. glass or silicon dioxide[6]) was patterned by lift-off with arrays of gold binding sites. The gold sites were functionalized with thiol-ended self-assembled monolayers (SAMs) [55] to selectively make them hydrophobic. Thanks to the wetting contrast between binding sites and native surfaces, lenses of hydrophobic methacrylate adhesive were then selectively dip-coated on the sites while sliding the substrate into water [4]. The interfaces between such lenses and water constituted the receptors for the parts to be assembled. The very thin parts were patterned on a single side with gold pads, whose shapes reproduced those of the binding sites. The pads were also functionalized by SAMs. For the assembly, the parts were pipetted into water toward the substrate. Minimization of interfacial energy, involving substitution of adhesive/water with lubricant/SAM interfaces, led the parts to maximize the overlap of their pads with the binding sites, thus reaching highly-accurate placement and orientation. Defects (e.g. unbound or interstitial parts, multiple parts crowding the same site) were removed by gentle flows of water. Finally, permanent bonding was obtained by cross-linking the adhesive either by thermal treatment, or by exposure to UV radiation [84]. The assembly of micromirrors of unprecedented flatness over an array of actuators for adaptive optics was demonstrated [83]. Pose errors as low as 0.2 μm for placement and of 0.3° for angular displacement were claimed.

Howe and colleagues later enhanced their original technique to assemble high-performance RF inductor chips over CMOS integrated circuits [80]. An initial capillary self-alignment step coarsely localized the chips onto the sites; fine chips registration and electrical connection to the substrate were then achieved by solder reflow. The chips and the binding sites were endowed with a pair of matching metal pads. They were both coated with resist, made hydrophobic by exposure to CF_4 plasma, while leaving the pads untouched. Solder was applied to the site pads, and the whole surface of the binding sites was dip-coated with hexadecane while the substrate was slid into water. The hexadecane film hid the solder bumps, and aided the coarse assembly of the chips to the binding sites. Later the substrate was extracted from water and dried. This allowed the pads on the chips to get in contact with the solder bumps on the sites. Thermal reflow finalized the self-aligned metallic bonding.

Several failure modes pertaining to the above mentioned technique were studied by Mastrangeli et al. [61]. Particularly, mutual parts aggregation and fluid coating disruption by part removal and improper agitation were evidenced as serious yield killers. Aggregates can form upon direct contact of multiple parts, as mediated by air bubbles in the carriers or by the same interactions driving the SA process (e.g. hydrophobic interactions, wetting of metal pads by exposed solder bumps). Such three-dimensional assemblies represent an impertinent interference with the intended two-dimensional SA process, and may involve a significant number of components. Moreover, impacts of parts on the binding sites as well as the accidental removal of parts partially adhering to the receptors may distort their fluid coating. While the

[6] Besides exploiting native surface properties, a host of methods are available to tune surface chemical properties of substrates, mainly including plasma treatments and derivatization with tailored moieties. Thorough surface cleaning is mandatory, as well.

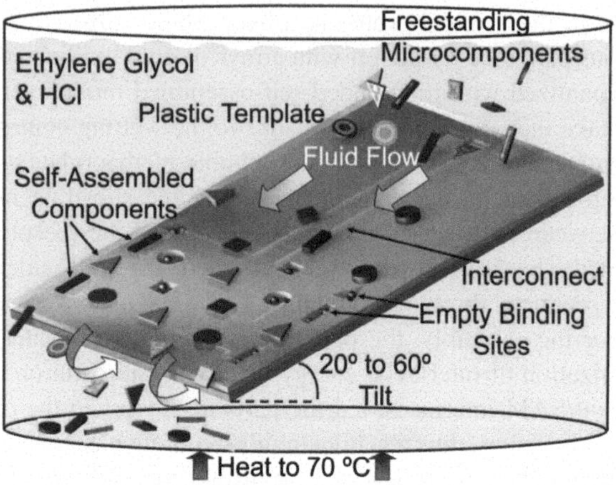

Fig. 12.4 Gravity-driven, stochastic fluidic SA of multiple batches of devices onto a templated substrate aided by shape-matching and molten solder (Reprinted with permission from [86], © (2006) National Academy of Sciences, U.S.A.)

former failure mode makes the involved components useless and unrecyclable, the latter induces false positive assemblies, partial part-to-site overlaps or useless sites altogether. The use of surface chemistries switching as a function of local chemical parameters (e.g. pH), and of recessed binding sites [62] were respectively proposed as possible solutions to the issues. Importantly, these failure modes pertain only to a fully-stochastic capillary approach to SA, where the parts can meet the substrate or other parts several times and with random orientations.

Böhringer's team at University of Washington made use of water-coated hydrophilic sites surrounded by hydrophobic surfaces as receptors for the assembly of MEMS resonator chips in air [51]. Assembly conducted in a (semi)dry environment helped avoiding stiction problems eventually induced in MEMS by a fluid carrier. Besides, the air/water interface is more energetic than the water/oil one. The patterned target substrate was mounted upside down over a vibrating diaphragm, which imparted enough kinetic energy to the parts to reach the receptors while apparently not damaging the devices. The method showed reasonably-high yield, self-correction of parts pose due to kinetic impact energy, and compatibility with multi-batch assembly by means of masked sites. A similar substrate pre-conditioning was employed by the group to assemble piezoelectric driving elements for micropumps over Pyrex substrates in air [26]. In this case, part feeding was manual and their in-plane alignment was refined by orbitally-shaking the target substrate.

Stauth and Parviz devised a gravity-driven process, combining parts-to-site geometrical shape recognition with the capillary binding forces of molten solders,[7] to achieve SA of microdevices onto non-standard planar substrates (e.g. plastic and glass) [86] (Fig. 12.4). In the process, a thick resist is lithographically patterned over a plastic substrate containing electrical interconnects. Recessed binding sites are cleared in the resist in coincidence with metal pads, later dip-coated with solder. The geometry of these traps is precisely tailored to the shape of the devices, so

[7] The process can be thought of as an evolution of the original *fluidic SA* by Yeh and Smith [97].

that multiple batches of parts can also be sequentially assembled (see Sect. 12.4.1.1). The substrate is immersed and tilted in a warm, acidic water solution, which melts the solder and continuously removes its superficial oxide layers. Once dispersed in the solution, the devices approach the substrate and eventually enter the shape-matching sites by gravity. Once in place, the molten solder refines their registration and finalizes their electrical and mechanical bonding. A laminar fluid flow and the controlled agitation of the substrate facilitate the sliding of devices across the substrate and the removal of incorrectly assembled devices (e.g. devices set upside down). A recycling apparatus was developed to automate the process, allowing unbound devices to undergo repeated assembly runs [42]. As theoretically expected, filling of all sites is achieved only asymptotically through multiple passes [42]. Finite element simulations were used to estimate and compare competing fluidic and capillary forces involved in the process [85]. While the claimed yield for the mechanical assembly step was as high as 97 % within 5 min, the overall yield was sensibly lower (62 %) due to failure of electrical interconnects [76]. Studies on the dimension, metallurgy and electrical performance of robust solder-based binding sites for SA purposes were consequently conducted [64, 74]. Parviz's group applied their SA process to the integration of a variety of functional systems, including single-crystal silicon devices onto large plastic substrates [86], optoelectronic devices onto glass and plastic [76], and fluorescence detection systems [43].

12.4.1.1 Programmability

Stochastic fluidic die-to-substrate SA allows for the massively parallel processing of multiple parts at once. Given specific constraints on parts and substrates, multiple batches of parts can be assembled on the same substrate, as well. The key enabler is the selective (de)activation of the binding sites—that is, the possibility of choosing in advance which sites could (not) host the assembly of parts.

Böhringer's team extended Srinivasan's methodology into multi-batch mode [94]. Selective site deactivation was achieved through the electrochemically-induced desorption of the thiol-ended SAMs from the chosen binding sites (Fig. 12.5a). In so doing, these sites lost their hydrophobicity by exposing anew a hydrophilic gold surface. This prevented them from being coated by the adhesive during the subsequent dip-coating step. The mechanical bond after the assembly was still provided by cross-linked polymers. The team addressed the issue of the electrical connection of the devices to the substrate, as well [94]. To this purpose, external vias were electroplated to link the electrodes of the substrate to assembled LEDs. However, the use of electrodesorption and of electrodeposition complicated the layout of the substrate, since it required all the sites to be externally accessible by two independent sets of electrodes and to host metal seeds—all structures to be used one time only.

Jacobs and colleagues demonstrated the electrothermally-programmable heterointegration of microdevices on substrates [21]. In their approach, copper binding sites were dip-coated with low-melting-point solder over a polyimide sheet. The solder bumps were individually activated (i.e. molten) by resistive heaters integrated

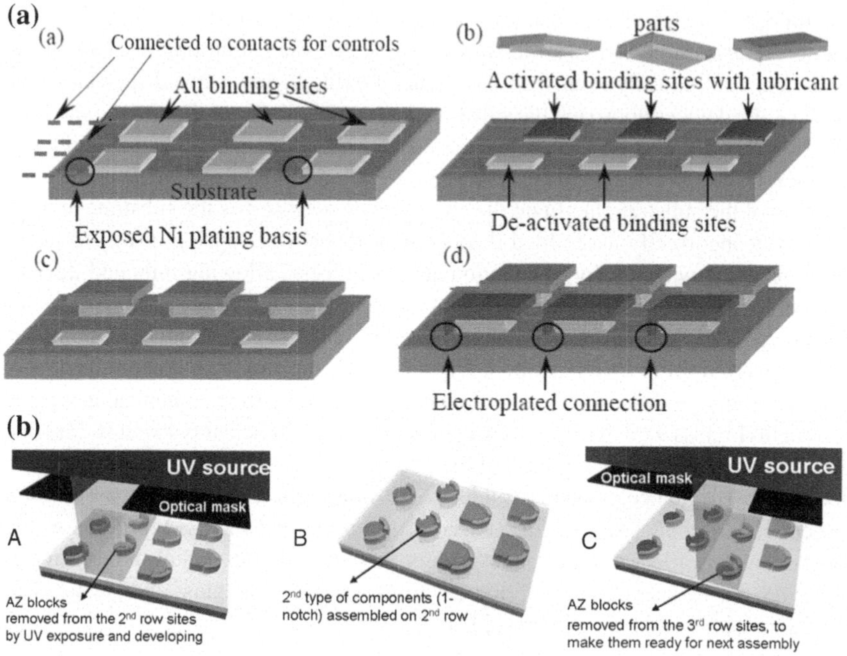

Fig. 12.5 Examples of programmable capillary die-to-substrate SA of microdevices (Reprinted with permission). **a** Electrochemical (Xiong et al. [93], with permission from the author) **b** Optical (Saeedi et al. [75], © (2011) Cambridge University Press)

on the backside of the flexible support. The generated heat power was calibrated to induce melting of solder on specific sites while avoiding both stationary and transient cross-talk to neighbouring sites. Applying the technique to the assembly of a multicolour LED-based flexible display, the substrate was wrapped inside a cylindrical drum, whose periodic motions allowed the tumbling of the LEDs in the fluid. LEDs of three different colours were assembled in subsequent steps. After this high-yield assembly of LEDs, a transparent layer patterned with metallic interconnections was self-aligned on top of them exploiting the reflow of solder bumps, providing the second electrical contact to the LEDs. Working arrays of several hundred devices were obtained with close to 100 % yields. The main cause of defects was the unprecise control over the solder volume dip-coated on large substrates.

Liu et al. sequentially activated multiple receptor sites according to the melting point of their solder bumps [54]. As a demonstration, two solders with different melting points (70 °C for solder A and 47 °C for solder B) were deposited on selected binding sites by covering the rest of the sites with resist. The liquid hosting the assembly was then heated up to the melting point of solder B, and the first batch of parts was introduced and assembled. After that, the temperature of the liquid environment was increased until it reached the melting point of solder A. At this point, the second batch of parts was introduced and assembled.

In an approach roughly complementary to the previous, Sharma utilized the thermally-induced gelling behavior of a hosting fluid to selectively prevent parts from reaching the binding sites on a substrate [81]. Hydrophobic sites were fabricated on the parts and on the assembly template, and the template sites were covered with hexadecane. The hosting fluid formed a gel around the sites when they were heated up: this prevented the assembly of parts onto undesired sites. Consequently, by individually heating up (i.e., deactivating) the sites, several part types could be assembled on a single template. Local gelling was also exploited by Krishnan et al. to control the fluidic assembly of blocks [49].

Stauth and Parviz made use of the inherent and strong selectivity of three-dimensional shape-matching between parts and recessed binding sites to enable multibatch assembly. They used sets of orthogonal shapes, so that parts of one shape (e.g. circles) could not fit inside the sites designed for a different shape (e.g. rectangles) [86]. This way, multiple batches could in principle be assembled also at the same time, besides sequentially (Fig. 12.4).

Finally, Parviz's group selected the binding sites to be activated for assembly by means of photolithography [75]. All binding sites on the substrate, defined by chaperones, were initially protected by plugs of resist. The chosen plugs could be selectively removed by exposure to light and subsequent development before the assembly process (Fig. 12.5b).

12.4.1.2 Unique Die Assembly Pose

The number of equivalent poses a thin (i.e. two-dimensional) die has access to once on its binding site of identical shape is geometrically determined by the in-plane symmetry of its shape. For instance, a square and a rectangle have four and two equivalent poses, respectively. This ambiguity is at odd with the electrical functionality of the die: among the possible poses, only a single one matches the electrical polarity of multiport circuits. This issue do not necessarily pertain to pick-and-place assembly, where the devices can be predictably pre-oriented before assembly.

Such important issue was already investigated empirically by Srinivasan et al. in their pioneering work [84]. It may indeed be thought that adopting shapes without any in-plane symmetry for the dies should avoid the problem. This intuition turns out to be not correct [58].

Böhringer's team has since early on extensively worked on the modelling of capillary die-to-substrate SA to aid binding site design. A first-order model predicting the energy landscape and consequent capillary forces acting on the parts was proposed [53]. The model is valid for lateral part displacements larger than the thickness of the lubricant film. The model makes use of the two-dimensional convolution of the geometrical shapes of interest to estimate their spatial overlap as a function of mutual orientation and displacement. The model helps quickly discovering local energy minima (i.e. sub-optimal poses) and estimating their relevance as compared to the global minimum, and is consistently supported by Surface Evolver (SE) simulations [12]. Optimal design for parts and sites was addressed by (i) deriving an

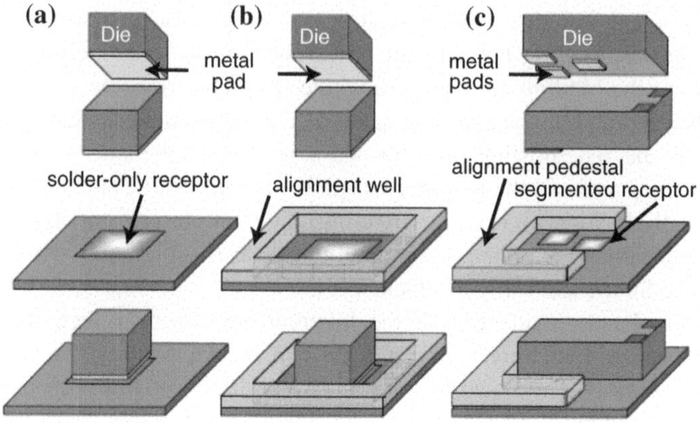

Fig. 12.6 Evolution of receptor sites for capillary die-to-substrate SA, after Knuesel et al. **a** Fluid-only receptor does not avoid assembly of multiple dies, **b** fluid receptor with alignment well allows single die assembly with multiple poses, and **c** segmented fluid receptor with alignment pedestal enforces single die assembly with unique pose (Adapted with permission from [45], © (2007) NSTI)

analytical, sufficient condition for the existence of continuous paths leading a part toward its minimum energy placement [96], and (ii) describing a class of patterns theoretically driving a die from any initial position toward a unique, arbitrary pose over the binding site [52]. Experimental evidence of the efficacy of this geometrical solution for actual die-to-substrate SA is nonetheless not convincing [58]. Moreover, the *inverse* assembly problem—i.e., finding a site shape generating a given energy landscape for a given part shape—has found no general solution so far [95].

Geometrical analysis is not the only way to cope with the problem at hand. At least three alternative solutions were demonstrated, all exploiting three-dimensional out-of-plane features.

The first, partial solution dates back to the fluidic SA process developed by Yeh and Smith at UC Berkeley [97] and is implicit in the use of receptor sites recessed into the substrate. L- or T-shaped sites with V-grooved cross-sections can enforce unique in-plane poses of geometrically-complementary parts more efficiently than their two-dimensional projections. However, this solution imposes the use of non-standard device shapes, and it does not resolve the geometrical pose ambiguity of typical device geometries (e.g. circular, rectangular).

More recently, Jacobs' group proposed a simpler method to achieve highly-accurate die-to-substrate assembly with unique in-plane pose [100]. The method makes use of pedestals surrounding three of the four sides of the binding sites, and of an asymmetric placement of solder bumps within the sites to further select the pose of the dies (Fig. 12.6). The pedestals protruding from the substrate exclude beforehand a number of wrong poses of the dies, besides multiple dies crowding a single site. Then, once a die has reached the correct alignment among the pedestals, the single working orientation out of the possible two is selected depending on whether the metal pads on the die and the solder bumps on the site are thereby superposed. This happens in the correct orientation only, as confirmed by the establishment of the metallic interconnections. Uncorrectly posed dies are continuously removed during the process by means of fluidic tumbling.

One more solution exploits complementary mechanical features patterned on both dies and substrates. It was devised by Fang and Böhringer in an attempt to extend the process to full-wafer scale, as described below.

12.4.1.3 Wafer-Level Assembly

Fang and Böhringer proposed a few methods for achieving high-density mounting and unique angular orientation of microcomponents at full-wafer scale. In these methods, the stochastic die motion is still dominant but aided by mechanical filtering of poses to increase reproducibility and shorten the time-to-assembly. Final alignment of dies to substrate sites is refined by means of capillarity.

In the first process [24], thin square parts, having one hydrophobic side, were distributed over a plate patterned with an array of apertures, and thereby randomly agitated till they fell into the apertures. The yield of this feeding and mechanical filtering step [38] can be demonstrably enhanced by introducing mechanical *catalysts*, in analogy with chemical reactions [3]. Then, the parts vertically-standing over a temporary carrier were projected onto the water-coated binding sites of a palletizing plate. When the aperture plate was removed, capillary torques and substrate agitation drove the parts to lie flat over the sites with their hydrophobic side facing upwards. After water evaporation, a bonding plate was aligned and brought into contact with the parts before performing a reflow step.

Fang and Böhringer also demonstrated two feature-directed processes to achieve unique die pose [25]. They exploited micropegs extruding from the bottom side of the dies and fitting into complementary-shaped holes of the substrates.

In the semi-dry method, one of the major sides of the square dies was made hydrophobic; the opposite side featured a circular peg offset from the centre. The hosting substrate was prepared with square sites featuring holes complementary to the pegs. Once palletized, the parts were released to the hosting substrate, which was shaken to accomplish the registration of the dies. The final alignment accuracy could be enhanced by capillarity.

A fully-dry assembly method was also demonstrated, and later developed by Ong et al. to include working electrical interconnections [66, 67].

12.4.1.4 Energy Cascade Method

A breakthrough in high-throughput and accurate die-to-substrate SA was recently reported by Knuesel and Jacobs [46]. It ingenuously combines preferential adsorbtion of preconditioned parts at a water/silicone oil interface (as seen in Sect. 12.2) with fast transfer of the parts from the interface to a target substrate (Fig. 12.7).

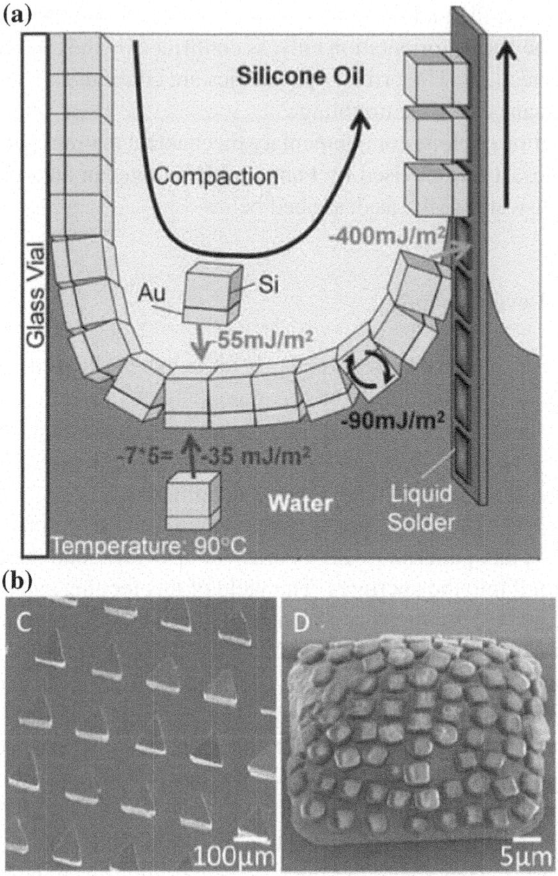

Fig. 12.7 **a** The energy cascade SA method at fluid/fluid/solid interface. **b** Assembly examples (Reprinted with permission from [46], © (2010) National Academy of Sciences, U.S.A.)

The sub-100 μm cuboidal parts are fabricated out of single-crystal Si. A single face of the parts exposes an Au pad derivatized with a hydrophilic SAM, while all other Si faces are derivatized with an orthogonal SAM to be hydrophobic. The substrate is templated with arrays of binding sites coated with solder. During the process, the pH of water is brought to about 2 and the bath is heated above the melting point of the solder. This way, the surface of molten solder is constantly freed from oxide layers. The parts are immersed in the bath, and when crossing the horizontal interface, they preferentially reorient to have the single hydrophilic face pointing downward to be wet by water. At this point, the substrate is vertically slid through the fluid interface, cyclically alternating downward and upward translations. This step reminds of the Langmuir-Blodgett transfer of monolayers of moieties onto substrates [69]. During the upward motion of the substrate, the water/oil interface is bent upward, and while receding across the substrate it pushes the parts in direct contact with the substrate.

Thanks to the above mentioned preferential preorientation, the parts approach the substrate with their Au-coated side. They can thereby selectively bind and self-align to the molten solder bumps. Conversely, during the downward motion, the fluid meniscus is inverted and the parts are repelled from the substrate. Unassembled or uncorrectly bound parts are thus removed and set back at the water/oil interface. A site filling ratio higher than 90 % can be achieved in a single pass, while site saturation is achieved through multiple passes. Interestingly, the sequence of steps involved in the assembly process (i.e. part transfer from the bulk to the interface, the preferential orientation at the interface and the final transfer from the interface to a solder-coated site on the substrate) form a so-called *energy cascade*, as each of them demonstrably decreases the excess free energy of the system. Furthermore, the technique complies with an aggressive downscaling of part dimensions since it does not depend on gravity nor sedimentation. While a lower scaling limit for the transport of nanometric parts is set by thermal energy, experimentally the process is challenged by the stability of molten metal bumps.

By means of their method, Knuesel and Jacobs demonstrated an unprecedented combination of assembly throughput (up to 62500 parts assembled in 45 s with 98 % of yield)) and accuracy (0.9 μm and 0.14° for translational and rotational error, respectively) in the fabrication of flexible, segmented monocrystalline solar cells [46]. They later enhanced the technique to the continuous self-tiling of patterned plastic surfaces, whereby electrically connected domains (with less than 3 % of vacancies) of monocrystalline Si, 20 μm-thin tiles were fabricated and could work as microconcentrator solar cell modules [47]. Precise design of the tiling domains is required to achieve such vacancy-free capillary mosaics. They demonstrated an assembly rate of 10000 tiles per minute for 20 μm^2 tiles, which is 100 times better than allowed by state-of-the-art serial robotic chip assembly, and a minimal assembled tile size of 3 μm^2, exceeding what currently possible in robotic assembly again by a factor of 100.

12.4.2 Hybrid Microhandling

Recently, the combination of capillary self-alignment and robotic manipulation has been investigated as candidate for an ideal precision manufacturing technique.[8] The idea is to bring the part close to the fluid-coated receptor site by means of a robotic arm, and thereby release the part upon its contact with the fluid/air interface (Fig. 12.8). The resulting method, called *hybrid microhandling* (HM) [78], could couple the complementary advantages of both approaches.

The dexterity and rather general handling capabilities of servoed manipulation is so far unmatched by SA. A large variety of grippers are available to address parts of arbitrary geometries, and to coordinate them within assemblies of unrivalled sophistication [27]. Robotic manipulation can handle up to 10^5 parts per hour with

[8] See e.g. the European Hydromel Project, and the FAB2ASM European Project.

accuracies in the order of several microns. On the other hand, it is an intrinsically serial technique requiring feedback control or supervision, and its performance is limited by a trade-off between placement accuracy and speed of operation. Moreover, contact handling of micrometric parts is physically hard for manipulators because of adhesion issues between parts and grippers [65], which severely limits the compliance of the technique with current device downscaling.

Capillary self-alignment is instead unsupervised since, once properly set up, it relies on the minimization of the energy of the system and the proper injection of noise to overcome local energy barriers. It also achieves very-high placement accuracy, as earlier described, and involves a strong adhesion force between the parts and the receptor site once in contact. This force can be larger than the adhesion force between gripper and parts.

Hybrid microhandling may thus perform well beyond the accuracy/throughput trade-off of established pick-and-place assembly. In fact, the gripper could release the die as soon as partial overlap and contact between die and (the fluid coating of the) receptor site[9] is established, and move to fetch the next die while the previous is self-aligning. This partial parallelization of the process will boost its throughput while maintaining the great accuracy typical of capillary self-alignment.

Koyanagi's research group at Tohoku University was the first to propose and implement the integration of a capillary self-alignment step into a die-stacking, through-silicon VIAs-based fabrication process for three-dimensional electronic circuits [33]. Improved through successive extensions, their original approach consisted in patterning by reactive ion etching a temporary carrier (e.g. an oxidized Si wafer) with reliefs shape-matching the profile of the dies to be stacked. Drops of mild solutions of hydrofluoric acid were deposited on the reliefs, whose sharp vertical edges helped confining the drops [90] possibly in combination with specific surface treatments. Then multiple, centimeter-sized dies where brought all at once (by means of an ad hoc parallel vacuum gripper) with their oxidized backsides in partial contact with the drops. Upon release, the dies self-aligned to the reliefs, and ended up being strongly bound to the carrier upon evaporation of the solution, which this way created a chemical bond between the contacting oxide surfaces [29]. A sub-micrometric accuracy in the registration of the edges of the die to those of the reliefs was demonstrated by SEM imaging of the cross-sections. The carrier was then flipped upside-down and aligned to the target substrate so that the dies could be transferred from the carrier to the stack. Both reconfigured wafer-to-wafer and multichip-to-wafer stacking could be addressed. The effectiveness of this method was demonstrated through e.g. the assembly of functional 3D memory and microprocessor chips, retinal circuits [30], and of components featuring cavities [31].

Alternatively, the group used a similar approach to assemble dies to receptor sites in a face-down fashion—thus enhancing the work of Scott et al. discussed in Sect. 12.4. To this purpose, the group built on its long experience in both capillary self-alignment and fabrication of InAu microbumps by means of evaporation and

[9] In die-to-substrate capillary SA, the capture cross-section (i.e. the self-correction range) is of the order of the area of the die, as compared to that of the solder bumps as in flip-chip assembly.

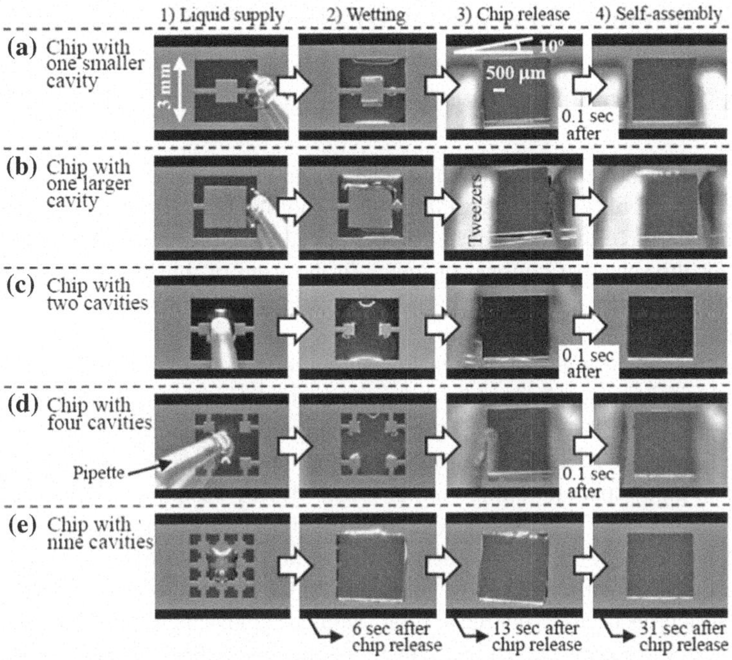

Fig. 12.8 Hybrid microhandling combining coarse (manual or robotic) pre-alignment with precise capillary self-alignment of components (Reprinted from [31], © (2011) MDPI under the Creative Commons Attribution License)

lift-off. They were able to assemble working, low-resistivity daisy-chains with pitches of few microns [32].

At about the same time, Zhou, Sariola and colleagues at Aalto University investigated the viability of hybrid microassembly in a more general assembly perspective [77]. They showed how capillarity can be successfully used to leverage both standard manipulation as well as hard assembly tasks, such as the construction of cantilevers and of hierarchical structures [79]. They also reported extensive statistical studies on the performance of the technique as applied to die-to-substrate assembly [77]. Its accuracy, of the order of lithographical tolerances, was confirmed; and its dynamics was proved to substantially depend on the tolerance in the vertical position relative to the drop and on the lateral misplacement of the part and of the drop as compared to the center of the receptor site. As a consequence, the dynamics of a self-aligning part can be subjected to large and unpredictable variations while still normally leading to the expected pose. Finally, the group also showed how hybrid microhandling can be used to successfully assemble electronic dies onto receptor sites consisting of non-connected patches [15], as earlier anticipated by Mastrangeli [61], and submillimetric dies [16].

12.5 SA of Three-Dimensional MEMS

12.5.1 Capillary Actuation of MEMS

Standard microfabrication techniques, whether additive (surface micromachining) or substractive (bulk), are planar and based on photolithography. As a consequence, the fabricated three-dimensional micromechanical structures result from the stacking and/or the selective removal of layers in the form of thin films. Though such structures can be rather sophisticated, they allow for a limited freedom in the choice of architectural solutions and designs.

Alternatively, devices and architectures composed by the interlocking of planar, patterned layers can be obtained by the out-of-plain deployment of such layers from the native substrate. The forces and torques arising from the minimization of an energetic fluid interface have been fruitfully exploited for this purpose. This approach to functional, high aspect-ratio three-dimensional MEMS was pioneered and recently reviewed by Syms at Imperial College [87]. The basic idea consists in depositing a controlled amount of reflowing material over the border between (at least) two adjacent, normally flat structures. After release from the native substrate, (at least) one of such structures needs to be free to rotate along the above mentioned border, either in presence or in absence of a hinge. The reflowing material, which can be e.g. glass, polymers or solder, needs to wet and strongly adhere to the underlying structural materials. Upon temperature increase the shrinkage of the reflowing material gives rise to capillary forces and particularly torques—discussed in Chaps. 3 and 4, respectively. Such transient capillary actuation can displace and rotate the structures. The final disposition of the structures can be engineered by the position of interlocking hinges, or predictably controlled by the amount of reflowing material deposited. A large variety of structures could be fabricated in this way, including optical scanners, diffraction lenses, out-of-plane inductors.

12.5.2 Self-Assembled Polyhedra

A key challenge for the full development of the potential of SA as a fabrication technique is how to best produce micro- and nanoscopic parts specifically tailored to undergo controlled aggregation into functional devices [7, 50]. Clearly, the extreme patterning accuracy achievable by lithography is to be reckoned with. On one hand, specific patterns on the surfaces bounding particles or devices are fundamental to encode information about the location, selectivity and directionality of the physical connections among the parts. On the other hand, it is hard to lithographically pattern surfaces extruding in three dimensions.

At the end of the 1990s the controlled self-folding of two-dimensional, patterned planar structures was proposed as a viable technique to face the challenge. By starting with its two-dimensional deployment defined onto a substrate, the desired three-dimensional structure could be fabricated by folding its planar components.

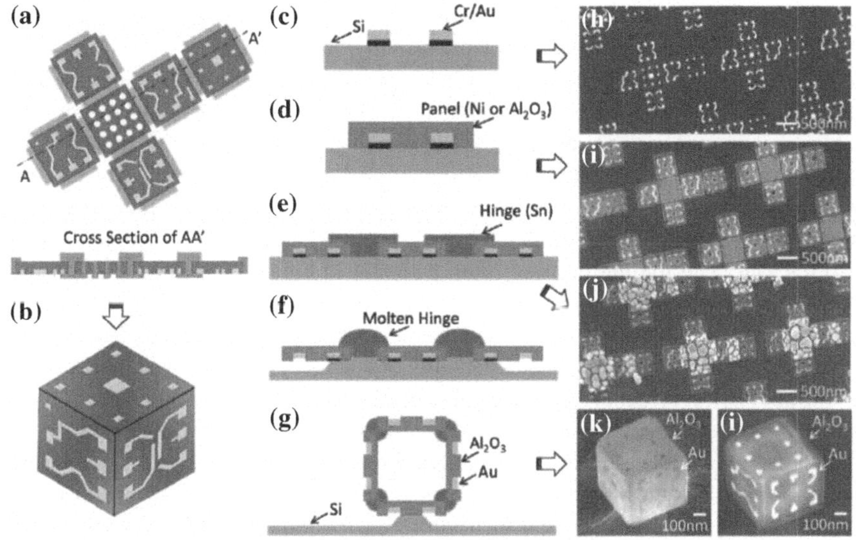

Fig. 12.9 Fabrication of patterned, multilayer sub-nanometric cubes for three-dimensional optical arrays by capillary self-folding (Reprinted with permission from [19], © (2011) John Wiley and Sons)

Among the proposed actuation mechanisms—which included also pre-stressed films and magnetic forces—capillarity emerged because of its ease of integration with standard machining techniques, and its favorable downscaling properties. This field of research was pioneered and extensively investigated by Gracias, first at the Whitesides Group at Harvard and later at Johns Hopkins University. Gracias mainly focused on polyhedral (typically Platonic) three-dimensional structures, whose decompositions into connected sets of planar panels (called *nets* in graph theory) are rather easy to fabricate. The capillary method adopted was inspired by the earlier work on capillary actuation of MEMS structures (see previous section). It relies on the capillary torque exerted by the reflowing materials (particularly tin, and polymers) deposited along the borders of adjacent panels to induce their folding. In this case, the reflowing material is additionally deposited on most of the edges of the panels, as it thereby induces the precise self-alignment of the panels upon contact and the final sealing of the structure, this way dramatically increasing the fabrication yield (Fig. 12.9). The location and sequence of concurrent folding steps could also be predefined. Ongoing research has so far notably achieved the fabrication of e.g. three-dimensional inertial sensors [18], cargos for biological samples [2], three-dimensional arrays with remarkable photonic properties [19], and more generally the first examples of nanoparticles featuring predetermined patterns all over the bounding surfaces [17].

Specific issues arising from downscaling include patterning accuracy, process parameters, and optimal choice of materials, surface patterns and nets. Electron beam lithography and lift-off were mostly used for patterning surface features with down to nanometric accuracy. Concurrently, the alignment between the subsequent masks integral to the fabrication became critical. As for the materials, the choice of the reflowing materials was not as much dependent on the surface tension value—because of scaling, surface tension dominates anyway over all other forces at nanoscale—as on the wetting with respect to the structural materials. Besides tin as reflowing material, Ni, Au and aluminum oxide were chosen for the latter. The control over the behavior of tin grain growth, surface mobility and coalescence was also relevant. The folding process could be initiated by the heat developed during the release of the panels, and was in general sensitive to its parameters as well to the geometrical dimensions of the actuated structures. Furthermore, a study on the influence of surface patterns on the assembly of crystal aggregates of polyhedra was performed [70]. It evidenced how, for cubic blocks, the global assembly performance is not only dependent on the energetics of one-to-one part assembly, but also on one-to-four and four-to-four ones. Also, the assembly error rate is sensitive to the binding patterns on the surface of the parts. Numerical and experimental evidence supported that the total area and the spatial distribution of the pattern affect the energy landascape associated to the assembly. Finally, heuristics were investigated to facilitate the choice of nets among the vast number of alternatives existing for each type of polyhedron [68]. Metrics defined over the graphs of the nets, generally prizing their compactness, were proposed to quantify the self-folding performance.

12.6 Conclusions

Surface tension forces can be fruitfully engineered to fabricate functional microsystems. Capillary SA can not only leverage standard assembly tasks, but also perform tasks currently precluded otherwise. Hybrid microhandling exemplifies the former instance. Capillary self-alignment is thereby decisive in overcoming the accuracy/throughput trade-off of standard pick and place assembly. The assembly rates demonstrated by the energy cascade method of Knuesel and Jacobs for dies as small as few tens of microns support conversely the latter claim. Besides, capillary self-folding has enabled the fabrication of the smallest patterned particles reported so far.

The further appreciation of the potentiality of capillary SA will possibly depend on more comprehensive characterization, modeling and standardization of the methods, and on its envisionable extension to higher volumes and large areas while unaffecting yield. The field is still burgeoning with creative and innovative approaches; yet further and extensive research is needed to systematically tackle such tasks.

References

1. F. Arai, D. Ando, T. Fukuda, Adhesion forces reduction for micro manipulation based on micro physics, in *IEEE 9th Annual Workshop on Micro Electro Mechanical Systems*, San Diego (CA, USA), 1996, pp. 354–359
2. A. Azam, K.E. Laflin, M. Jamal, R. Fernandes, D.H. Gracias, Self-folding micropatterned polymeric containers. Biomed. Microdevices **13**, 51–58 (2011)
3. R. Baskaran, J.H. Hoo, B. Cheng, K.F. Böhringer, Catalyst enhanced micro scale batch assembly, in *IEEE 21st International Conference on Micro Electro Mechanical Systems*, Tucson (AZ, USA), 2008, pp. 1069–1072
4. H.A. Biebuyck, G.M. Whitesides, Self-organization of organic liquids on patterned self-assembled monolayers of alkanethiolates on gold. Langmuir **10**, 2790–2793 (1994)
5. M. Boncheva, D.A. Bruzewicz, G.M. Whitesides, Millimeter-scale self-assembly and its applications. Pure Appl. Chem. **75**, 621–630 (2003)
6. M. Boncheva, D.H. Gracias, H.O. Jacobs, G.M. Whitesides, Biomimetic self-assembly of a functional asymmetrical electronic device. Proc. Natl. Acad. Sci. **99**, 4937–4940 (2002)
7. M. Boncheva, G.M. Whitesides, Making things by self-assembly. MRS Bull. **30**, 736–742 (2005)
8. N. Bowden, F. Arias, T. Deng, G.M. Whitesides, Self-assembly of microscale objects at a liquid/liquid interface through lateral capillary forces. Langmuir **17**, 1757–1765 (2001)
9. N. Bowden, I.S. Choi, B.A. Grzybowski, G.M. Whitesides, Mesoscale self-assembly of hexagonal plates using lateral capillary forces: synthesis using the 'capillary bond'. J. Am. Chem. Soc. **121**, 5373–5391 (1999)
10. N. Bowden, S.R.J. Oliver, G.M. Whitesides, Mesoscale self-assembly: capillary bonds and negative menisci. J. Phys. Chem. B **104**, 2714–2724 (2000)
11. N. Bowden, A. Terfort, J. Carbeck, G.M. Whitesides, Self-assembly of mesoscale objects into ordered two-dimensional arrays. Science **276**, 233–235 (1997)
12. K.A. Brakke, The surface evolver. Exp. Math. **1**, 141–165 (1992)
13. T.L. Breen, J. Tien, S.R.J. Oliver, T. Hadzic, G.M. Whitesides, Design and self-assembly of open, regular, 3d mesosctructures. Science **284**, 948–951 (1999)
14. A.H. Cannon, Y. Hua, C.L. Henderson, W.P. King, Self-assembly for three-dimensional integrations of functional electrical components. J. Micromech. Microeng. **15**, 2172–2178 (2005)
15. B. Chang, I. Routa, V. Sariola, Q. Zhou, Self-alignment of rfid dies on four-pad patterns with water droplet for sparse self-assembly. J. Micromech. Microeng. **21**, 095024 (2011)
16. B. Chang, V. Sariola, S. Aura, R.H.A. Ras, M. Klonner, H. Lipsanen, Q. Zhou, Capillary-driven self-assembly of microchips on oleophilic/oleophobic patterned surface using adhesive droplet in ambient air. Appl. Phys. Lett. **99**, 034104 (2011)
17. J.-H. Cho, A. Azam, D.G. Gracias, Three dimensional nanofabrication using surface forces. Langmuir **26**, 16534–16539 (2010)
18. J.-H. Cho, S. Hu, D.G. Gracias, Self-assembly of orthogonal three-axis sensors. Appl. Phys. Lett. **93**, 043505 (2008)
19. J.-H. Cho, M.D. Keung, N. Verellen, L. Lagae, V.V. Moshchalkov, P. van Dorpe, D.H. Gracias, Nanoscale origami for 3d optics. Small **7**, 1943–1948 (2011)
20. I.S. Choi, M. Weck, B. Xu, N.L. Jeon, G.M. Whitesides, Mesoscopic, templated self-assembly at the fluid-fluid interface. Langmuir **16**, 2997–2999 (2000)
21. J. Chung, W. Zheng, T.J. Hatch, H.O. Jacobs, Programmable reconfigurable self-assembly: parallel heterogeneous integration of chip-scale components on planar and nonplanar surfaces. J. Microelectromech. Syst. **15**, 457–464 (2006)
22. S.E. Chung, W. Park, S. Shin, S.A. Lee, S. Kwon, Guided and fluidic self-assembly of microstructures using railed microfluidic channels. Nat. Mater. **7**, 581–587 (2008)
23. T.D. Clark, J. Tien, D.C. Duffy, K.E. Paul, G.M. Whitesides, Self-assembly of 10μm-sized objects into ordered three-dimensional arrays. J. Am. Chem. Soc. **123**, 7677–7682 (2001)
24. J. Fang, K.F. Böhringer, Parallel micro components-to-substrate assembly with controlled poses and high surface coverage. J. Micromech. Microeng. **16**, 721–730 (2006)

25. J. Fang, K.F. Böhringer, Wafer-level packaging based on uniquely orienting self-assembly (the duo-spass processes). J. MIcroelectromech. Syst. **15**, 531–540 (2006)
26. J. Fang, K. Wang, K.F. Böhringer, Self-assembly of pzt actuators for micropumps with high process repeatability. J. Microelectromech. Syst. **15**, 871–878 (2006)
27. G. Fantoni, M. Porta, in *A Critical Review of Releasing Strategies in Microparts Handling*, ed. by S.M. Ratchev, S. Koelemeijer. International Federation for Information Processing (Springer, Berlin, 2008), pp. 223–234
28. M. Fialkowski, K.J.M. Bishop, R. Klajn, S.K. Smoukov, C.J. Campbell, B.A. Grzybowski, Principles and implementations of dissipative (dynamic) self-assembly. J. Phys. Chem. B **110**, 2482–2496 (2006)
29. T. Fukushima, E. Iwata, T. Konno, J.-C. Bea, K.-W. Lee, T. Tanaka, M. Koyanagi, Surface tension-driven chip self-assembly with load-free hydrogen fluoride-assisted direct bonding at room temperature for three-dimensional integrated circuits. Appl. Phys. Lett. **96**, 154105 (2010)
30. T. Fukushima, H. Kikuchi, Y. Yamada, T. Konno, J. Liang, K. Sasaki, K. Inamura, T. Tanaka, M. Koyanagi, New three-dimensional integration technology based on reconfigured wafer-on-wafer bonding technique, in *International Electron Devices Meeting*, Washington, D.C. (USA), 2007, pp. 985–958
31. T. Fukushima, T. Konno, E. Iwata, R. Kobayashi, T. Kojima, M. Murugesan, J.-C. Bea, K.-W. Lee, T. Tanaka, M. Koyanagi, Self-assembly of chip-size components with cavity structures: high-precision alignment and direct bonding without thermal compression for hetero integration. Micromachines **2**, 49–68 (2011)
32. T. Fukushima, Y. Ohara, M. Murugesan, J.-C. Bea, K.-W. Lee, T. Tanaka, M. Koyanagi, Self-assembly technologies with high-precision chip alignment and fine-pitch microbump bonding for advanced die-to-wafer 3d integration, in *Electronic Components and Technology Conference*, Lake Buena Vista (FL, USA), 2011, pp. 2050–2055
33. T. Fukushima, Y. Yamada, H. Kikuchi, M. Koyanagi, New three-dimensional integration technology using self-assembly technique, in *International Electron Devices Meeting*, Washington, D.C. (USA), 2005, pp. 348–351
34. L.S. Goodman, Geometrical optimization of controlled collapse interconnections. IBM J. Res. Dev. **13**, 251–265 (1969)
35. D.H. Gracias, M. Boncheva, O. Omoregie, G.M. Whitesides, Biomimetic self-assembly of helical electrical circuits using orthogonal capillary interactions. Appl. Phys. Lett. **80**, 2802–2804 (2002)
36. D.H. Gracias, J. Tien, T.L. Breen, C. Hsu, G.M. Whitesides, Forming electrical networks in three-dimensions by self-assembly. Science **289**, 1170–1172 (2000)
37. B.A. Grzybowski, N. Bowden, F. Arias, H. Yang, G.M. Whitesides, Modeling of menisci and capillary forces from the millimeter to the micrometer size range. J. Phys. Chem. **105**, 404–412 (2001)
38. J. Hoo, R. Baskaran, K.F. Böhringer, Programmable batch assembly of microparts with 100% yield, in *15th International Conference on Solid-State Sensors, Actuators and Microsystems*, Denver (CO, USA), 2009, pp. 829–832
39. K. Hosokawa, I. Shimoyama, H. Miura, Dynamics of self-assembling systems analogy with chemical kinetics. Artif. Life **1**, 413–427 (1995)
40. K. Hosokawa, I. Shimoyama, H. Miura, Two-dimensional micro-self-assembly using the surface tension of water. Sens. Actuators A **57**, 117–125 (1996)
41. H.O. Jacobs, A.R. Tao, A. Schwartz, D.H. Gracias, G.M. Whitesides, Fabrication of a cylindrical display by patterned assembly. Science **296**, 323–5 (2002)
42. S.S. Kim, E. Saeedi, J.R. Etzkorn, B.A. Parviz, Large scale self-assembly of crystalline semiconductor microcomponents onto plastic substrates via microfluidic traps, in *IEEE 4th International Conference on Automation Science and Engineering*, Washington, D.C. (USA), 2008, pp. 967–970
43. S.S. Kim, E. Saeedi, D.R. Meldrum, B.A. Parviz, Self-assembled heterogeneous integrated fluorescence detection system, in *IEEE 2nd International Conference on Nano/Micro Engineered and Molecular Systems*, Bangkok (TH), 2007, pp. 927–931

44. E. Klavins, in *Toward the Control of Self-Assembling Systems*, ed. by A. Bicchi, H. Christensen, D. Prattichizzo. Control Problems in Robotics (Springer, Berlin, 2003), pp. 153–168
45. R. Knuesel, S. Bose, W. Zheng, H.O. Jacobs, Angular orientation-specific directed self-assembly and integration of ultra small dies, in *NSTI-Nanotechnology Conference*, Santa Clara (CA, USA), 2007, pp. 112–115
46. R.J. Knuesel, H.O. Jacobs, Self-assembly of microscopic chiplets at a liquid-liquid-solid interface forming a flexible segmented monocrystalline solar cell. Proc. Natl. Acad. Sci. **107**, 993–998 (2010)
47. R.J. Knuesel, H.O. Jacobs, Self-tiling monocrystalline silicon; a process to produce electrically connected domains of si and microconcentrator solar cell modules on plastic supports. Adv. Mater. **23**, 2727–2733 (2011)
48. P.A. Kralchevsky, K. Nagayama, Capillary interactions between particles bound to interfaces, liquid films and biomembranes. Adv. Colloid Interface Sci. **85**, 145–192 (2000)
49. M. Krishnan, M.T. Tolley, H. Lipson, D. Erickson, Hydrodynamically tunable affinities for fluidic assembly. Langmuir **25**, 3769–3774 (2009)
50. T.G. Leong, A.M. Zarafshar, D.G. Gracias, Three-dimensional fabrication at small size scales. Small **6**, 792–806 (2010)
51. S.-H. Liang, K. Wang, K.F. Böhringer, Self-assembly of mems components in air assisted by diaphragm agitation, in *IEEE 18th International Conference on Micro Electro Mechanical Systems*, Miami (FL, USA), 2005, pp. 592–595
52. S.-H. Liang, X. Xiong, K.F. Böhringer, Toward optimal designs for self-alignment in surface tension driven micro-assembly, in *IEEE 17th International Conference on Micro Electro Mechanical Systems*, Maastricht (NL), 2004, pp. 9–12
53. J. Lienemann, A. Greiner, J.G. Korvink, X. Xiong, Y. Hanein, K.F. Böhringer, Modelling, simulation and experiment of a promising new packaging technology parallel fluidic self-assembly of microdevices. Sens. Update **13**, 3–43 (2003)
54. M. Liu, W.M. Lau, J. Yang, On-demand multi-batch self-assembly of hybrid mems by patterning solders of different melting points. J. Micromech. Microeng. **17**, 2163–2168 (2007)
55. J.C. Love, L.A. Estroff, J.K. Kriebel, R.G. Nuzzo, G.M. Whitesides, Self-assembled monolayers of thiolates on metals as a form of nanotechnology. Chem. Rev. **105**, 1103–1169 (2005)
56. C. Mao, V.R. Thallati, D.B. Wolfe, S. Whitesides, G.M. Whitesides, Dissections: Self-assembled aggregates that spontaneously reconfigure their structures when their environment changes. J. Am. Chem. Soc. **124**, 14508–14509 (2002)
57. B.R. Martin, D.C. Furnage, T.N. Jackson, T.E. Mallouk, T.S. Mayer, Self-alignment of patterned wafers using capillary forces at a water-air interface. Adv. Funct. Mater. **11**, 381–386 (2001)
58. M. Mastrangeli, Enabling capillary self-assembly for microsystem integration. PhD thesis, Arenberg Doctoral School, Katholieke Universiteit Leuven, 2010
59. M. Mastrangeli, S. Abbasi, C. Varel, C. van Hoof, J.-P. Celis, K.F. Böhringer, Self-assembly from milli- to nanoscale: methods and applications. J. Micromech. Microeng. **19**, 083001 (2009)
60. M. Mastrangeli, G. Mermoud, A. Martinoli, Modeling self-assembly across scales: the unifying perspective of smart minimal particles. Micromachines **2**, 82–115 (2011)
61. M. Mastrangeli, W. Ruythooren, J.-P. Celis, C. Van Hoof, Challenges for capillary self-assembly of microsystems. IEEE Trans. Compon. Packag. Technol. **1**, 133–149 (2011)
62. M. Mastrangeli, W. Ruythooren, C. Van Hoof, J.-P. Celis, Conformal dip-coating of patterned surfaces for capillary die-to-substrate self-assembly. J. Micromech. Microeng. **19**, 045015 (2009)
63. M. Mastrangeli, C. van Hoof, R. Baskaran, J.-P. Celis, K.F. Böhringer, Agent-based modeling of mems fluidic self-assembly, in *IEEE 23rd International Conference on Micro Electro Mechanical Systems*, Wanchai, Hong Kong (SAR China), 2010, pp. 476–479
64. C.J. Morris, B.A. Parviz, Micro-scale metal contacts for capillary force-driven self-assembly. J. Micromech. Microeng. **18**, 015022 (10pp) (2008)

65. C.J. Morris, S.A. Stauth, B.A. Parviz, Self-assembly for microscale and nanoscale packaging: steps toward self-packaging. IEEE Trans. Adv. Packag. **28**, 600–611 (2005)
66. Y.Y. Ong, Y.L. Lim, L.L. Yan, E.B. Liao, V. Kripesh, Dry self-assembly and gang bonding of micro-components from silicon carrier to substrate wafer, in *IEEE/CPMT International Electronics Manufacturing Technology Conference*, Petaling Jaya (MY), 2006, pp. 486–491
67. Y.Y. Ong, Y.L. Lim, L.L. Yan, S. Vempati, E.B. Liao, V. Kripesh, S.U. Yoon, Self-assembly of components using shape-matching, in *Electronics Packaging Technology Conference*, Singapore, 2007, pp. 826–829
68. S. Pandey, M. Ewing, A. Kunas, N. Nguyen, D.G. Gracias, G. Menon, Algorithmic design of self-folding polyhedra. Proc. Natl. Acad. Sci. **108**, 19885–19890 (2011)
69. M.C. Petty, *Langmuir-Blodgett Films: An Introduction* (Cambridge University Press, Cambridge, 1996)
70. J.S. Randhawa, L.N. Kanu, G. Singh, D.G. Gracias, Importance of surface patterns for defect mitigation in three-dimensional self-assembly. Langmuir **26**, 12534–12539 (2010)
71. P.W.K. Rothemund, Using lateral capillary forces to compute by self-assembly. Proc. Natl. Acad. Sci. **97**, 984–989 (2000)
72. P.W.K. Rothemund, Folding DNA to create nanoscale shapes and patterns. Nature **440**, 297–302 (2006)
73. P.W.K. Rothemund, N. Papadakism, E. Winfree, Algorithmic self-assembly of DNA Sierpinski triangles. PLOS Biol. **2**, e424 (2004)
74. E. Saeedi, S. Abbasi, K.F. Böhringer, B.A. Parviz, Molten-alloy driven self-assembly for nano and micro scale system integration. Fluid Dyn. Mater. Process. **2**, 221–246 (2007)
75. E. Saeedi, J.R. Etzkorn, B.A. Parviz, Sequential self-assembly of micron-scale components with light. J. Mater. Res. **26**, 268–276 (2011)
76. E. Saeedi, S. Kim, B.A. Parviz, Self-assembled crystalline semiconductor electronics on glass and plastic. J. Micromech. Microeng. **18**, 075019 (7pp), 2008
77. V. Sariola, M. Jääskeläinen, Q. Zhou, Hybrid microassembly combining robotics and water droplet self-alignment. IEEE Trans. Rob. **26**, 965–977 (2010)
78. V. Sariola, Droplet Self-Alignment: High-Precision Robotic Microassembly and Self-Assembly. PhD thesis, Aalto University, 2012
79. V. Sariola, Q. Zhou, H.N. Koivo, Three dimensional hybrid microassembly combining robotic microhandling and self-assembly, in *IEEE International Conference on Robotics and Automation*, 2009
80. K.L. Scott, T. Hirano, H. Yang, H. Singh, R.T. Howe, A.N. Niknejadk, High-performance inductors using capillary based fluidic self-assembly. J. Microelectromech. Syst. **13**, 300–309 (2004)
81. R. Sharma, Thermally controlled fluidic self-assembly. Langmuir **23**, 6843–6849 (2007)
82. J.S. Smith, High density, low parasitic direct integration by fluidic self assembly (fsa), in *International Electronics Devices Meeting*, San Francisco (CA, USA), 2000, pp. 201–204
83. U. Srinivasan, M.A. Helmbrecht, C. Rembe, R.S. Muller, R.T. Howe, Fluidic self-assembly of micromirrors onto microactuators using capillary forces. IEEE J. Sel. Top. Quantum Electron. **8**, 4–11 (2002)
84. U. Srinivasan, D. Liepmann, R.T. Howe, Microstructure to substrate self-assembly using capillary forces. J. Microelectromech. Syst. **10**, 17–24 (2001)
85. S.A. Stauth, B.A. Parviz, Modeling of fluidic self-assembly for integration of silicon components on plastic, in *IEEE 19th International Conference on Micro Electro Mechanical Systems*, Instanbul (TR), 2006, pp. 194–197
86. S.A. Stauth, B.A. Parviz, Self-assembled single-crystal silicon circuits on plastic. Proc. Natl. Acad. Sci. **103**, 13922–13927 (2006)
87. R.R.A. Syms, E.M. Yeatman, V.M. Bright, G.M. Whitesides, Surface-tension powered self-assembly of microstructures the state of the art. J. Microelectromech. Syst. **12**, 387–417 (2003)
88. A. Terfort, N. Bowden, G.M. Whitesides, Three-dimensional self-assembly of millimetre-scale components. Nature **386**, 162–164 (1997)

89. A. Terfort, G.M. Whitesides, Self-assembly of an operating electrical circuit based on shape complementarity and the hydrophobic effect. Adv. Mater. **10**, 470–473 (1998)
90. C.G. Tsai, C.M. Hsieh, J.A. Yeh, Self-alignment of microchips using surface tension and solid edge. Sens. Actuators A **139**, 343–349 (2007)
91. M.R. Tupek, K.T. Turner, Submicron aligned wafer bonding via capillary forces. J. Vac. Sci. Technol. **25**, 1976–1981 (2007)
92. D.B. Wolfe, A. Snead, C. Mao, N.B. Bowden, G.M. Whitesides, Mesoscale self-assembly: capillary interactions when positive and negative menisci have similar amplitudes. Langmuir **19**, 2206–2214 (2003)
93. X. Xiong, Y. Hanein, J. Fang, D.T Schwartz, K.F. Böhringer, Multi-batch self-assembly for microsystem integration, in *3rd International Workshop on Microfactories (IWMF '02)*, Minneapolis (MN, USA), 2002, pp. 25–28
94. X. Xiong, Y. Hanein, J. Fang, Y. Wang, W. Wang, D.T Schwartz, K.F. Böhringer. Controlled multibatch self-assembly of microdevices. J. Microelectromech. Syst. **12**, 117–127 (2003)
95. X. Xiong, Y. Hanein, W. Wang, D.T. Schwartz, K.F.Böhringer, Multi-batch micro-selfassembly via controlled capillary forces, in *IEEE/RSJ International Conference on Intelligent Robots and Systems*, Maui (HI, USA), 2001, pp. 1335–1342
96. X. Xiong, S.-H. Liang, K.F. Böhringer, Geometric binding site design for surface-tension driven self-assembly, in *IEEE International Conference on Robotics and Automation*, Barcelona (ES), 2004, pp. 1141–1148
97. H.-J.J. Yeh, J.S. Smith, Fluidic assembly for the integration of gaas light-emitting diodes on Si substrates. IEEE Photonics Technol. Lett. **46**, 706–709 (1994)
98. W. Zheng, J. Chung, H.O. Jacobs, Fluidic heterogeneous microsystem assembly and packaging. J. Microelectromech. Syst. **15**, 864–870 (2006)
99. W. Zheng, H.O. Jacobs, Fabrication of multicomponent microsystems by directed three-dimensional self-assembly. Adv. Funct. Mater. **15**, 732–738 (2005)
100. W. Zheng, H.O. Jacobs, Self-assembly process to integrate and interconnect semiconductor dies on surfaces with single-angular orientation and contact-pad registration. Adv. Mater. **18**, 1387–1392 (2006)

Chapter 13
Surface Tension Driven Actuation

Ruba T. Borno and Michel M. Maharbiz

Abstract The ability of plants to extract work from evaporation is the inspiration behind all of the work in this chapter. Transpiration mechanisms are explored as a method for driving mechanical and electrical energy conversion. We present two types of energy scavenging devices that were inspired by nature, and specifically, by different transpiration mechanisms used in plants. The first type of device consists of a new class of mechanical actuators that operate via the evaporation of water. The second type of device uses evaporation-driven flow to generate electrical power using a charge-pumping mechanism.

13.1 Introduction

The term *biomimetics* was coined in 1969 by Otto Schmitt and is used to describe the significant body of work—ranging from engineering to architecture and art—which studies natural systems and components to inform the design of materials, machines and structures [1–3]. The research presented here is inspired by, and mimics components of, transpiration in plants to scavenge mechanical energy for actuation and to generate electrical energy.

There are only a few examples [4–7] of researchers making microscale devices which scavenge energy from evaporation. Even so, there are many environments favorable for transpiration; environments need only a humidity gradient to cause evaporation to occur. Evaporation is driven by changes in multiple parameters including temperature, relative humidity, pressure, and air flow. A change in any of those

R. T. Borno · M. M. Maharbiz (✉)
Electrical Engineering and Computer Science Department,
University of Michigan, Ann Arbor, USA
e-mail: rborno@gmail.com

M. M. Maharbiz
e-mail: maharbiz@eecs.berkeley.edu

P. Lambert (ed.), *Surface Tension in Microsystems*, Microtechnology and MEMS,
DOI: 10.1007/978-3-642-37552-1_13, © Springer-Verlag Berlin Heidelberg 2013

ambient conditions could be exploited by an evaporation harvesting technology. The continuous harvesting of energy would require either a constantly replenished volume of water or, equivalently, cyclical condensation and evaporation phases during operation.

Plants have evolved methods for extracting work from evaporation and the surface tension of water. Trees are capable of moving large volumes of water from the ground through their vascular networks, without the use of active pumps. *Evapotranspiration* is a term used to denote processes occurring at the combined plant and soil evaporating surfaces, where evaporation occurs at the soil-air interface, and transpiration occurs at the plant-air interface, namely leaves [8]. In describing these processes, a *cohesion-tension theory* has been formulated to describe the passive pumping of water from root to leaves, driven primarily via evaporation at microscale stomatal pores [8–11]. Cohesion-tension theory posits that plants utilize a negative pressure difference across water contained under tension within the plant vasculature—extending from root to leaf. Cohesion-tension theory will be discussed further later in this chapter. The ability of plants to extract work from evaporation is the inspiration behind all of the work in this chapter. Transpiration mechanisms are explored as a method for driving mechanical and electrical energy conversion. We present two types of energy scavenging devices that were inspired by nature, and specifically, by different transpiration mechanisms used in plants. The first type of device consists of a new class of mechanical actuators that operate via the evaporation of water (Sect. 13.2). The second type of device uses evaporation-driven flow to generate electrical power using a charge-pumping mechanism (Sect. 13.3).

13.2 Transpiration Actuation

This section presents the design, fabrication and characterization of large displacement, distributed-force polymer biomimetic actuators driven only by the surface tension of water. The devices were inspired by the hygroscopic spore dispersal mechanism in fern sporangia. The microdevices were fabricated through a single mask process using a commercial photo-patternable silicone polymer to mimic the mechanical characteristics of plant cellulose. An analytical model for predicting the microactuator behavior was developed using the *principle of virtual work*, and a variety of designs were simulated and compared to empirical data. Fabricated devices experienced tip deflections of more than 3.5 mm and angular rotations of more than 330° due to the surface tension of water. The devices generated forces per unit length of 5.75–67.75 $\mathrm{mNm^{-1}}$ and were dependent on device geometry. The actuation mechanism presented in this work may provide a robust method for embedding geometry-programmable and environment-scavenged force generation into common materials.

13.2.1 Fern Sporangium

As discussed above, the ability to extract work from both environmental conditions and the surface tension of water is a hallmark of plant evolution [8–11]. In addition to pumping water through their vasculature using evaporative processes, ferns also make use of water evaporation within specialized microstructures to obtain fast motion and high forces for spore dispersal [11]. As part of their reproductive processes, ferns grow specialized vessels called sporangia which house and disperse spores during certain times of the year. As they dry, these millimeter scale sporangia open violently to release microscale spherical spores into the air. This mechanism is illustrated in Fig. 13.1. Each sporangium is surrounded by a single layer of dead, water-filled cells called an annulus. Each cell is comprised of two rib-like structures filled with water. As the water inside the annulus cells dries, surface tension between the water and the cell wall gives rise to high forces causing a deflection along the outer edge of the annulus. The combined deflection of each wall straightens the entire annulus structure and tears open the spore sac.

The discovery of the fern sporangium spore dispersal mechanism is credited to Ursprung and Renner who, in 1915, independently used the sporangium to try and measure the tensile strength of water [11, 13, 45]. They placed sporangia in

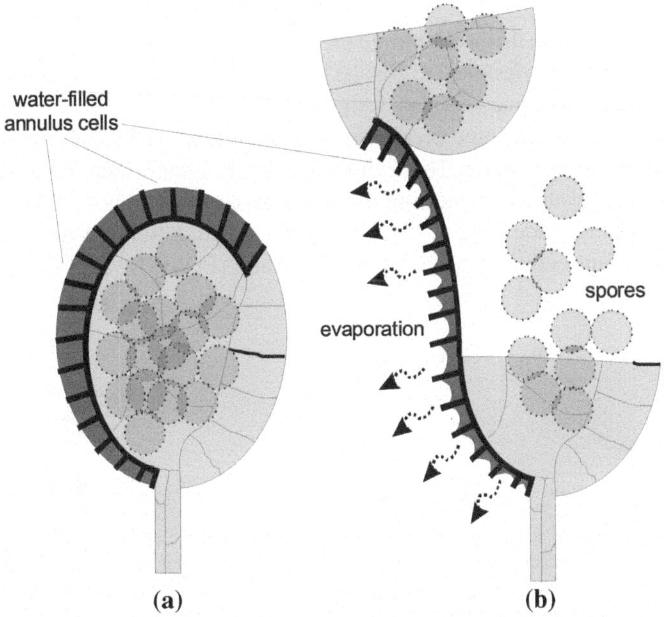

Fig. 13.1 Illustration of the actuation mechanism of fern sporangium: **a** before and **b** after evaporation induced opening of annulus to release spores (adapted from [11]). Reprinted with permission from [12]. Copyright 2006 Institute of Physics

controlled test chambers in which the relative humidity was regulated by a concentrated solution. Using a microscope, they noted that the sporangia structures opened when enough water evaporated within each annulus cell to cause mechanical distortion of the entire annulus. However, they also noted that the sporangium snapped shut, to effectively throw the spores out into the air, at a solution osmotic potential of −35 MPa. Ziegenspeck, in 1928, and Haider, in 1954, both showed that the sporangium snapped closed once enough water evaporated from each annulus cell (approximately 10–20 μm wide) to 'break' the water and form a vapor bubble inside the cell [11, 13, 14]. Interestingly, Ziegenspeck and Haider found that once a single annulus cell underwent cavitation, or generated a water vapor bubble, it promoted almost instantaneous cavitation of the rest of the annulus cells.

The issue of sporangium actuation was also investigated in the quest to understand the general theory behind the passive pumping of water in trees and plants [15]. However, since the early 1950s, relatively few studies of fern sporangium mechanics have been undertaken [13]. As a whole, it should also be noted that few, if any, micro- and nanomechanical platforms exist for testing water-pumping and actuation hypotheses for these natural structures.

13.2.2 Design and Modeling

The devices designed in this work mimic the geometries of fern sporangia, as show in Fig. 13.2. Devices were designed to consist of a curved spine that straightened during device operation. The spine was also designed around the physics of a slender beam. Ribs extending from the spine create multiple individual cells with two sidewalls which can be filled with water. These ribs act as levers upon which forces due to the surface tension of water act to deform the spine, as shown in Figs. 13.2b–c. Both the spine and ribs were made from a silicone polymer. In order to mimic the mechanical properties of plant cellulose, polydimethylsiloxane (PDMS) was used because of

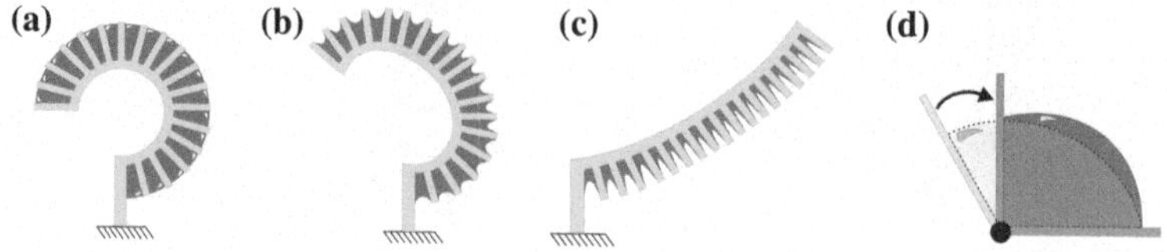

Fig. 13.2 Illustration of the actuation mechanism of microfabricated devices inspired by fern sporangium: **a** Water is placed between ribs to wet device. **b** Immediately after wetting, the surface tension of water pulls on the ribs at the meniscus edge. **c** As the water seeks to minimize its surface energy, the force due to surface tension pulls on the ribs as though they were levers and causes the spine to deform. **d** A similar example of energy minimization has been used to rotate hinges using molten solder as presented in [17] as Fig. 13.3d. The device in **a**–**c** multiplies this energy minimization effect by incorporating multiple ribs to divide the volume of liquid. Reprinted with permission from [12]. Copyright 2006 Institute of Physics

its comparable Young's modulus. Cellulose, the material that comprises plant cell walls, has a Young's modulus of 120–500 MPa [16]. The Young's modulus of the photo-patternable silicone employed in this work was 160 MPa, as reported by the manufacturer.

The behavior of these devices can be compared to previous work on melted solder-driven self-assembling plates. There has been much research on surface tension driven actuation utilizing solder at high temperatures [17–20]. As illustrated in Fig. 13.2d, when the solder melts, the liquid minimizes its energy by reducing the interfacial surface area. The surface tension forces at the solder-plate interface cause the hinged plate to rotate as the liquid solder's surface energy is minimized. The microactuators in the work presented in this paper employ a distributed energy minimization mechanism by including multiple liquid volumes that seek to reduce their surface energies. As will be explained, this distribution of actuation is useful for achieving additional control such as the time response and for embedding force-generation profiles into the actuation with device geometry.

This section shows the use of the principle of virtual work as the basis of an analytical model that predicts the mechanical energy scavenging device performance based on geometric parameters and material properties. First, the principle of virtual work is described, followed by an analysis of the various energy components in the presented system. Then, the strain energy due to deformation of the spine is derived, as well as the surface energy of the liquid between the device's ribs. The derived energy values are then used to find the minimum total potential energy.

13.2.2.1 Principle of Virtual Work

The task of predicting the exact nature of device deformations using force balance methods can be quite challenging for even the simple curved geometries presented in this work. Although it is becoming increasingly possible to use finite element models and commercial solvers to approach problems including surface-tension effects, large deformation analysis, and non-linear material properties, these simulations are typically computationally expensive if there are no obvious symmetries to exploit at the global level and the finite element model simulations do not always converge. In this light, energy methods based on the principle of virtual work are particularly useful and provide an efficient means of device modeling. Such methods have been used extensively to model MEMS devices [21]. The model presented here provides an approximate solution to the maximum achievable device deformation based on geometric parameters and material properties. In the present context, the principle of virtual work states that when a body in equilibrium experiences external forces that cause deformation, the energy added to the deformed body is solely due to those applied forces [21, 22]. In this analysis, all forces other than surface tension are neglected. The total potential energy of the microactuators presented here is the sum of the deformation energy of the device spine and the surface energy of the water columns between the ribs. The developed analytical model calculates the total energy of various resting states of the device when the ribs are completely filled with water

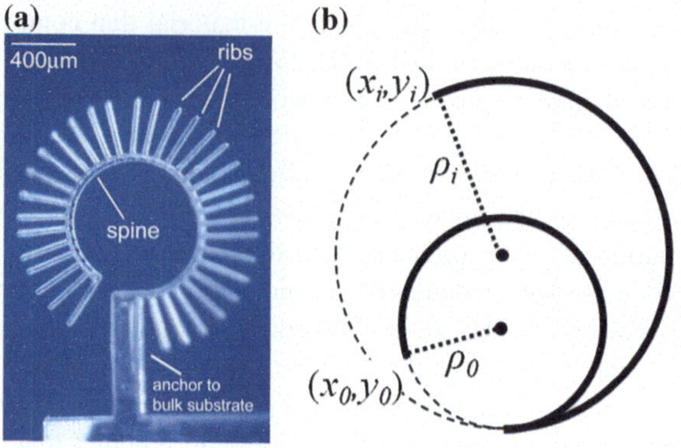

Fig. 13.3 Transpiration based microactuator: **a** A micrograph of a device in its initial rest state. **b** An illustration of how the strain energy of the spine is calculated, using a constant arc length with an increasing radius of curvature ρ. Tip deflections, δ, are calculated as $\delta = \sqrt{(x_i - x_0)^2 - (y_i - y_0)^2}$. Reprinted with permission from [12]. Copyright 2006 Institute of Physics

and seeks to find the state of minimum potential energy. In the ideal case, there are no fabrication defects, the ribs are distributed evenly along the spine, each of the cells comprised of two ribs is completely filled with water, and the evaporation rate for each cell is the same at any given time. The model seeks to find the equilibrium state of the system. Transient response is not being modeled. The model is further based on the following statements:

- Devices are designed to have an initial stress free shape that is curved like a circle. A micrograph of an actual device is shown in Fig. 13.3a. The energy of the stress free curve is zero;
- The model assumes no shear forces and no out-of-plane distortion;
- The spine was designed to have uniform radius of curvature in its stress-free shape. The assumption of no shear implies that any deformed shape of the spine will also have uniform radius of curvature. An illustration of how the strain energy of the spine is calculated is shown in Fig. 13.3b. A constant arc length with an increasing radius of curvature, ρ, is used to calculated tip deflections, δ, using (13.1):

$$\delta = \sqrt{(x_i - x_0)^2 - (y_i - y_0)^2} \tag{13.1}$$

- Ribs are assumed to be rigid so that they do not store any deformation energy.

As the device deforms due to surface tension the strain energy in the spine increases. Equilibrium is reached when the total potential energy is at a minimum. The total system energy for a device experiencing various spine deformations, or curves, like those in Fig. 13.3b, is calculated and given in Fig. 13.4a. Each curve number corresponds to a different uniformly curved arc, ρ, for which both the strain energy of the spine and the surface energy of the water within the ribs was calculated and summed to find the total potential energy of the actuator. The curve with

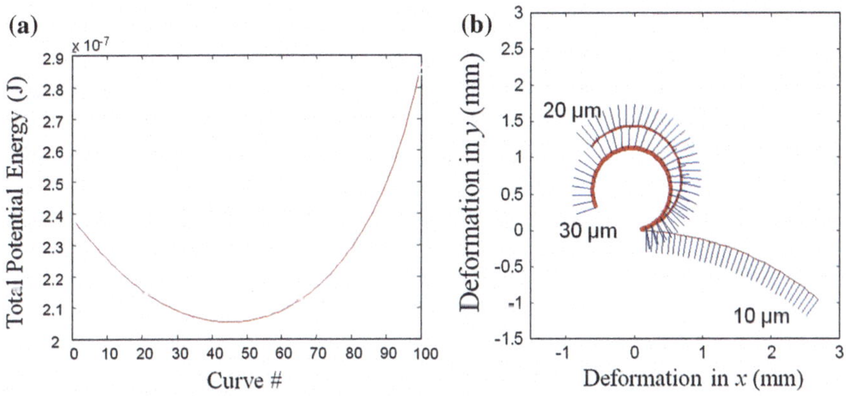

Fig. 13.4 System energy and deformation simulation results for microactuator: **a** Simulated total system energy for a device experiencing various spine deformations represented simply as *curves*. Each curve number corresponds to a different uniformly curved arc for which both the strain energy of the spine and the surface energy of the water within the ribs was calculated and summed to find the total potential energy in the actuator. The modeled spine width was 10 μm with a Young's modulus of 60 MPa. The curve with the lowest potential energy indicates the curve for the equilibrium state. **b** Simulated deformation curves at the point of equilibrium between the strain energy of the spine and the surface energy of the water for devices with varying spine thicknesses (10, 20 and 30 μm) and a rib length of 400 μm. As the spine thickness is decreased, equilibrium is reached at higher deflections. Reprinted with permission from [12]. Copyright 2006 Institute of Physics

the lowest potential energy indicates the curve for the equilibrium state. Simulated deformation curves are shown in Fig. 13.4b for the point of equilibrium between the strain energy of the spine and the surface energy of the water for devices with varying spine thickness or 10, 20, and 30 μm and a rib length of 400 μm. As the spine thickness is decreased, equilibrium is reached at higher deflections.

13.2.2.2 Strain Energy of the Spine

The strain energy is defined as the integral of a scalar effort, modeled by stress, over a scalar displacement, modeled by strain [21]. The spine of the microactuator presented here can be modeled as a continuous curved beam. The strain energy density of the spine is modeled by $\hat{W}$ in (13.2) where E is the Young's modulus and ε is the strain.

$$\hat{W}(x, y, z) = \int_{0}^{\varepsilon(x,y,z)} E\varepsilon d\varepsilon = \frac{1}{2}E\varepsilon^2(x, y, z) \tag{13.2}$$

For large deformations in polymers, nonlinear expressions from neo-Hookean and Mooney-Rivlin models can be used and include additional non-linear terms to the axial stress-strain relationship [23, 24]. However, the maximum strains of the geometries presented in this work do not exceed ≈ 0.1 for spine thicknesses $<100\,\mu\text{m}$. Due to the small strains of the device geometries, a linear axial stress-strain relationship based on Hooke's Law, can be used [23, 25].

The strain energy is based on the physics of a straight cantilever. This is the opposite of the actual actuator geometry which is an initially curved spine. Although there are cases in which an initially curved beam cannot be modeled using the altered equations for a straight beam, it is acceptable in this case since the calculated neutral axis radius of the straight spine is within 0.05 % of the geometrically calculated curved spine value using spine thicknesses, h $< 50\,\mu$m. Therefore, the energy present in an initially curved beam that is straightened during actuation is essentially equivalent to the strain energy in an initially straight beam that is then curved during actuation.

In order to calculate the energy stored in the beam, the expression must be integrated over the volume of the segment of the spine that is being analyzed. A uniform spine can be modeled by a number of small segments, each with an angle of deformation from the previous segment. The energy of segment can be found as the product of a constant and the square of the segment's angle. The total energy of the spine is the sum of all the segment energies. A sum of squares is at its minimum when all of the elements are equal. Thus, the energy of the spine would be at its minimum when all of the angles are equal, or equivalently, when the shape is in the form of a circular arc. The strain energy, W_{strain}, of a spine at deformed curve radius, ρ, radius of rest shape curve, ρ_0, Young's modulus of the spine material, E, length of spine, L, and rectangular cross-section of the spine defined by w and h, is given in (13.3) [26].

$$W_{\text{strain}} = \frac{Eh^3wL}{24}\left(\frac{1}{\rho} - \frac{1}{\rho_0}\right)^2 \tag{13.3}$$

As shown in Fig. 13.5a, w is defined as the depth of the device into the page and l is the in-plane thickness of the spine. The stress-free curvature, ρ_0, is given by (13.4), where θ_{initial} is the angle of the arc swept by the curved spine. The bending stiffness, EI, or the product of Young's modulus, E, and the moment of inertia of the spine, I, is given by (13.5), where h and w are as defined above.

$$\frac{1}{\rho_0} = \frac{\theta_{\text{initial}}}{L} \tag{13.4}$$

$$EI = \frac{Eh^3w}{12} \tag{13.5}$$

13.2.2.3 Surface Energy of Water

As previously mentioned, the equilibrium state of the device deformation is when the total energy in the system, comprised of strain energy and surface energy, is at a minimum. The surface of the volume of water filling in the region between adjacent ribs is actually a complex three-dimensional shape. In the presented analysis, this surface has been approximated as set of trapezoids for the pair of side areas which touch both the adjacent ribs and the spine, as shown in Fig. 13.5a, Area A, and as a partial cylinder of appropriate radius for the top areas which touch just the adjacent ribs, as shown in Fig. 13.5a, Area B. The sum of these two regions provides an approximation of the surface area of the volume of water present between the ribs. The work done by the surface tension of water, W_{surface}, for a given spine deformation is expressed by (13.6) where γ is the surface tension of water, A is the area of the side meniscus on both sides of the device, B is the area of the top meniscus between the tips of a set of ribs, and n is the number of ribs.

$$W_{\text{surface}} = 2\left(\sum_{0}^{n} A\right) + \sum_{0}^{n} B \tag{13.6}$$

The area of the side meniscus, A, can be approximated by a trapezoid that depends on the length of the rib, l, as given by (13.7). The curvature of the surface due to the meniscus of the water is neglected for cases where the depth of the rib, w, is much greater than the distance between the tips of the ribs, H.

$$A = l(H - \tan\frac{\phi}{2}) \tag{13.7}$$

The length of the meniscus, L_{m}, within each cell from rib-tip to rib-tip can be solved using the water-to-rib contact angle, θ_{c}, and a geometric construction presented in (13.8–13.10), where r is the radius of curvature of the meniscus between two ribs, ϕ is the angle between two ribs, H is the distance between the tip of the ribs, and

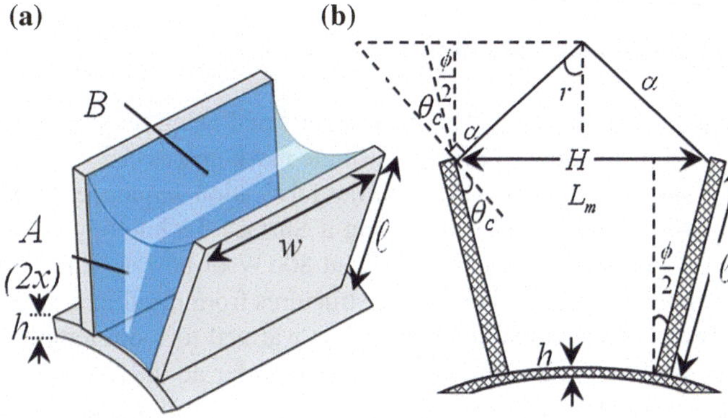

Fig. 13.5 Definition of geometric variables used in the calculation of strain energy and surface energy of the actuator: **a** The two components in the total surface energy of water: Area A indicates the side area of the meniscus which depends on rib length, l, ($\times 2$) to account for the front and back areas and Area B indicates the area of the meniscus above the ribs which depends on the device depth, w. The spine thickness is represented by h. **b** Variables used in the calculation of the length of the meniscus L_{m}: r is the radius of curvature of the meniscus between two ribs, H is the distance between the tip of the ribs, θ_{c} is the contact angle between water and rib, 2α is the angle sweeping the arc length of the meniscus, and ϕ is the angle between two ribs. Reprinted with permission from [12]. Copyright 2006 Institute of Physics

w is the depth of the ribs. The partial cylinder used to approximate the top area, B, takes into account the menisci generated by the liquid-solid interface and is given by (13.11).

$$\alpha = \frac{\pi}{2} - \frac{\phi}{2} - \theta_c \tag{13.8}$$

$$r = \frac{H}{2 \sin \alpha} \tag{13.9}$$

$$L_m = 2\alpha r \tag{13.10}$$

$$B = L_m w - 2(\alpha r^2 - r^2 \sin \alpha \cos \alpha) \tag{13.11}$$

The approximation for B subtracts the area that does not contain water due to the concave shape of the meniscus along the depth of the ribs, w. The calculation for B also subtracts the area due to the concave meniscus at the tip of the ribs coming from the side of the volume of water. In 13.11 the unadjusted partial cylinder area is given in the first term, $L_m w$, and the subtracted areas due to the menisci are given in the second term, $2(\alpha r^2 - r^2 \sin \alpha \cos \alpha)$. These approximations result in a value for the total surface area of the water contained within a set of ribs which takes into account the contact angle, θ_c, of water to the ribs.

13.2.3 Experimental Method

13.2.3.1 Fabrication and Assembly

The microactuators in this work were batch fabricated using the photo-patternable silicone WL-5150 from Dow Corning in a single mask process.

Figure 13.6 shows the process steps in the fabrication sequence. A passivation layer was deposited on a clean wafer using a Surface Technology Systems ICP-DRIE tool for 1 min with 85 sccm of C_4F_8 at 800 W of power. This hydrophobic layer facilitates the removal of the silicone structures from the silicon substrate post fabrication. The photo-patternable silicone was warmed to room temperature, and was then deposited and spun onto the passivated wafer at 300 rpm for a depth, or vertical thickness, of 80 μm on the substrate. The coated wafer was then soft baked on a hotplate at 110 °C for 2 min to remove carrier solvents. The patternable silicone layer was exposed to 1000 mJ cm^{-2} of UV-light using a GCA Autostep 200 i-line stepper to activate polymerization. The exposed silicone layer was then baked on a hotplate for 2 min at 150 °C to achieve full polymerization of the exposed regions. Stoddard solvent, which was also obtained from Dow Corning, was used to remove the unexposed silicone during a 1-minute puddle development. Fresh developer was deposited onto the wafer with a pipette for 10 s as it was spun at 200 rpm. Isopropyl alcohol was then squirted onto the wafer while it as spun at 500 rpm to remove

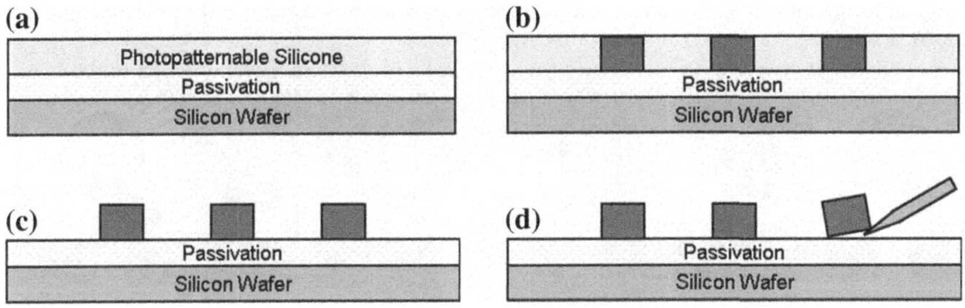

Fig. 13.6 Fabrication sequence of lithographically patterned polymer actuators: **a** A thin passivation layer is deposited on a bare wafer and photopatternable silicone. **b** Negative photolithography is used to activate polymerization of the silicone. **c** The silicone is developed. **d** The patterned devices are removed from the wafer. Reprinted with permission from [12]. Copyright 2006 Institute of Physics

solvents and any undeveloped silicone. Subsequently, the wafer was spun dry for 30 s at 500 rpm. The final photo-patterned silicone features were cured on a hotplate at 250 °C for 10 min. The microactuator devices were separated from their carrier wafers using a razor blade. Next, they were immersed in methanol and dried on a clean silicon wafer. Following this, the devices were lifted from the wafer and mounted at their base anchor for testing. Finally, a 50 W, 2-minute O_2 plasma treatment at 250 mTorr was used to make the structures hydrophilic prior to testing.

13.2.3.2 Testing

Following the O_2 plasma step, water was applied onto the hydrophilic surfaces using a needle and syringe. Evaporation took place at room temperature. Device actuation was observed and recorded using a Nikon Coolpix 5 Megapixel digital camera mounted on a compound microscope using 10 times magnification. In some cases, 10 μM fluorescein (Sigma-Aldrich Corporation) was dissolved in deionized water to help visualize the water and the meniscus.

13.2.4 Experimental Results

Device performance was quantified in two ways:

- Device tip deflection, which is measured as the distance that the entire device tip at the last rib moved from its initial position to the point of equilibrium;
- Angular rotation, which measured the angle swept by the device tip from the starting position to the point of equilibrium.

In order to test the derived analytical model and understand scaling phenomena, a family of devices with parametrically varied geometries was fabricated and assembled. The varied parameters of rib length, rib spacing, spine thickness and device

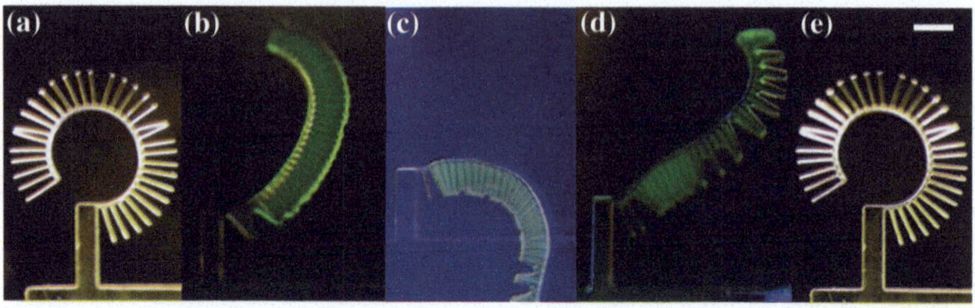

Fig. 13.7 Micrographs of an actuator device unfurling: **a** Completely dry device before filling. **b** Wetted device 0.5 s after filling. **c** Device at full deflection 2 s after filling. **d** Device drying at 30 s after filling. **e** The device returns to its original profile once water evaporates completely. 10 μM of fluorescein was added to the water to help visualize meniscus. Scale bar indicates 400 μm. Reprinted with permission from [12]. Copyright 2006 Institute of Physics

depth, enabled the investigation of the effect of linear scaling on the magnitude of total device tip deflection, angular rotation, as well as the geometry of deformation and geometry controlled time response to evaporation-driven actuation.

Typical as-drawn geometric parameters of the fabricated devices include: 300, 400, and 500 μm rib lengths, constant 75 μm rib spacing, 10, 20 and 30 μm spine thicknesses, and 80 μm device depth (the thickness of the spun photopatternable silicone).

Figure 13.7 shows a time-lapsed typical device movement during actuation. Once wetted, devices quickly moved to their equilibrium position, as shown in Fig. 13.7b–c. As water evaporated, the devices began to relieve the strain in the spine. Devices returned to their undeformed state once dry.

13.2.4.1 Device Performance as Function of Spine Thickness

Devices with the smallest feature size of 10 μm spine thicknesses experienced resultant tip deflections in excess of 4 mm while unfurling 300 °C, as shown in Fig. 13.8.

Devices exhibited forces per unit length ranging from $5.75\,\mathrm{mN\cdot m^{-1}}$ for 10 μm spine thicknesses to $67.67\,\mathrm{mN\cdot m^{-1}}$ for 30 μm spine thicknesses. Decreasing the spine thickness predictably increased the deflection of the device, but also increased the spread of the measured data as the device became more sensitive to test environment conditions. The data in Fig. 13.8a show an increase in the standard deviation as the spine thickness is decreased. This is partly due to test environment limitations. Since the test environment of an optical microscope for image capture cannot be guaranteed to provide evenly distributed light, devices with narrower spine geometries expressed the variation much more noticeably than wider devices which did not deflect as much. The variation among a given spine thickness is rather constant relative to total angular rotation and displacement.

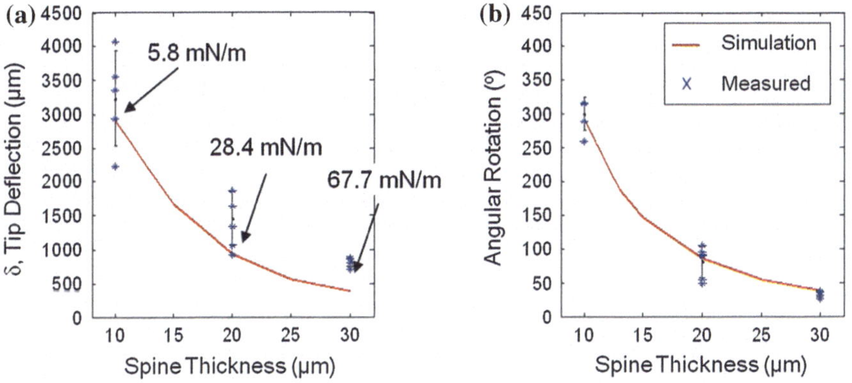

Fig. 13.8 Measurement and simulation results of device performance versus spine thickness: **a** Tip deflection, δ defined in (13.1), at the tip of device as a function of spine thickness (riblength $= 400\,\mu m$). **b** Angle of rotation of as a function of spine thickness (riblength $= 400\,\mu m$). The results of the simulation are given in red for a Young's Modulus value of 60 MPa. Reprinted with permission from [12]. Copyright 2006 Institute of Physics

13.2.4.2 Drying of Liquid Phase Meniscus as a Function of Device Geometry

Additional visual observations were made as a result of the varied geometric parameters. This information is not captured by the analytical model since the model seeks to find the point of equilibrium and not the transient response. The drying of the liquid phase was found to exhibit a varied response to rib length. For shorter rib lengths of 300 µm, the meniscus was observed to dry from the rib tips towards the spine, as can be seen in Fig. 13.9a. At a certain time point, many devices underwent a breaking of symmetry due to varied light intensity in the test environment, as can be seen in Fig. 13.9b–c. However, it is interesting to note that devices in which the meniscus dried toward the spine slowly returned to their starting position at a rate proportional to the evaporation of water.

For devices with longer rib lengths of 500 µm, the actuation is so extreme that the ribs are forced into contact, as captured by Fig. 13.10a–b. Note that, unlike the typical MEMS stiction of two cantilevers [27], the liquid is observed to dry towards the point contact at the tip of the ribs and away from the spine, as seen in Fig. 13.10c. Since the liquid dries towards the tips of the ribs, the device remained completely deformed at its point of equilibrium until all the liquid evaporated. At that point the ribs immediately separated and the microactuator snapped back to its starting position. The takeaways from Figs. 13.9 and 13.10 are that the meniscus can be geometrically programmed to dry towards the spine using shorter rib lengths, or to dry towards the tip of the ribs, using longer rib lengths that touch at the tip during actuation. This also allows for control of the speed with which a device returns to its original position. Devices with shorter rib lengths returned to their original position in a gradual manner, while those with longer rib lengths snapped back to their original position when the last drop of liquid holding two rib tips together dried.

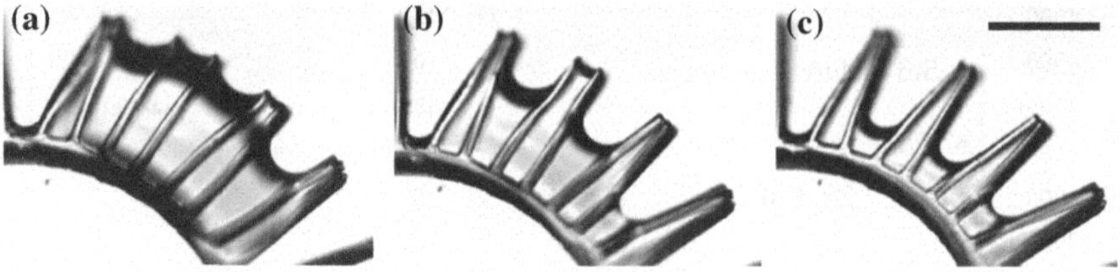

Fig. 13.9 Observations of drying phenomena in between ribs due to uneven test environment: **a** Drying of liquid phase for device whose ribs are not forced into contact. **b–c** At a certain time point devices undergo a breaking of symmetry due to uneven test environments which cause the water within some cells to evaporate faster than in others (i.e., non-uniform light intensity). Images are two seconds apart. Scale bar indicates 300 μm. Reprinted with permission from [12]. Copyright 2006 Institute of Physics

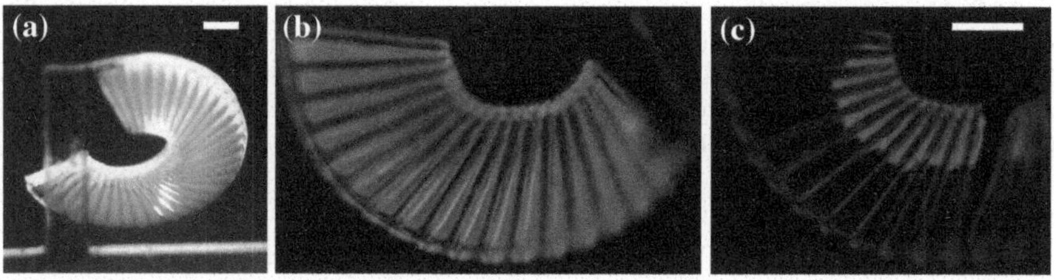

Fig. 13.10 Geometric control of the way that the device returns to its original position: **a** A photograph of a device whose deflection is so extreme that the ribs are forced into contact. **b** Close-up of the device drying to show the liquid phase. The liquid is observed to dry towards the tip of the ribs. **c** Due to the direction of the receding meniscus during drying, the device remained completely deformed at its maximum possible actuated position until all the liquid evaporated. Scale bar indicates 250 μm. Reprinted with permission from [12]. Copyright 2006 Institute of Physics

13.3 Evaporation Driven Fluid Flow in Plants

The transpiration actuation devices presented in the previous section demonstrated that evaporation can be exploited to generate mechanical power. This section also explores the possible use of evaporation to generate electrical power. Many scenarios and environments exist where evaporation of water, and the corresponding formation of humidity gradients, allow for energy to be scavenged. Examples include the surface of the skin due to perspiration, near the surface of bodies of water, and the soil-air interface which is utilized to do work via transpiration by the vast majority of plants on earth. This section will explore the use of evaporation driven flow to generate electrical power in plant-like fluidic networks. Therefore, it is important to present an introduction to how plants use transpiration efficiently for their own purposes. This will include the cohesion-tension mechanism used by plants to passively pump water through their vascular network, an overview of the energy requirements of evaporation, and how plants maximize their hydraulic flow by obeying Murray's Law.

13.3.1 Transpiration in Plants and the Cohesion-Tension Theory

Vascular plants have evolved to take advantage of the process of transpiration in order to obtain an internal one-way fluidic transport network as well as cooling. Transpiration is evaporative-based pumping driven by some form of external energy source, which in the outdoors is almost exclusively solar energy. The process relies on the combined effects of both the evaporation of water from an air-water interface and the cohesive-adhesive properties of water. A plant's water requirements directly parallel the amount that is exchanged during transpiration. In fact, other water requirements for plant growth and photosynthesis pale in comparison to the ninety-five percent or more that is lost to transpiration through the plant's leaves [28]. Water flow in plants is used to transport minerals throughout the plant, and the transpiration-driven flow of water from the roots to the leaves, provides a method of doing so at minimal energy cost [8].

The mechanism that drives the transpiration stream in plants is known as the cohesion-tension theory, or simply the cohesion theory, which explains that water in plant microvasculature is under tension generated at the leaves as water evaporates [23]. This driving force is a result of the automatic coupling between evaporation and the negative pressure achieved by capillary forces within plant vascular cell walls. The negative values of pressure indicate that the water in the plant system has a lower potential than the reference potential of water at sea level and standard temperature and pressure (zero bar) [29].

A plant minimizes its vascular maintenance costs by flowing water through high conductance tubes of dead cells, known as xylem. Xylem is developed by programmed death of protoplast and a compression-resistant cell wall to avoid collapse by negative pressure within the capillaries [27]. Thus, the cost of transpiration is reduced to water-flow side effects of the construction of the microvasculature networks, flow-induced water stress or negative apoplast pressure, and the layers of living tissues with hydrophilic cell walls that filter the water into the xylem from the soil [30]. As a result, plants allow transpiration to passively pump water through their microvasculature without expending much energy to actively pump the water themselves.

Modern microfabrication techniques allow for the production of microchannels of appropriate size to allow capillary action to become one of the dominant forces and to approach the scale at which the process of transpiration in plants can take place [32]. The movement of water in the microchannels of a transpiration-based energy conversion system can do work. However, in order to produce flow, water molecules must be removed from a leaf-like evaporating area. Therefore, energy is used in order to evaporate each molecule of water.

13.3.2 Energy and Rate of Evaporation

In order to derive the energy required to evaporate a given weight of water, E_{w}, two actions must be taken into account. First, the given volume of water must be heated to 100 °C, and then the latent heat of vaporization, $\Delta_{\nu}H$, must be applied to the volume of water. This can easily be calculated using (13.12), where c is the specific heat of water and is equal to 4.186 J, $\Delta_{\nu}H$ is the latent heat of vaporization of water and is equal to 2261 J g^{-1}, and T_{w} is the temperature of the volume of water to evaporate, 100 indicates the temperature in degrees Celsius to which the water has to be brought, and g_{w} is the weight of the water in grams. Therefore, the energy required to evaporate a given amount of water can be reduced by increasing the operating temperature.

$$E_{\mathrm{w}} = g_{\mathrm{w}}[(100 - T - \mathrm{w})c + \Delta_{\nu}H] \tag{13.12}$$

The attempt to quantify the rate of evaporation of a fluid is the topic of much research [33, 34]. One theory has predicted accurate flow rates and is given by (13.12) [35], where Q_{evap} is the rate of total evaporation, R is the radius of the water surface, D_{w} is the diffusion coefficient of water in air which is approximately 0.25 cm^{2}s^{-1}, n_{w} is the is the water vapor density at the surface of the water, and n_0 is the water vapor density of the background environment [24].

$$Q_{\mathrm{evap}} = 4RD_{\mathrm{w}}(n_{\mathrm{w}} - n_0) \tag{13.13}$$

13.3.3 Synthetic Leaf Design

A microfabricated synthetic 'leaf' was designed to leverage evaporation, scavenge and store useful electrical power from evaporation-driven flow. The device consists of a porous microfluidic evaporator ('leaf'), as shown in Fig. 13.11a (central image). This 'leaf' converges onto a central channel, or 'stem' as shown in Fig. 13.11b. The stem is embedded with thin-film metal plates and slugs of a mobile non-aqueous phase; one plate is passivated to prevent electrolysis. Gas bubbles were used as the mobile phase to demonstrate the proof of concept.[1] Each pair of capacitor plates is connected into a rectifying charge pump, shown in Fig. 13.11f, similar to that used in vibration scavenging work [36]. The stem is connected to a non-porous water reservoir. A given dielectric slug transits between the capacitor plates at the speed of the evaporation-induced flow, Fig. 13.11c, generating a capacitance change, dC/dt as shown in Fig. 13.11d at each dielectric-water interface. This capacitance change induces a current, shown in Fig. 13.11e into and out of an electric circuit represented

[1] Bubbles were generated using an input stream of dry air introduced into a liquid channel at a T-junction so as to create air bubbles interspersed with water.

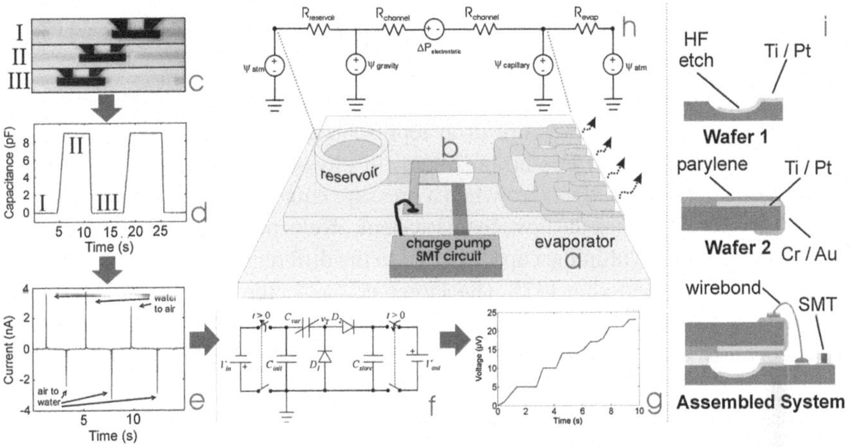

Fig. 13.11 Synthetic leaf design, fabrication and test results. **a** Illustration of porous microfluidic evaporator 'leaf' designed using Murray's Law for the bifurcating branches. The stem is connected to a water reservoir. **b** Central channel, or 'stem' embedded with thin-film metal plates and slugs of a mobile gas bubbles. **c** Gas bubbles move through capacitor plates via evaporative driven flow. Numerals I, II and III indicate sequential time points spaced several seconds apart in plots **c** and **d**. **d** Gas bubble movement through capacitor plates generates capacitance change, dC/dt, at each dielectric-water interface. **e** Capacitance change induces a current. **f** Rectifying circuit utilized to convert current to pump up the voltage on a storage capacitor. **g** Voltage generated by bubbles transiting through capacitor plates via evaporative driven flow. **h** A lumped model representation of the fluidic network of the transpiration based energy converter. The four sources of pressure shown are atmospheric (ψ_{atm}), gravitational (ψ_{gravity}), capillary ($\psi_{\text{capillary}}$), and electroosmotic pressure ($\psi_{\text{electrostatic}}$) due to the non-zero voltage on the capacitor plates. At the microscale, the evaporative pressure drop is dominated by the $\psi_{\text{capillary}}$ (i.e. $\psi_{\text{capillary}} \gg \psi_{\text{atm}}$ and ψ_{gravity}). **i** An illustration of the fabrication process of the transpiration based energy conversion devices. Reprinted with permission from [31]. Copyright 2009 American Institute of Physics

by Fig. 13.11f; with the proper rectification, current moving into the circuit can be used to pump up the voltage, as shown on Fig. 13.11g on a storage capacitor, thus harvesting and storing a portion of the energy in the flow.

13.3.4 Electrical Parameter Model

Energy scavenging from evaporation-driven flow necessitates a trade-off between energy density and power density for a given design. Consider the lumped element model of fluid flow in the microsystem shown in Fig. 13.11h. Ignoring the effect of voltage-induced pressure ($\psi_{\text{electrostatic}}$) for the moment [21], the pressure drop across the entire microfluidic system is set by the difference between the pressure generated by the evaporative pores on one end ($\psi_{\text{capillary}}$) and atmospheric pressure at the other (ψ_{atm}). The surface tension-based pressure at the evaporator scales with the radius of the pores:

$$\psi_{\text{capillary}} = \frac{-2\gamma \cos\theta}{d} \tag{13.14}$$

where γ is liquid surface tension, θ is the liquid-solid contact angle, and d is the pore diameter. In a sense, this pressure drop represents the volumetric energy density per pore extractable from this arrangement. In physical terms, liquid flow inside the microchannels will fall to zero when a pressure equal or greater in magnitude to $\psi_{\text{capillary}}$ is exerted against the flow. In this system, an electrostatic force arises against the motion of a slug within the capacitor due to the difference in dielectric constants between the water and slug [37]. The larger $\psi_{\text{capillary}}$, the larger the electrostatic pressure required to stop the flow. Thus, more energy can be extracted. However, as the pore size decreases, the absolute evaporation rate at each pore decreases. For example, at 20 °C, 50 % relative humidity, liquid in a 6 nm pore is in equilibrium with the vapor phase and no net evaporation occurs [29].

At frequencies below 1 MHz, water molecules and aqueous mobile ions arrange to form a layer of charge across any immersed conductor with a potential across it. This well-known double layer effectively looks like a large capacitor (in the range 0.1–1 pFm^{-2} [38, 39]). As evaporative flow forces the mobile dielectric phase between the plates, displacing the water, the effective capacitance of the two plates changes dramatically (Fig. 13.11d). Because $Q = CV$ for any capacitor, a change in capacitance (dC/dt) must result in either a change in voltage (dV/dt) or a current (dQ/dt). The conversion circuit (Fig. 13.11f) keeps the capacitor voltage constant while using diodes to allow current flow from the variable capacitor into the circuit as the dielectric enters the plates, but prevents the trailing edge from extracting current. In this way, work is extracted from the kinetic energy of the fluid flow and can be stored (Fig. 13.11g).

13.3.5 Fabrication

Devices were microfabricated on glass wafers (Fig. 13.11i). Microchannels were etched isotropically in HF to achieve 75 µm deep channels. Ti/Pt electrodes were evaporated and patterned onto both wafers and then the wafers were diced. Cr/Au was sputtered around the edge of Wafer 2 to create an electrical contact to the back side of the die. Wafer 2 dice were then coated in 1.2 µm of Parylene C (SCS Coatings, Indianapolis, IN). Holes were manually drilled into Wafer 2 using a press for fluidic connection to a fluid reservoir. The two substrates were aligned using a stereo microscope then bonded using Sylgard 184 PDMS (Dow Corning). The surface mount components for the circuit were then soldered onto the Ti/Pt pads on Wafer 1. Finally, the electrical contact of Wafer 2 was wire-bonded to Wafer 1. If used, porous ceramic evaporator discs (Soil Moisture Equipment Corp) were placed in contact with open microchannels and held with epoxy dots.

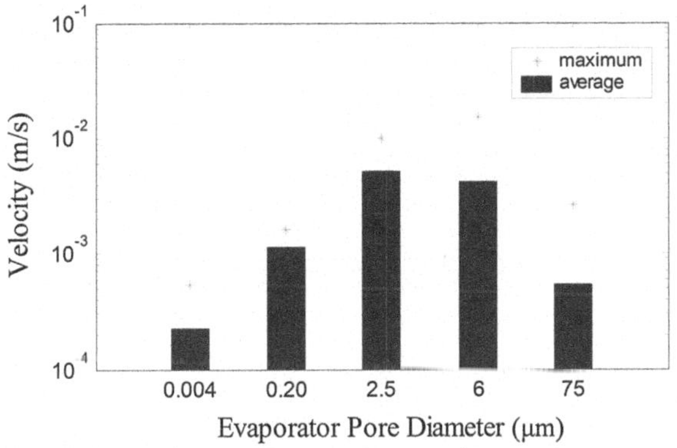

Fig. 13.12 Evaporation driven velocity versus evaporator pore diameter. The average and maximum evaporation-driven velocity for each evaporator pore diameter tested: 4, 200 nm, 2.5, 6, and 75 μm. Channel height, $h = 45\,\mu\text{m}$ and channel width, $w = 500\,\mu\text{m}$

13.3.6 Testing Results: Evaporation Driven Flow

Figure 13.12 presents fluid velocities generated by devices connected to evaporators with different diameter pores. The maximum fluid velocity measured at STP was $1.5\,\text{cms}^{-1}$ (6 μm pores); for 3 mm long plates, at this velocity, the frequency bandwidth associated with a gas-liquid capacitive change is 5 Hz.[2] Since the effective value of the double layer capacitance is frequency dependent, direct measurements of dC/dt as a bubble transited were taken at various frequencies and extrapolated to DC. Figure 13.11d shows $dC/dt = 8$–10 pF for a single bubble taken at 1 MHz. In order to directly measure the current extracted with each bubble, one stem capacitor was wired in parallel with a surface mount capacitor (1 μF) and an ammeter placed between the two components. Both capacitors were then pre-charged to $1\,\text{V} @ 100\,\mu\text{L min}^{-1}$, the leading edge of the bubble produced a 3–4 nA current into the SMT capacitor which flowed back into the stem capacitor on the trailing edge (Fig. 13.11e). This continued as long as there were bubbles in the channel. As expected, the current at each transition was proportional to dC/dt which was proportional to fluid velocity (not shown).

13.3.7 Testing Results: Energy Scavenging Circuit

The full scavenging circuit in Fig. 13.11f uses the current generated by the leading edge of the bubble to incrementally increase the voltage of a storage capacitor, C_{store}. Diodes prevent current flow out of the storage capacitor, so each bubble effectively 'pumps up' the energy stored in C_{store}. C_{var} is the microfabricated capacitor, whose

[2] $f = v/x = 1.5\,\text{cms}^{-1}/0.3\,\text{cm}$.

value ranges from C_{max} (water) to C_{min} (dielectric). Initially, $C_{var} = C_{max}$; both C_{var} and C_{init} are pre-charged to $V_{in} = -5$ V. The total energy in the system is given by (13.15) and (13.16). V_{out} on C_{store} starts at 0 V.

$$E_{tot} = E_{init} + E_{var} + E_{store} \tag{13.15}$$

$$E_{tot} = \frac{1}{2}C_{init}V_{in}^2 + \frac{1}{2}C_{var}V_{in}^2 + \frac{1}{2}C_{store}V_{out}^2 \tag{13.16}$$

$$\Delta_{cycle} = \frac{1}{2}C_{store}V_{out}^2 \tag{13.17}$$

When the dielectric enters, C_{var} changes its value from C_{max} to C_{min}; this ΔC_{var} results in an increase of V_{out} on C_{store}. The positive current due to a decrease in the capacitance of C_{var} results in a positive current across D_2 to C_{store}. When a bubble exits, C_{var} changes its value from C_{min} to C_{max} but there is no voltage change at V_{out} (due to the diodes). At this point, a full energy conversion cycle has occurred and the circuit has a net positive energy gain, Δ_{cycle}, as given by (13.17).

To characterize this circuit, the water reservoir was replaced with tubing loaded with alternating sections of water and air bubbles to provide a large number of sequential interfaces. A syringe pump drove the bubble train into the device at a flow rate of 100 μL min^{-1}. The circuit was pre-charged to −5 V and source was disconnected. V_{out} was measured (Fig. 13.11g) and video of the bubbles traversing was synchronized with V_{out}. V_{out} decays slowly due to the finite input impedance of the voltmeter (10 GΩ) when bubbles are not transiting and generating power (Fig. 13.11d, Regions I and III). During operation (Fig. 13.11d, Region II), each leading bubble edge results in an increase in output voltage (ΔV_{out}) of approximately 2–5 μV (Fig. 13.11g). It is important to note that this circuit works in the face of leakage parasitics in all the SMT components ($R_{parallel} = 10$ GΩ).

As long as the energy scavenged is a small fraction of the energy in the fluid, each interface generates dC/dt of approximately the same size, forcing a fixed ΔV_{out} on C_{store}. However, as the energy on C_{store} increases, raising the voltage by ΔV_{out} requires increasing amounts of energy (Fig. 13.11g).

$$\Delta E = \frac{1}{2}C_{store}(V + \Delta V)^2 - \frac{1}{2}C_{store}V^2 \tag{13.18}$$

Thus, the average power generated increases with time (up to a maximum dictated by the flow velocity). As a comparison for the values in Fig. 13.11b, a minimum-sized slug of liquid (two menisci facing each other) moving at 1.5 cm s^{-1} has 1.2 pJ of kinetic energy for a slug 500 μm wide, 75 μm high, and 150 μm long (measured curvature of meniscus, not shown). The maximum volumetric power density of this device can be calculated by envisioning a channel composed of trains of minimally-sized slugs moving over arrays of metal plates which are one meniscus wide; this scenario maximizes the interfacial area transiting over plates at any given time.

$$
\begin{aligned}
\Delta E &= 1.2 \times 10^{-12}\,\mathrm{J} \\
t &= \frac{300\,\mu\mathrm{m}}{1.5 \times 10^{-2}\,\mathrm{m\,s^{-1}}} = 20\,\mathrm{ms} \\
P_{\max} &= \frac{\Delta E}{t} = 60\,\mathrm{pW} \\
N &= \frac{1\,\mathrm{cm}^3}{300\,\mu\mathrm{m} \times 500\,\mu\mathrm{m} \times 75\,\mu\mathrm{m}} = 88.9 \\
p &= = N \cdot P_{\max} = 5\,\mu\mathrm{W\,cm^{-3}}
\end{aligned}
\tag{13.19}
$$

where ΔE is the kinetic energy of a slug, d is the length, and v is the velocity of each slug, $P_{\max}$ is the maximum power extractable from one slug, N is the number of slugs in $1\,\mathrm{cm}^3$ and p is the maximum power density per cm^3. Adding the ceramic evaporator volume ($1.414\,\mathrm{cm}^3$) drops power density to $2\,\mu\mathrm{W\,cm^{-3}}$. This compares well with vibration scavenging efforts given that the existing device can be significantly optimized [40].

The extraction efficiency is limited by the circuit's resistive parasitics. Leakage on Cstore burns away stored energy. Leakage on C_{init} eventually prevents the circuit from operating. Ultra-low power electronics for periodic refreshing a pre-charge condition are a topic of ongoing investigation [41, 42]. The consumption of the overhead electronics needs to be less than $1\,\mu\mathrm{J}$ per refresh cycle in order to breakeven if $C_{\mathrm{store}} = 20\,\mathrm{nF}$. For comparison, a single $0.25\,\mu\mathrm{m}$ MOSFET running at 5 V would consume about $0.5 \times C_{\mathrm{gate}} V^2 = 62.5\,\mathrm{fJ}$.

13.3.8 Future Improvements

This technology is the first, to our knowledge, to exploit the same mechanism used by plants to scavenge energy from spatial and temporal humidity gradients and should prove useful as the basis for a cyclically renewable source of power. However, several improvements are needed. For a realistic scavenger, the bubbles need to be replaced with a circulating (e.g. beads) or oscillating dielectric phase. Further, the pore size of the evaporator and the fluidic resistance matching between evaporator and main channel need to be optimized; parameters were set by commercially available ceramic disks and any improvement will immediately increase power density. A daily pre-charge circuit is needed to keep the scavenger active indefinitely. Lastly, micro-fabricated condensers need to be added to collect water droplets during the condensation cycle.

13.4 Conclusion

A number of efforts, often biomimetic, have focused on the efficient condensation and collection of atmospheric water, whether by specific geometries or surface characteristics (such as patterns of alternating hydrophobic and hydrophilic areas)

[43, 44]. Given this, it is possible to consider continually renewable systems which, via condensation at night and evaporation during the day, can extract small amounts of power regularly and repeatedly. Moving forward, two central challenges need to be solved. First, large area evaporative networks with properly matched fluidic channels ranging from cm-scale down to μm-scale (and from there to porous surfaces) need to be developed in a material that is cheap, light-weight, mechanical robust and mechanically compliant. Second, given a properly matched fluidic network, the basic scheme used above needs to be refined to eliminate bubbles and maximize the interfacial dielectric-water area being shuttled across the conductors.

References

1. Y. Bar-Cohen, Biomimetics - using nature to inspire human innovation. Biosinspiration Biomimetics **1**, 1 (2006)
2. O.H. Schmitt, Some interesting and useful biomimetic transforms, in *Proceedings of the 3rd International Biophysics Congress*, Boston (MA), 1969
3. B. Bhushan, Y.C. Yung, Natural and biomimetic artificial surfaces for superhydrophobicity, self-cleaning, low adhesion, and drag reduction. Prog. Mater. Sci. **56**, 1–108 (2011)
4. V. Namasivayam, R.G. Larson, D.T. Burke, M.A. Burns, Transpiration-based micropump for delivering continuous ultra-low flow rates. J. Micromech. Microeng. **13**, 261 (2003)
5. T. Vestad, D.W.M. Marr, J. Oakey, Flow control for capillary-pumped microfluidic systems. J. Micromech. Microeng. **14**, 1503 (2004)
6. J.M. Walker, D.J. Beebe, An evaporation-based microfluidic sample concentration method. Lab Chip **2**, 57–61 (2002)
7. N. Goedecke, J. Eijkel, A. Manz, Evaporation driven pumping for chromatography application. Lab Chip **2**, 219–223 (2002)
8. R.O. Slatyer, Movement of water in the vascular system, in *Plant-Water Relationships* (Academic Press, New-York, 1967) pp. 211–215
9. J. Sutcliffe, Mechanisms of water movement, in *Plant-Water Relationships* vol 14, The Institute of Biology's Studies in Biology. (Edward Arnold (Publishers) Ltd, London, 1968) pp. 72–76
10. D.H. Trevena, *Cavitation and Tension in Liquids* (Adam Hilger, Philadelphia, 1987)
11. M.T. Tyree, M.H. Zimmerman, *Xylem Structure and the Ascent of Sap* (Springer, New York, 2002)
12. R.T. Borno, J.D. Steinmeyer, M.M. Maharbiz, Transpiration actuation: the design, fabrication and characterization of biomimetic microactuators driven by the surface tension of water. J. Micromech. Microeng. **16**, 2375–2383 (2006)
13. V. Raghavan, *Developmental Biology of Fern Gametophytes* (Cambridge University Press, New York, 1989)
14. K. Haider, Zur Morpholegie und Physciologie der Sporangien leptosporangiater Farne. Planta **44**, 370–411 (1954)
15. E. Steudle, Acquisition of water by plant roots. Annu. Rev. Plant Physiol. Plant Mol. Biol. **52** (2001)
16. E. Chanliaud, K. Burrows, G. Jeronimidis, M. Gidley, Mechanical properties of primary plant cell wall analogues. Planta **215**, 989–996 (2002)
17. R.R.A. Syms, E.M. Yeatman, V.M. Bright, G.M. Whitesides, Surface tension-powered self assembly of microstructures—the state-of-the-art. J. Microelectromech. Syst. **12**, 387 (2003)
18. P.W. Green, R.R.A. Syms, E.M. Yeatman, Demonstration of three-dimensional microstructure self-assembly. J. Microelectromech. Syst. **4**, 170 (1995)
19. M. Boncheva, G.M. Whitesides, Templated self-assembly: formation of folded structures by relaxation of pre-stressed, planar tapes. Adv. Mater. **17**, 553 (2005)

20. D.H. Gracias, J. Tien, T.L. Breen, C. Hsu, G.M. Whitesides, Forming electrical networks in three dimensions by self-assembly. Science **289**, 1170 (2000)
21. S.D. Senturia, *Microsystem Design* (Kluwer, Boston, 2000), pp. 240–244
22. J.W. Weaver, S.P. Timoshenko, D.H. Young, *Vibration Problems in Engineering*, 5th ed. (Wiley, New York, 1990), pp. 518–520
23. D.L. Bower, *An Introduction to Polymer Physics* (Cambridge University Press, New York, 2002), pp. 164–176
24. C.E. Maneschy, M. Massoudi, V.R. Velloso, Dynamic elastic solutions in neo-Hookean and Mooney-Rivlin materials. Int. J. Non-Linear Mech. **28**, 531 (1993)
25. J.M. Gere, in *Bending Stresses in Beams in The Engineering Handbook*, ed. by R.C. Dorf (CRC Press, Inc., Boca Raton, FL, 1996), pp. 52–59
26. S. Goyal, N.C. Perkins, C.L. Lee, Nonlinear dynamics and loop formation in Kirchoff rods with implications to the mechanics of DNA and cables. J. Comput. Phys. **209**, 371–389 (2005)
27. R. Maboudian, R.T. Howe, Critical review: adhesion in surface micromechanical structures. J. Vac. Sci. Technol. B Microelectron. Process. Phenom. **15**, 1 (1997)
28. J.S. Sperry, Evolution of water transport and xylem structure. Int. J. Plant Sci. **163**, 115–127 (2003)
29. J.C.T. Eijkel, A.v.d. Berg, Water in micro- and nanofluidics systems described using the water potential. Lab Chip **5**, 1202–1209 (2005)
30. M.J. Canny, The transpiration stream in the leaf apoplast: water and solutes. Philos. Trans. Biol. Sci. **341**, 87–100 (1993)
31. R.T. Borno, J.D. Steinmeyer, M.M. Maharbiz, Charge-pumping in a synthetic leaf for harvesting energy from evaporation-driven flows. Appl. Phys. Lett. **95**(1), 013705 (2009)
32. P. Gravesen, J. Branebjerg, O.S. Jensen, Microfluidics-a review. J. Micromech. Microeng. **3**, 168 (1993)
33. K. Yasuoka, M. Matsumoto, Y. Kataoka, Evaporation and condensation at a liquid surface. I. Argon. J. Chem. Phys. **101**(9), 7904–7911 (1994)
34. C.A. Ward, G. Fang, Expression for predicting liquid evaporation flux: statistical rate theory approach. Phys. Rev. E **59**(1), 429–440 (1999)
35. K. Hisatake, M. Fukuda, J. Kimura, M. Maeda, Y. Fukuda, Experimental and theoretical study of evaporation of water in a vessel. J. Appl. Phys. **77**(12), 6664–6674 (1995)
36. S. Roundy, Energy scavenging for wireless sensor nodes with a focus on vibration to electricity conversion. Ph.D. Dissertation, University of California-Berkeley, 2003
37. C. Carnero, J. Carretero, A study on the physical behaviour of a dielectric slab inserted into a parallel-plate capacitor. Eur. J. Phys. **17**, 220–225 (1996)
38. J.O'M. Bockris, A.K.N. Reddy, *Modern Electrochemistry* (Plenum, New York, 1970)
39. L. Mateu, F. Moll, Review of energy harvesting techniques and applications for microelectronics, in *Proceedings of the SPIE Microtechnologies for the New Millenium*, Sevilla, Spain, pp. 359–373, 2005
40. K.A. McCulloh, J.S. Sperry, F.R. Adler, Water transport in plants obeys Murray's Law. Nature **27**, 939–942 (2003)
41. A. Host-Madsen, J. Zhang, Capacity bounds and power allocation for wireless relay channels. IEEE Trans. Inf. Theory **51**, 2020–2040 (2005)
42. X. Cheng, B. Narahari, R. Simha, M.X. Cheng, D. Liu, Strong minimum energy topology in wireless sensor networks: NP-completeness and heuristics. IEEE Trans. Mob. Comput. **2**(3), 248–256 (2003)
43. M. Helmer, Biomaterials: dew catchers. Nature, **463**(7281), 618 (2010)
44. R.P. Garrod, L.G. Harris, W.C. Schofield, J. McGettrick, L.J. Ward, D.O. Teare, J.P. Badyal, Mimicking a Stenocara beetle's back for microcondensation using plasmachemical patterned superhydrophobic-superhydrophilic surfaces. Langmuir **23**(2), 689–693 (2007)
45. A. Ursprung, Über die Kohäsion des Wasser in Farnannulus. *Berichte der Deutsch Botanischen Gesellschaft* **33**, 152–162 (1915)

Chapter 14
Capillary Forces in Atomic Force Microscopy and Liquid Nanodispensing

Thierry Ondarçuhu and Laure Fabié

Abstract Capillary forces are an important issue when performing atomic force microscopy (AFM) in air. The formation of a water meniscus between the tip and the surface often leads to a deterioration of the tip or sample which may affect the spatial resolution of the instrument. The development of dynamic modes allowed decreasing these effects but artefacts due to capillary forces are still present. The presence of a small liquid bridge between the tip and the surface has also been used to locally modify the surface. Several tip-assisted lithography techniques have been developed and may find a wide range of applications in nanociences. In particular, a liquid nanodispensing technique (NADIS) using specific AFM tips allows to deposit directly on the surface liquid droplets with volumes in the femto- to attoliter range. In this chapter, we describe the effect of capillary forces on these AFM-based techniques. After a review of the influence of capillary forces in AFM in static and dynamic modes, we focus on the NADIS technique which provides a unique way to study nanomenisci with controlled size.

14.1 Introduction

Soon after its invention in 1986 [12], atomic force microscopy (AFM) has become a widespread characterization technique in research and industry. This technique allows the formation of images of sample surfaces with nanometric resolution, in ambient environment, without requiring specific sample preparation as for electron microscopy. The image is formed using a microfabricated tip with a radius of

T. Ondarçuhu (✉)
Nanosciences group, CEMES-CNRS,
29, rue Jeanne Marvig, 31055 Toulouse cedex 4, France
e-mail: ondar@cemes.fr

L. Fabié
Nanosciences group, CEMES-CNRS and Université de Toulouse,
29, rue Jeanne Marvig, 31055 Toulouse cedex 4, France

P. Lambert (ed.), *Surface Tension in Microsystems*, Microtechnology and MEMS,
DOI: 10.1007/978-3-642-37552-1_14, © Springer-Verlag Berlin Heidelberg 2013

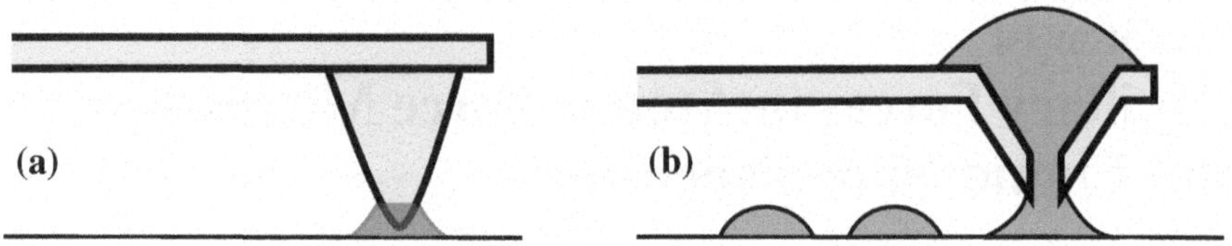

Fig. 14.1 Sketch of the techniques described in this chapter: **a** In AFM, a liquid meniscus condensates during scanning of the tip in contact with the surface; **b** In NADIS, the liquid solution is brought to the surface from a reservoir droplet deposited on the cantilever, through a nanochannel milled at the tip apex

curvature of several nanometers at its apex, which is scanned in contact (or very close from the surface in dynamical modes) with the sample surface. Since it has been known for a long time, that two solid surfaces brought in close contact lead to the formation of a liquid bridge [1, 36], even for low humidity, capillary effects are expected to play an important role in the imaging process. Indeed, the presence of a liquid meniscus formed by capillary condensation at the tip-sample contact (Fig. 14.1a) is often detrimental for AFM imaging as it leads to strong adhesive forces which may damage the tip or the solid surface under inspection. The effect of the meniscus is also crucial in the case of the oscillating AFM modes which were developed in order to limit the effect of friction between tip and surface. The formation and rupture of the liquid bridge during each oscillation of the cantilever may lead to important errors in height measurements and makes difficult the quantification of the mechanical properties of the surface which is available with this technique. The precise understanding of capillary adhesion is therefore crucial to extract quantitative information from AFM data. By its ability to measure tiny forces [19], AFM has been intensively used as a model system to study capillary forces at nanometer scale. These studies are relevant not only to understand in detail the AFM imaging process but also in domains, such as tribology, where capillary adhesion is involved. Indeed, it is recognized that macroscopic friction is ruled by tribology processes occurring, in parallel, at many micro- or nano-sized contacts [4, 13]. AFM therefore provides a unique method to probe the effect of capillarity at the level of a single nano-asperity, the AFM tip being considered as a model nanocontact.

In addition to its application for probing surfaces with high resolution, AFM can be used to modify and pattern surfaces at nanometer scale. It was shown that the tip of an AFM (or STM) can modify a surface by mechanical, thermal or electrical action [71]. A major development was brought by Jaschke and Butt [41] who demonstrated that the AFM tip can be used to add locally spots of molecules to the surface. Later on, this method was fully developped by Piner et al. [59], and the so-called "dip pen nanolithography (DPN)" technique became a scientific and technological reference in the domain [65]. The method relies on the diffusive transfer of molecules prealably adsorbed on the tip, to the surface during the tip-surface contact time. Provided experimental conditions are carefully controlled, spots of self-assembled molecules can be reproducibly patterned with typical dimensions of the order of 100 nm in routine, and can be downscaled to an ultimate resolution of 15 nm. However, by nature, DPN is highly molecule-dependent. Specific tip and surface functionalizations are often needed for a given molecule-surface system [76].

Recently, a nanodispensing (NADIS) technique was developed in order to extend the field of applications of scanning probe lithographies. The idea of NADIS is to use an AFM tip to directly deposit and manipulate ultrasmall amounts of liquid solutions [50]. Doing so, it can be applied to any soluble material, independently of its chemical nature and size. Contrary to DPN, NADIS requires the fabrication of specific AFM tips with a nanochannel drilled at their apex. This channel is used to transfer the liquid solution from a reservoir droplet deposited on the cantilever, to the surface (Fig. 14.1b). This technique allowed patterning of surfaces with an ultimate resolution of 70 nm [30], comparable with the one of DPN. Since it combines the resolution of the scanning probe lithography methods with the versatility of liquid lithography in terms of deposited molecules, NADIS is therefore an alternative to DPN and may find many applications.

In addition to its applications in nanopatterning, the NADIS technique is a unique tool for a systematic study of capillary forces and mechanics of nanomenisci with controlled dimensions, in a wide range of situations including the case of a meniscus which is connected to a reservoir [28]. This will be the main topic of this chapter.

After giving some elements on the computation of capillary forces, we discuss the influence of capillary force in atomic force microscopy in static and dynamic modes. We then focus on the study of capillary forces using the NADIS technique.

14.2 Computation of Capillary Forces

Due to their implication in many natural and industrial processes, the calculation of capillary forces has attracted a lot of interest. Many models have been developed as detailed in this book and in [18, 46]. In this section, we simply recall the main methods allowing to compute capillary forces, restricting our discussion to the ones which will be used in this chapter. We consider a liquid meniscus bridging two solid surfaces, namely a flat surface and a conical tip. This system is sketched in Fig. 14.2a where the main parameters are represented. Given the sizes and velocities used in AFM, we neglect gravity effects and dynamic capillary forces. Two main methods have been used to compute the capillary force (Chap. 2):

- In the energetic approach, the force results from the derivation of the total energy of the system with respect to the tip-surface distance z.

$$F_{\text{cap}} = -\frac{dW_{\text{tot}}}{dz} \tag{14.1}$$

The W_{tot} energy comprises two terms: one term corresponds to the energy of the various solid/liquid, solid/vapor and liquid/vapor interfaces and a second term is

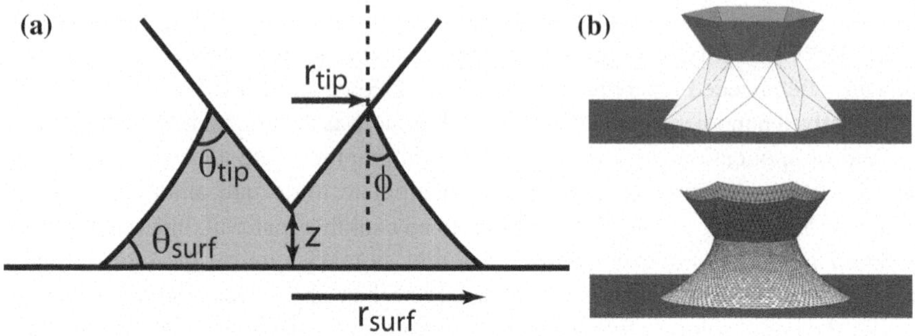

Fig. 14.2 **a** Sketch of the studied configuration. z is the tip-sample distance, r_{tip} (r_{surf}) and θ_{tip} (θ_{surf}) are the radius of the contact line and contact angle at the tip (surface), respectively, and ϕ is the angle of the liquid interface with respect to the vertical, at the contact line; **b** shape of the meniscus between a conical tip and a surface before and after optimization using the Surface Evolver software. Adapted from [28]; Copyright 2010 American Chemical Society

given by the bulk energy $W_{\text{bulk}} = \int -P.dV$ where P is the pressure inside the liquid and V its volume.

- In the geometric approach, the force is derived from the shape of the meniscus taking into account two different components : the Laplace term F_{L} due to the pressure P inside the meniscus, and the tension term F_{T} which results from the liquid surface tension acting on the contact line. In the situation represented in Fig. 14.2a, these two terms write:

$$F_{\text{cap}} = F_{\text{L}} + F_{\text{P}} = P.\pi r_{\text{tip}}^2 - 2\pi r_{\text{tip}}\gamma \cos\phi \tag{14.2}$$

The equivalence between these two approaches was demonstrated by Lambert et al. [47].

Both methods require the knowledge of the liquid interface shape, which verifies the Laplace Eq. (1.1). This equation can be solved analytically in the case of a zero pressure, leading to a catenoïd profile of the form:

$$r = C\cosh\left(\frac{z - z_0}{C}\right) \tag{14.3}$$

where C and z_0 are two parameters allowing to match the boundary conditions on tip and substrate. This profile is often approximated by a circular or parabolic equation in order to simplify the calculations (Sect. 2.2.4). In the cases where the radius of curvature R of one of the solid surfaces is large compared to its separation with a plane substrate z, considering that the tension force is negligible, this approximation leads to a simple expression for the capillary force [36]:

$$F_{\text{cap}} = \frac{4\pi R\cos\theta}{1 + \frac{z}{d}} \tag{14.4}$$

where θ is the contact angle of the liquid on both surfaces and d the radius of the liquid bridge. Since the deviation of this solution compared to numerical solutions is smaller than 5 % [24], (14.4) is used in many cases. However, it fails to describe situations where the radius of curvature of the tip is of the order of the tip-sample separation, as encountered in AFM. This case requires numerical simulations. For axisymmetric systems, the Laplace Equation (1.1) can be solved numerically, using a double iterative method for meniscus with constant volume [48] (Sect. 2.2.3). The force is then calculated using the geometrical method.

For more general situations, including non axisymmetric systems, the Surface Evolver software allows to compute the 3D shape of the meniscus by energy minimization with defined boundary conditions and constraints [16]. Starting from a simple initial surface, the mesh is refined and optimized using a gradient descent method (Fig. 14.2b). The boundary conditions describe the behaviour of the contact lines, on both the tip and the surface. Depending on the situations, two different conditions are applied: when the contact line is pinned on the solid (mainly for hydrophilic surface), a constant radius (r_{tip} or r_{surf}) is applied; on more hydrophobic surfaces, the contact line can move with given contact angle (θ_{tip} or θ_{surf}) used as boundary condition. Constraints apply to the liquid body during the minimization and can be of two kinds: a constant volume constraint is generally used for non volatile liquids whereas a constant pressure is typically involved in the case of capillary condensation. Once the profile is optimized, the capillary force is derived numerically from the total energy according to Eq. 14.1 [25, 26]. The process is iterated to compute the whole force curve.

The choice of the method used depends on the system under study and on the precision required on the results. In the following, we will use approximated equations (for example Eq. 14.4) in the case of capillary forces involved in AFM because the precise shape is not known experimentally. On the contrary, for investigations with NADIS tips which lead to many different situations and for which the non axisymmetric pyramidal shape of the tip may play a role, we used computations based on the Surface Evolver software, which is the only method able to handle all the encountered situations. More details on the method are given in Sect. 14.4.1.

14.3 Capillary Forces in Atomic Force Microscopy

In this section, we give a brief description of the effect of capillary forces in AFM. Rather than an exhaustive review of the large number of experiments dedicated to this subject, we chose examples emphasizing on nanoscale aspects and on the artefacts that may appear when making quantitative measurements of surface topography and properties. We first describe the contact mode and, in a second part, the oscillating AFM mode.

14.3.1 Contact Mode

In contact mode, the tip is maintained in contact while scanning the surface. In standard ambient condition, a nanomeniscus is present at the tip-surface contact, leading to a strong adhesive force. The meniscus may form from liquid films adsorbed on tip or substrate surface, or by capillary condensation. The resulting capillary force dominates over other interactions such as van der Waals or electrostatic forces [57]. The presence of this meniscus does not directly alter the imaging process. In particular, in the case of hard materials, the height measurements are not modified by the capillary force. However, this force noticeably increases the tip-surface friction which may lead to a rapid deterioration of the tip or surface, and consequently of the image resolution. The nanomeniscus also makes difficult the study of very soft samples, such as biological materials, by contact AFM in air. In the latter case, imaging is usually performed in liquid environment which, in addition to preserve the native environment of the objects, allows to get rid of the liquid meniscus [21, 31].

Due to its ability to measure interaction forces, AFM was intensively used to study capillary condensation and the properties of nanomeniscus. Most of the studies focused on the influence of relative humidity (RH) on capillary force. However, it is difficult to extract a general trend from the experimental results since the same system (for example, a Si_3N_4 tip on a mica surface) may lead to different dependencies. The force is found to increase monotonically [34, 43] or step-like [11, 68], or present a maximum as a function of RH [35, 77]. A counter-intuitive decrease of adhesion force even at low relative humidity, was also reported recently [44].

These contradictory results were reconciled by Kober et al. [44] by a careful examination of the AFM tip shape by Scanning Electron Microscopy (SEM). The influence of humidity (see Fig. 14.3a for two examples of variation of F_{cap} with RH) could be correlated to the precise nanoscale shape of the tip apex. The fact that tip nano-geometry is not controlled in routine experiments may thus explain the reported problems of reproducibility. The results could be interpreted by computing the force between a flat surface and a tip whose shape could be changed from a perfectly sharp cone, to a cone with a round apex and a truncated cone. The results obtained by minimizing the total energy are reported in Fig. 14.3b. A large variety of behaviours which correspond to reported experimental results are evidenced. The main trends can be simply obtained if one considers the Laplace pressure component of the capillary force, thus neglecting the tension term. Using this approximation, the force can be written as:

$$F \approx P_{\text{Laplace}} \pi r_{\text{tip}}^2 \approx \frac{\gamma}{r_{\text{K}}} \pi r_{\text{tip}}^2 \tag{14.5}$$

where r_{tip} is the radius of the wetted part of the tip and r_{K} the (negative) Kelvin radius i.e. the radius of curvature of the condensed meniscus (the absolute value of r_{K} increases with RH following $r_{\text{K}} = \frac{\gamma V}{RT log(RH)}$, where V is the molar volume and R is the gas constant) [1, 36]. In the case of a perfect cone, one expects that r_{tip}

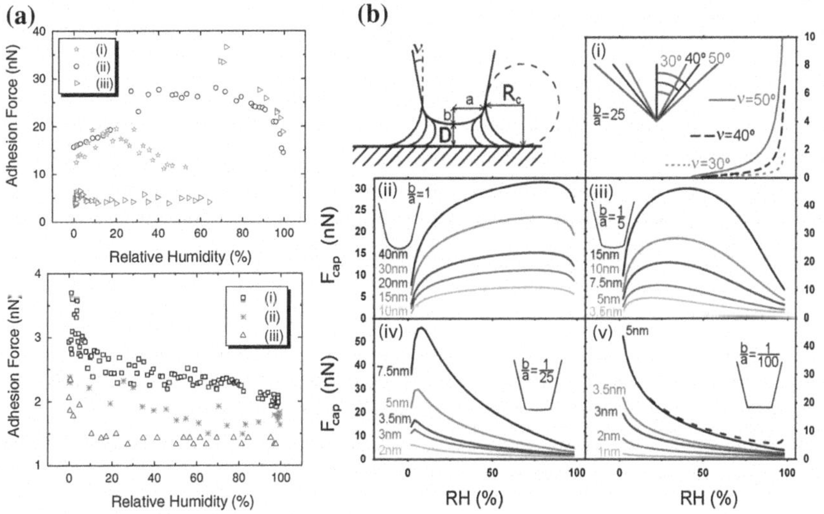

Fig. 14.3 **a** Adhesion force versus RH curves for three different tips on a flat mica surface. (*top*) The adhesion force exhibits a maximum with RH. SEM images of the tip revealed damaged tips with large radii at the apex; (*bottom*) Experiments performed with a maximum applied load of 10 nN. In this case, the adhesion force decreases with RH. Within the SEM resolution the nominal tip radius was preserved after all three experiments. **b** Calculation of capillary forces versus RH curves, for different geometries, represented in the insets. Reprinted figure with permission from [44]. Copyright 2010 by Wiley

and r_K increase similarly with RH, which, according to Eq. 14.5, leads to an increase of F (see Fig. 14.3b(i)). If, on the contrary, the apex of the tip is broken with a flat apex (Fig. 14.3b(v)), as may occur with a monocrystalline material, and provided the tip radius is small enough to be comparable with r_K, the contact line is pinned at the tip edges during most of the separation process: if r_{tip} remains constant while r_K increases, F is expected to decrease monotonically with RH. For rounded tips or broken tips with larger radius, intermediate situations with a maximum are put into evidence. This is the most frequent situation in experiments.

This example illustrates the interest of AFM to study the influence of capillary forces on nanocontact. It shows that the control of the precise shape of the asperities is essential to control the capillary force. The design of well-defined asperities may lead to original structures for which capillary adhesion decreases with humidity.

AFM can also provide information on the kinetics of neck formation which is crucial for dynamic studies. The characteristic time for water condensation was estimated in the millisecond range by AFM friction measurements [69]. Very recently, Noël et al. developed the so-called "circular AFM" mode that allows measuring adhesion forces while the probe is sliding on the surface at continuous and constant velocities [54]. This original experimental setup opens the way to systematic studies of the dynamics of meniscus formation.

14.3.2 Oscillating Modes

The amplitude-modulation (AM-AFM) mode, also named *Tapping* mode or *Intermittent contact* mode, was developed in order to minimize the friction which could damage both tip and surface in contact mode [32]. It rapidly became the most extensively used mode for routine sample imaging in air, on hard and soft materials. In the AM-AFM mode, the cantilever is oscillated close to its resonance frequency. An interaction with the surface leads to a shift and/or a widening of the cantilever resonance peak which results in a variation of the amplitude of oscillation A and a phase shift φ between the cantilever excitation and its response. The amplitude signal A is used as feedback signal to measure the topography of the surface while the phase signal φ leads to images which can exhibit a contrast when different materials are present on the surface. This latter image contains information on the surface itself (stiffness) and on the tip-sample interaction (adhesion, dissipation). In particular, the average dissipated power can be estimated using [22]:

$$P_{\text{diss}} = \frac{kA^2\omega}{2Q}\left(\frac{A_0}{A}sin\varphi - 1\right) \tag{14.6}$$

where k is the cantilever spring constant, Q its quality factor and ω the pulsation of the excitation.

The tip-surface interaction plays a crucial role in dynamic AFM mode. Since it strongly depends on imaging conditions (oscillation amplitude, setpoint used in the close loop), a precise understanding of the imaging process is essential to perform quantitative topographical and sample properties measurements. In this framework, capillarity plays a major role in ambient conditions. The formation of a liquid bridge at the tip-surface contact, at each oscillation, can drastically change the images and lead to important artifacts.

Two distinct interaction regimes have been evidenced both experimentally and theoretically. In the attractive regime, usually observed for large tip-surface distances, the tip experiences an overall attractive force during oscillation. This results in a positive phase shift. When the tip comes closer to the surface, a sudden transition can be observed to a repulsive regime where the tip is in stronger contact with the surface during one part of the cycle and is subject to a repulsive force in average. This leads to a negative phase shift. The phase signal which is monitored in real time in AFM measurements is therefore a useful way to know exactly in which regime the AFM is operating. These two behaviours could be reproduced theoretically solving the motion equation of the cantilever, considered as a point mass, submitted to an oscillating driving force. In addition, the cantilever elastic restoring force, a viscous damping by air and the tip-sample interaction forces are used. By taking into account in the latter term, long range van der Waals attractive force and a repulsive force depending on substrate elastic properties (Young modulus and Poisson coefficient) of the tip and the sample, Garcia et al. could reproduce the two different regimes [32]. Zitzler et al. [80] introduced an additional term taking into account

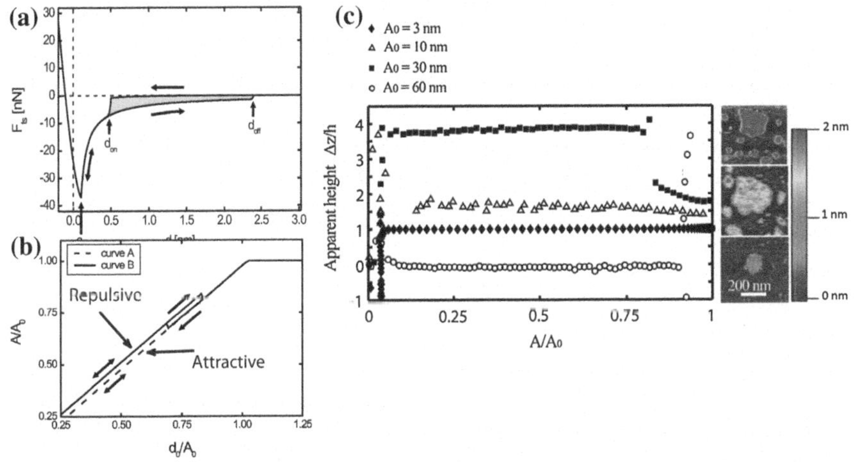

Fig. 14.4 **a** Tip-surface interaction curve used in the model. Capillary forces introduce a hysteresis between $d = 2h$ and $d = d_{\text{off}}$; **b** Calculated Amplitude-distance curve for $A_0 < A_C$ (curve A) and $A_0 > A_C$ (curve B); Reprinted with permission from [80]. Copyright 2002 by the American Physical Society. **c** Calculated apparent height of water patches on $BaF_2(111)$ as a function of imaging conditions A and A_0 and AFM images of such patches. Reprinted with permission from [67]. Copyright 2011 by the IOP Publishing

capillary force. Neglecting capillary condensation and assuming that the meniscus volume comes from the water films adsorbed on the surface, they used the approximated expression of the capillary force reported Eq. 14.4. This force applies from $d = 2h$ where h is the height of the adsorbed water film to $d = d_{\text{off}}$, distance at which the meniscus breaks [75]. The resulting total interaction force is reported in Fig. 14.4a. The formation and rupture of the meniscus introduces an hysteresis in the curve which leads to a dissipated energy W_{diss} given by the area shaded grey. Once introduced in the motion equation, the two different regimes are reproduced as evidenced in the amplitude-distance curve shown in Fig. 14.4b. For free amplitude A_0 smaller than a certain free amplitude A_C, the system remains in the attractive regime in the whole range of tip-surface distances (curve A). For $A_0 > A_C$, the system switches in the repulsive regime when approaching to the surface (curve B). The switching distances are different in the approach and retract curves, which leads to a hysteresis in the force-distance curve characterized by a bistable regime. The influence of relative humidity observed in the experiments is also captured by this model. These results show that imaging conditions (A and A_0 values) and local tip-sample interaction (dominated by capillary effects in ambient conditions) play a crucial role in the oscillation regime in which the AFM is operating. This has two important consequences:

- the quantitative measurement of surface topography requires a good control of imaging parameters. In particular, if materials with different elastic or wetting properties are present on the surface, for a given setpoint, the system may be working in different regimes on both systems which leads to *imaging artefacts* [53]. An example is given in Fig. 14.4c where the calculated apparent height of water patches on a $BaF_2(111)$ sample is plotted as a function of the ratio A/A_0 for different A_0 values [67]. Erroneous height measurements (ranging from fractions of the true height in one extreme to values which are as large as four times the real height in the other) are predicted and observed experimentally (see AFM images in Fig. 14.4c) due to a difference of imaging regime in wet and dry parts of the substrate.

- Since the capillary force curve and consequently the dissipated energy strongly depend on the size of the nanomeniscus condensing at the tip apex, a *mapping of the hydrophobicity of the surface* can be envisioned. This is of particular interest in biological applications where species recognition is often based on hydrophilic/hydrophobic properties. However, as shown above, the dissipation contrast is a complex interplay between the dissipated energy W_{diss} and the bistable motion of the cantilever which may switch between attractive and repulsive regimes. It was shown that P_{diss} given by Eq. 14.6 can be used to quantify hydrophobicity, but this is valid only if the system is operating in repulsive mode on the different parts of the heterogeneous surface under inspection [64].

As a conclusion, AM-AFM mode is a complex imaging mode which requires a lot of attention as soon as quantitative measurements are involved. In this context, capillary effects play a major role. A precise understanding of static and dynamics of meniscus at nanometer scale is therefore needed to interpret the data. Under well-controlled conditions, AM-AFM provides a unique way to access and map a large variety of sample properties.

14.4 Capillary Forces in Liquid Nanodispensing

The liquid nanodispensing (NADIS) method is an AFM-based technique developed recently to deposit ultrasmall liquid quantities on a surface. It brings more versatility in terms of deposited molecules than the "Dip Pen Nanolithography" (DPN) technique which is the reference in the domain of scanning probe lithography (see [9, 33, 45, 65] for reviews on this technique). In DPN, the molecules diffuse from the tip to the surface with the help of the water nanomeniscus which condensates at the tip contact. This rather complex mechanism is not totally elucidated. Spot size is found to increase as the square root of contact time, in a diffusive-like behaviour [63] which strongly depends on relative humidity. However, by nature, DPN is very molecule-dependent. It has been intensively developed in the case of a thioalkane or thiolated oligonucleotides which self-assemble on a gold surface. The obtained patterns are then used as templates to attach, in a second step, the molecules or nano-objects of interest by specific interaction. Direct patterning of large objects such as nanoparticles or proteins may result difficult, and requires specific tip and surface functionalization [76].

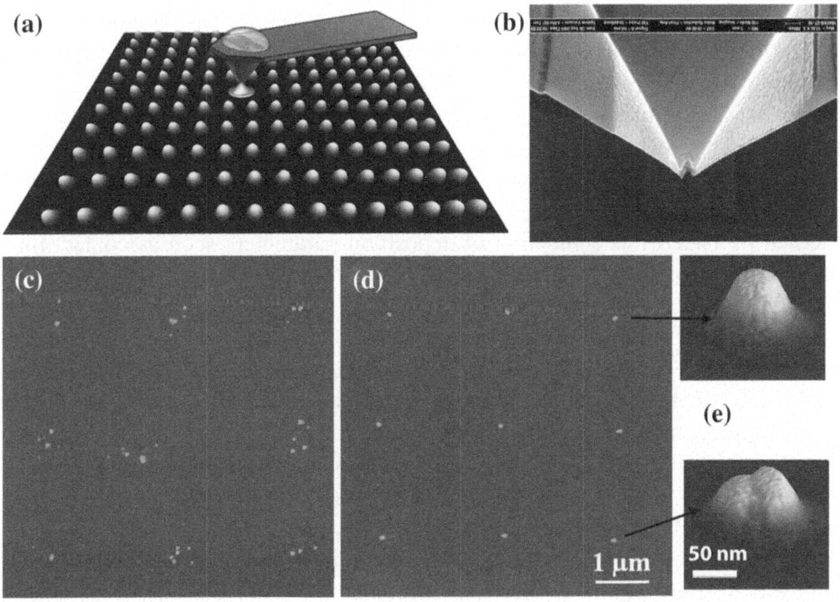

Fig. 14.5 **a** AFM image of a 1 µm pitch array of ionic liquid droplets with a diameter of 600 nm deposited with a NADIS tip (represented schematically). **b** SEM side view of a NADIS tip with a 120 × 70 nm aperture. **c** Array of spots of 25 nm polystyrene nanoparticles obtained with a 300 nm aperture NADIS tip; **d** same with a 130 nm aperture tip; all spots comprise only one or two nanoparticles as shown in the zooms reported in (**e**). This result is rather surprising since a more statistical repartition of nanoparticles, with missing dots, was expected. A concentration of nanoparticles at the tip apex, due to evaporation, may explain this result

NADIS was developed in order to overcome this limitation [50]. Specific AFM tips are used to manipulate and deposit directly liquid solutions (Figs. 14.1b and 14.5a). Doing so, it can be applied to any soluble material, independently of its chemical nature and size. NADIS requires the fabrication of specific AFM tips with a nanochannel drilled at their apex (Fig. 14.5b). The liquid solution is transferred from a reservoir droplet deposited on the cantilever, to the surface through this aperture. Arrays of liquid droplets (or lines) with sizes ranging from 70 nm to several microns were fabricated changing parameters such as the size of the aperture, the tip surface properties and the contact time [28, 29]. Interestingly, for a given tip and surface association, the spot size is only determined by the solvent properties, regardless of the nature of the solute, which is an important point when considering applications. In particular, nano-objects such as nanoparticles or proteins, which are difficult to deposit with DPN, can be patterned routinely with NADIS. Moreover, given the small volumes manipulated by NADIS (in the atto- to femto-litre range), the droplets contains, for standard dilutions a small number of active molecules, thus opening the way to the deposition of individual nano-objects (see Fig. 14.5c–e).

In addition to its application in nanopatterning, NADIS is also a unique tool to perform more fundamental studies on the behaviour of liquids at sub-micron scale. By its ability to manipulate ultra-small liquid quantities, it allows to perform model experiments at sub-micron scale, a range of size which is hardly accessible with other techniques. This was illustrated by a study of the evaporation of micron and sub-micron sessile droplets (femto-droplets) deposited by NADIS on ultrasensitive mass sensors [5]. A recent investigation of the dimension of spots and lines dimensions patterned by NADIS as a function of deposition parameters such as contact time or speed also provided a method to investigate spreading of liquid at sub-micron scale [29].

One important point with the NADIS technique is that the cantilever used to deposit liquid on the surface can also measure the force exerted on the tip by the liquid meniscus. Indeed, the capillary force which is monitored during the whole process (approach and retraction of the tip) is the only real time available signal. Since the shape of the force curves contains information on the deposition process, it is important to analyse them in detail to get some insight on the mechanism of liquid transfer and optimise the method. Conversely, NADIS provides a way to study meniscus with well controlled dimensions, in original situations, and is therefore a useful way to study capillary forces and mechanics of nanomeniscus, at nanometer scale. In the following, we report in detail a study of the capillary forces measured during nanodispensing.

14.4.1 Experimental and Calculation Methods

The experimental details of the NADIS technique have been reported elsewhere [30]. We only mention here the main steps of the procedure. NADIS tips are fabricated starting from hollow pyramidal and sharpened AFM tips (Olympus, OMCL-RC800). These silicon nitride tips are coated by a 50 nm-thick layer of gold. The nanochannel is drilled at the tip apex by Focused Ion Beam (FIB) in two steps: the tip wall is first thinned from the top before the milling of the aperture, performed from the tip side in order to optimize the centering. Channels with diameter as small as 35 nm could be fabricated using this procedure. If needed, the tip outer wall surface properties is tuned using surface functionalization of the gold layer by thiol molecules, in vapor phase.

Using micromanipulation techniques, the tips are loaded with a droplet of about 30 μm used as a reservoir, deposited at the end of the cantilever. The small size of the (open) reservoir limits the range of usable liquids. Only solvents with very low volatility such as glycerol (or glycerol/water mixture) or dibutylphtalate, can be used. Cantilevers with closed channel cantilevers were recently developed in order to address this issue [51]. The loaded tip is then mounted on an AFM. Spots were deposited in force spectroscopy mode, which allows to approach the tip from the surface until a preset contact force is applied, remain in contact for a given contact time, and withdraw the tip with a constant velocity. In the case of line deposition, we used a dedicated setup incorporating a nanopositioning table in the sample holder [29, 55]. After contact with the surface, the tip remains at a fixed position and the substrate is moved at constant velocity (from 0.1 to 580 μm/s). The feedback loop of the AFM is used to maintain a constant force during the line writing. In the experiments reported below, it is important to know precisely the size of the meniscus created during the deposition. With that aim, we used highly concentrated 16 mM solutions of a ruthenium complex [74] in a 90:10 v:v glycerol-water mixture. Previous experiments showed that, in this situation, the patterns observed by standard AFM tapping mode imaging, after solvent evaporation, matched the initial liquid shape [30].

As discussed in Sect. 14.2, force calculations were performed using the Surface Evolver software [16]. Using a well-defined situation with analytical solution, the calculation parameters were optimized to provide a relative error smaller than 10^{-6} while keeping reasonable computation times [28]. All results presented below were then obtained refining the mesh until getting more than 20 000 facets and using a variation of height $\delta z = 10^{-3}R$ for the energy derivation (Eq. 14.1). The method was also assessed in a simple case which consists on a zero pressure liquid meniscus connecting two pads with radius 1 μm separated by a distance z. The result obtained by Surface Evolver with $r_{\text{tip}} = r_{\text{surf}} = 1$ μm as boundary conditions and $P = 0$ as constraint was compared to the one deduced from the catenoïd profile (Eq. 14.3) which is the exact solution at zero pressure. In the latter case the two parameters C and z_0 were calculated for each z value and the force was deduced as the only tension term F_{T} (see Eq. 14.2) with the ϕ value determined from the derivation of the profile at the contact line. The maximum deviation between both curves obtained using energetic (Surface Evolver) and geometric (catenoïd) approaches, respectively, is smaller than 0.4 %, which validates our approach (Fig. 14.6).

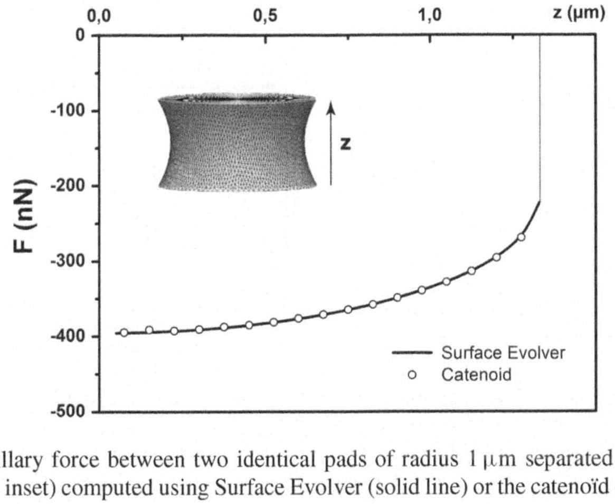

Fig. 14.6 Capillary force between two identical pads of radius 1 μm separated by a distance z (sketched in the inset) computed using Surface Evolver (solid line) or the catenoïd equation (empty symbols)

14.4.2 Examples of Capillary Force Curves

In Fig. 14.7a are reported the force curves measured during the approach and retraction of a loaded NADIS tip at a given point of the surface. The value of the force F is deduced from the cantilever deflection d using $F = kd$ where k is the cantilever spring constant. The actual tip-distance z between tip and surface is also monitored. In the approach phase (a), the tip does not experience any force until a jump to contact is observed. The tip is then pressed on the surface until a preset force (30 nN in the example of Fig. 14.7a) is obtained (b). When withdrawn from the surface, at the beginning, the contact is maintained between tip and surface until an attractive force of 80 nN for which the tip detaches from the substrate. The tip is then retracted with the meniscus bridging tip and surface (c). Note that this large attractive force, which extends over large tip separation distances, is not observed with standard AFM tips on a dry surface. For a separation of about 300 nm, the force abruptly drops to zero, due to the rupture of the liquid bridge.

In the separation phase, the force experienced by the tip is due to the capillary force exerted by the liquid meniscus. van der Waals forces between tip and surface, responsible for the small peak observed at the separation from the surface, are negligible. Viscous forces may also play a role. An estimate of this force leads to $F_v \approx \frac{6\pi\mu R^2}{t_s}$ where μ is the dynamic viscosity of the liquid, R the radius of the contact zone and t_s the separation time [20]. In typical conditions used in NADIS experiments with glycerol ($R = 50$ nm, $t_s = 0.2\,s$ and $\mu = 1.5$ Pa s) this leads to a viscous force in the order of 10^{-13} N which can be neglected in comparison with capillary forces.

Contrary to the approach curve which do not contain a lot of information, the shape of the retraction curve varies considerably as a function of the experimental conditions. The three major types of curves observed in the experiments are reported in Fig. 14.7b.

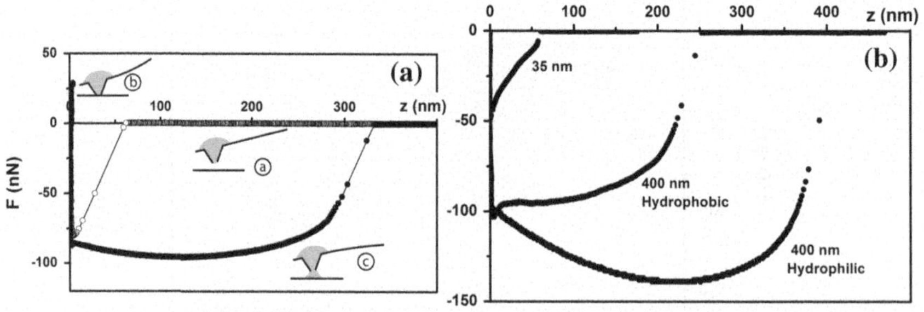

Fig. 14.7 **a** Experimental force curve obtained during the deposition of droplets with a NADIS tip with a 280 nm diameter and sketch of three different stages of the process; Approach and retraction curves are plotted in empty and filled symbols, respectively; **b** Retraction curves obtained with three different tips, the characteristics of which being given on the figure. Adapted from [28]; Copyright 2010 American Chemical Society

- In the case of hydrophilic tip with a channel aperture larger than 200 nm, the amplitude of the force slowly increases and then decreases before the snapoff.
- The same tip, functionalized with dodecanethiol in order to make its outer wall hydrophobic, leads to a different behaviour: the force decreases continuously and then jumps to zero for a tip-surface separation much smaller than in the previous case.
- Curves obtained with a tip with a 35 nm aperture show a different shape, with a concavity which is the opposite of the previous examples.

These examples show that the retraction curves contain a lot of information on the mechanism of liquid transfer. It is therefore important to perform modelisation of these curves in order to get insights on the processes involved in the deposition.

14.4.3 Interpretation of Experimental Force Curves

Liquid nanodispensing using the NADIS method provides a way to study capillary forces in a large variety of experimental conditions, depending on parameters such as the aperture size and tip or substrate surface properties. In order to put into evidence the role of the nanochannel feeding the nanomeniscus which is the originality of NADIS, we first describe a model experiment with standard AFM tips, which is used to assess the force calculation method. Hydrophilic and hydrophobic NADIS tips are then described, emphasizing on the boundary conditions and constraints to be used in the model to describe the experimental results.

14.4.3.1 Standard AFM Tips

We first consider the case of a hydrophilic standard AFM tip (with no channel) dipped in preexisting glycerol droplets with diameter ranging from 1 to a few μm, deposited by NADIS on a hydrophilic surface. The retraction curves obtained with a given tip, withdrawn from three different droplets are given in Fig. 14.8a (black dots).

In this model situation, boundary conditions and constraint can be defined without any ambiguity. Since both tip and surface are hydrophilic (after cleaning by piranha solution), the contact lines are assumed to remain fixed on the solid because there is no dewetting during the retraction. We therefore used fixed contact line radii ($r_{\rm tip}$ and $r_{\rm surf}$) as boundary conditions. Moreover, given the short acquisition time of the force curve (less than 1 s), the volume of the liquid meniscus can be considered as constant, giving the constraint to use in the model.

The results of calculations using these conditions are reported on Fig. 14.8a. We find a good agreement using as parameters $r_{\rm tip} = 1, 0.89$ and $0.63\,\mu$m, $r_{\rm surf} = 2, 1.8$ and $1.26\,\mu$m, $V = 8, 5.64$ and 2 fL, from large to small droplet, respectively. A positive shift in z comprised between 30 and 50 nm is also introduced to take into account the sharpened tip apex (inset of Fig. 14.8a).

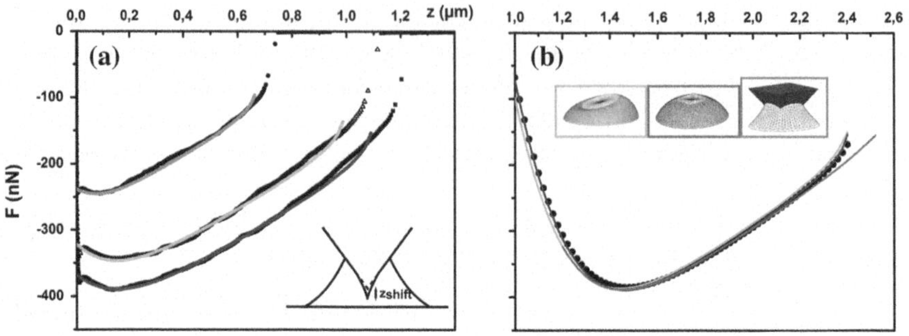

Fig. 14.8 **a** Experimental and calculated force curves obtained when dipping a standard AFM tip in three droplets with different diameters; **b** Force curves computed using various boundary conditions sketched in the insets. Adapted from [28]; Copyright 2010 American Chemical Society

In order to assess the robustness of the model to the precise shape of the AFM tip, we performed, in a given situation, calculations in different configurations: (i) a circular contact line as in the previous case; (ii) a circular contact line inclined by 10° (the inclination of the tip in our AFM setup); (iii) a square contact line (with the same perimeter as the circular one); (iv) a contact line on a pyramidal tip (the contact line was first optimized with a given contact angle at each point, and the resulting shape was then fixed during the retraction phase). The results reported in Fig. 14.8b, show small variations (less than 6 % in the major part of the curve) except at the very end of the separation process. In the following, we therefore considered the simple case of a circular horizontal contact line. This experiment validates our calculation method, which was then applied to more complex situations using NADIS tips.

14.4.3.2 Hydrophilic NADIS Tips

In order to put into evidence the effect of the channel feeding the liquid bridge, we first consider the case of hydrophilic NADIS tips in order to keep the same boundary conditions as in the previous section. We therefore used tips with different aperture diameters to pattern a hydrophilic surface. The case of patterning of hydrophobic surfaces is briefly mentioned at the end of this section.

Influence of Aperture Size

As shown in Fig. 14.7, the diameter of the channel has a strong influence on the shape of the retraction force curve. The two asymptotic behaviours are described in the next paragraphs.

NADIS Tips with Large Apertures (>200 nm)

NADIS tips with a channel diameter larger than 200 nm, lead to smooth curves with a strong attractive force which extends over large distances. An example of such curve obtained with an aperture of 280 nm is reported on Fig. 14.9. The shape of this curve notably differs from the one obtained with standard AFM tips presented in the previous section. Since the boundary condition are identical in both cases, the difference comes from the fact that with NADIS the meniscus is fed by the reservoir, through the channel. With NADIS, the volume of the liquid can not be considered anymore as a constant. For large apertures, we assumed that the flow in the channel is sufficient to insure that the total system is in equilibrium at all times. The pressure in the liquid bridge is therefore imposed by the Laplace pressure of the reservoir and remains constant during the separation process. The radius of curvature of the reservoir, estimated by optical microscopy, was of the order of 20 μm. This leads to a Laplace pressure $P = 6300$ Pa which was used as the constraint in the optimization process.

The results of the simulation reported in Fig. 14.9 show a very good agreement with the experimental data. Since the r_{surf} parameter is determined experimentally as the radius of the molecules spot let on the surface ($r_{surf} = 400$ nm), the only adjustable parameter is r_{tip}. We found $r_{tip} = 250$ nm. The tip is therefore wetted with a diameter of 500 nm much larger than the aperture size (280 nm), consistent with the fact that the tip is hydrophilic. As in the previous section, a shift in z is introduced to take into account the fact that the tip profile differs from a perfect cone. Contrarily to the case of standard AFM tips for which the z_{shift} parameter is positive, it is negative in NADIS case due to the truncated tip apex.

The good agreement observed between experimental and calculated force curves with only one adjustable parameter validates our approach and in particular the

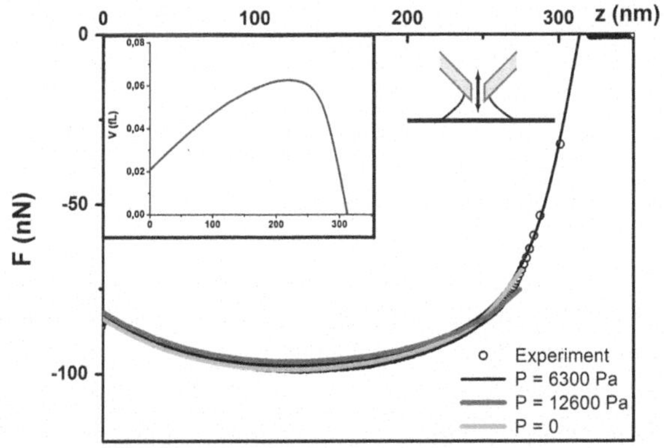

Fig. 14.9 Experimental (empty dots) and calculated force curves with $P = 0$ pressure (red solid line), $P = 6300$ Pa (black solid line) and $P = 12600$ Pa (green solid line) for a NADIS tip with a 280 nm aperture; Inset: Meniscus volume as a function of tip-surface separation z

constant pressure hypothesis. Large NADIS tips are therefore a way to study the capillary force resulting from a liquid meniscus with controlled pressure and size. In order to investigate the influence of the pressure, we compared the force curve with zero pressure (either by surface evolver or using a catenoïd), $P = 6300$ Pa and $P = 12600$Pa (Fig. 14.9a). The difference between these curves is less than 2 % except at the very end of the process. The lack of precise experimental estimate of the Laplace pressure therefore does not change the conclusions of the model. The effect of pressure may be more important in the case of Fluid-FM, for which an external pressure can be directly applied to the meniscus through close-channel cantilever [51].

The modelization can be used to compute the volume of the liquid bridge during the deposition process. The result is shown in the inset of Fig. 14.9. The volume increases by a factor of about 3 before decreasing before the separation. The flow rate in the nanochannel during the first part is of the order of 0.4 fL/s.

NADIS Tips with Small Apertures (<100 nm)

For aperture size smaller than 200 nm, the constant pressure hypothesis does not describe properly the experimental results. Due to the limitation of the flow in the channel by viscous effect the system is not anymore under hydrostatic equilibrium. An example of force curve obtained with a very small aperture (diameter 35 nm) is reported on Fig. 14.10. The force decreases rapidly with separation in contrast with the shape measured with large aperture. We assumed that for very small apertures, the flow in the channel is negligible, resulting in a meniscus with constant volume. We therefore performed the modelling with a fixed volume V which is not known experimentally and was used as an adjustable parameter. The adjustment of the experimental curve by the model is reported on Fig. 14.10. The parameters used were $r_{surf} = 95$ nm (determined experimentally), $r_{tip} = 65$ nm and $V = 0.4$ aL. Again, we find a very good agreement between model and experiments which validates our hypothesis of a negligible flow in the channel. A negative z_{shift} was also applied to take into account the truncated tip apex.

Transition Between Regimes

The significant difference between force curves as a function of the aperture size emphasizes the importance of the liquid flow in the channel. Two asymptotic behaviours were obtained: a free flow in the channel for large apertures, and a negligible flow for small ones. In order to validate this conclusion, we estimated the flow in the channel assuming a Poiseuille flow resulting from a pressure difference Δp between both extremities of the channel. The flow rate Q_P is given by $Q_P = \frac{\Delta p \pi a^4}{8 \mu L}$ where a and L are the channel radius and length, respectively. Assuming, in a rough approximation, a negligible pressure in the meniscus and a $P = 6300$ Pa pressure in the

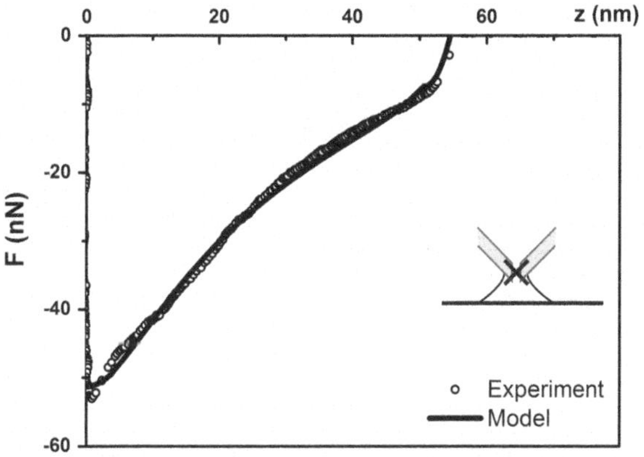

Fig. 14.10 Experimental (empty symbols) and calculated (solid line) force curves obtained with a NADIS tip with a diameter of 35 nm; inset: sketch of the condition used in the calculation. Adapted from [28]; Copyright 2010 American Chemical Society

reservoir, we obtain a maximum flow rate provided by the reservoir equal to $Q_P = 3$ fL/s for an aperture of 280 nm and $Q_P = 0.7$ aL/s for an aperture of 35 nm. As shown in Fig. 14.9 we can also estimate the flow rate needed to maintain a constant pressure in the meniscus. For the large aperture, we find $Q = 0.4$ fL/s, noticeably smaller than the available Q_P, consistent with the fact that the pressure can be maintained constant in the meniscus. On the contrary, for the 35 nm aperture, the flow rate required to maintain a constant pressure $Q = 5$ aL/s can not be provided by the Poiseuille flow ($Q_P = 0.7$ aL), justifying the constant volume hypothesis. Note that since the channel wall is hydrophilic, the effect of slippage should remain negligible [14] even in the smallest nanochannels, which justifies the use of Poiseuille equation in this estimation.

Components of the Capillary Force

In order to compare in greater detail the two different situations reported above, we plotted on Fig. 14.11, the calculated curves of Figs. 14.9 and 14.10 normalized by $2\pi\gamma r_{\text{tip}}$ for the force and by r_{tip} for z. In this representation, we observe that the constant volume situation leads to much higher forces over a smaller extension than the constant pressure case (see Fig. 14.11b for snapshots of meniscus profiles as function of tip height). As shown in Eq. 14.2, the capillary force can be decomposed in two components which can be extracted from the calculations by Surface Evolver and are reported on Fig. 14.11. For large apertures, the Laplace term is negligible (less than 5 % of the total force). On the contrary, it plays a major role in the constant volume regime. Indeed, it shows that a negative pressure builds in the meniscus and then decreases to zero. The magnitude of the Laplace force is comparable to the tension term and is predominant during a large part of the separation process. The

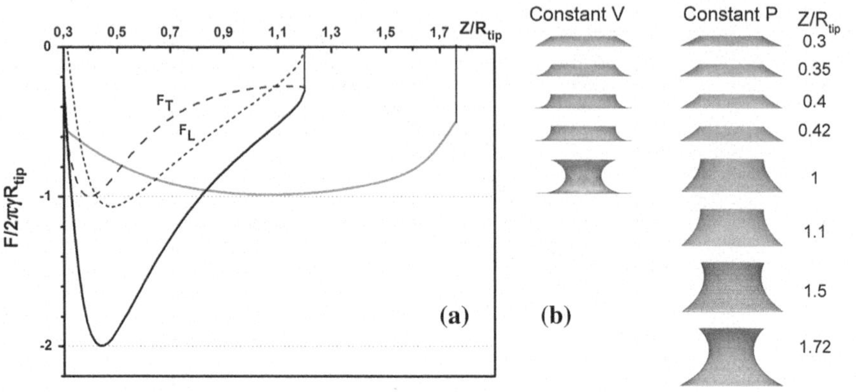

Fig. 14.11 **a** Normalized force curves calculated for constant pressure (grey solid line) and constant volume constraint (black solid line). In the latter case, the Laplace (black dotted line) and tension (black dashed line) components are also represented. **b** Shapes of menisci calculated with Surface Evolver in both conditions for several tip heights. Adapted from [28]; Copyright 2010 American Chemical Society

addition of these two components explains the large normalized forces observed for small apertures, an effect which may be of importance in the case of patterning of soft substrates.

Real-Time Monitoring of Droplet Size

As shown above, the capillary force depends on several parameters, among which the size of the meniscus on the surface r_{surf}. This parameter is important as it gives the size of the resulting spot. We therefore tried to correlate this parameter r_{surf} with the force curve which is the only available quantity during deposition. We chose the tip-surface separation z_{rupt} at meniscus rupture, which is an easily measured quantity. We studied the relationship between z_{rupt} and r_{surf} for a series of spots deposited with a large aperture NADIS tip, changing the contact time. It was shown that, in the regime of femtolitre droplets, the contact time plays an important role in the dimension of the spot [30]. A $R \approx t^{1/4}$ power law was evidenced at short contact times, relationship which could be interpreted as the result of the spreading of liquid on the surface under imposed pressure [29]. A series of droplets with diameters ranging from 250 to 500 nm, was obtained. For each spot, we measured the radius r_{surf} on the AFM image and the z_{rupt} value on the corresponding force curve as the tip altitude for the meniscus breakage (see Fig. 14.12). The collected data show that z_{rupt} increases by 50 % for r_{surf} increasing by a factor 2. In order to interpret this curve, we modelled force curves, with pressure constraint ($P = 6300$ Pa), for a given r_{tip} value, varying r_{surf}, and determined in each curve the z_{rupt} value. As shown in Fig. 14.12, a very good agreement is found with the experimental data, with r_{tip} as

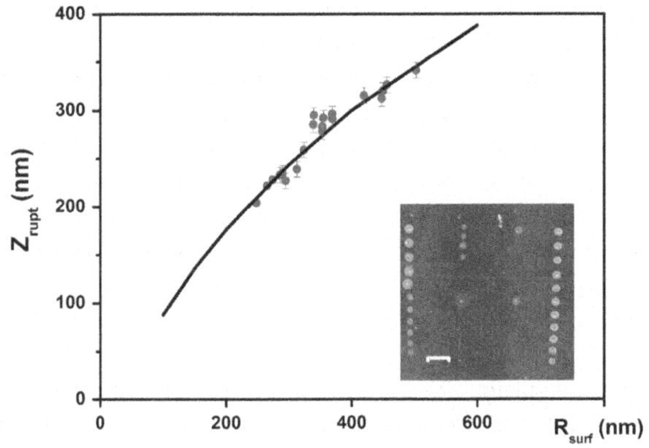

Fig. 14.12 Experimental (symbols) and computed (solid line) values of z_{rupt} as a function of r_{surf}; inset: AFM images of spots obtained with different contact times, used to determine the values of the radius r_{surf}. Adapted with permission from [28]; Copyright 2010 American Chemical Society

the only adjustable parameter. In the case of Fig. 14.12, a value $r_{tip} = 250\,\text{nm}$ was used.

The good agreement between experimental and simulated force curves therefore provides a simple method for real time monitoring of the deposition. From the value of z_{rupt} which can be measured instantaneously on the curves, the size of the droplet can be estimated, and the deposition parameters adjusted in order to obtain the desired spot size. This is an important point for applications.

Influence of Surface Properties

In the studies presented above, we considered hydrophilic surfaces. In this condition, we assumed that the contact line was fixed on the surface during the retraction of the tip. We also deposited droplets on hydrophobic surface ($\theta_{av} = 90°$) obtained by chemical functionalization by a fluorinated silane (1H1H2H2H-perfluorodecyl-trichlorosilane). The main characteristics are spot size smaller than the tip aperture and a force curve without any signature of a liquid bridge [30]: the tip detaches suddenly from the surface with no feature on the curve which could be attributed to the presence of a meniscus. In order to interpret these facts, we performed calculations using a constant contact angle θ_{surf} as boundary condition. The radius of the meniscus r_{surf} can therefore change. The other boundary condition (r_{tip} fixed) and constraint (constant P) are kept identical to previous cases. The resulting $F(z)$ and $r_{surf}(z)$ are reported on Fig. 14.13. We observe that for contact angle larger than 70°, the maximum of force magnitude disappears: the force decreases continuously, with a slope which increases with contact angle. For contact angles larger than 80°, this slope is larger than the cantilever spring constant. This leads to a mechanical instability which explains the sudden detachment of the tip from the surface observed in the

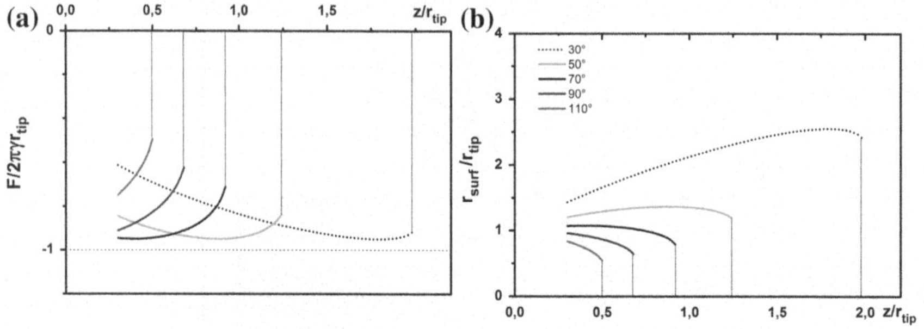

Fig. 14.13 **a** Computed normalized force curves for a hydrophilic tip patterning substrates with different equilibrium contact angles. **b** Plot of the radius of the meniscus on the substrate $r_{surf}(z)$ as a function of tip-surface separation z, both normalized by $r_{tip}(z)$

experiments. Stiffer cantilevers would be required to study the capillary force due to the meniscus in the latter cases.

The computation of $r_{surf}(z)$ reported in Fig. 14.13b shows that the size of the meniscus on the surface follows different trends depending on contact angle. For contact angles smaller than 50°, r_{surf} increases continuously and finally breaks for a value which may be 2.5 larger than r_{tip}. For $\theta_{surf} = 50°$, r_{surf} does not evolve a lot during retraction, while for values smaller than 50° it decreases continuously. Therefore, on hydrophobic surfaces, the final size of the deposit can be much smaller than the aperture size which is coherent with the experimental observations [30]. However, quantitative comparison are difficult since, in the ultimate stage of meniscus failure, dynamic effects play an important role [60, 72].

14.4.3.3 Hydrophobic Tips

In order to decrease the size of the droplets and reach ultimate dimensions, the tips are functionalized using dodecanethiol. This hydrophobic surface treatment prevents the spreading of liquid on the tip outer wall, thus confining the deposition at the tip apex. As shown on Fig. 14.7b, the tip functionalization also drastically changes the shape of the force curve. The force decreases monotonously before it jumps to zero for a z value much smaller than with the same hydrophilic tip. In order to model this situation for a large aperture tip (400 nm for the example of Fig. 14.7b), we changed the boundary condition on the (conical) tip. We used a fixed contact angle θ_{tip} instead of a fixed radius r_{tip}, for the hydrophilic tip. In this case, the liquid can dewet on the tip during the retraction, thus changing the r_{tip} value. Computed force curves when changing θ_{tip} are reported on Fig. 14.14a. We observe that the magnitude of the force decreases monotonously during the separation process, in qualitative agreement with the experimental observation. The slope is rather independent on θ_{tip} while z_{rupt} decreases significantly when increasing contact angle.

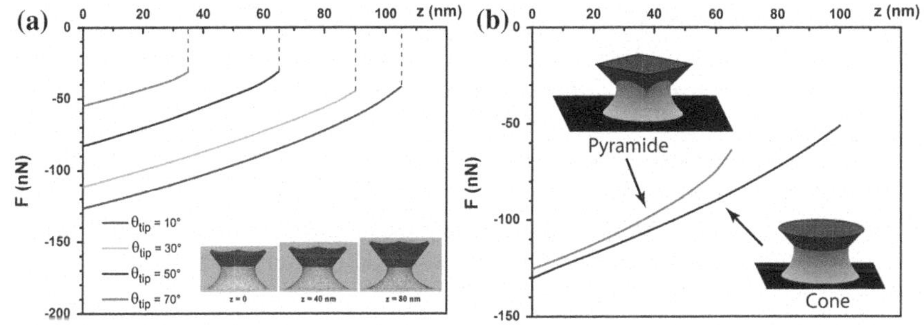

Fig. 14.14 **a** Calculated force curves obtained on a hydrophilic surface (contact line pinned on the substrate) changing the contact angle of the liquid on the tip; inset: Snapshots of meniscus profile at different heights, for $\theta_{tip} = 30°$; **b** Comparison of force curves obtained with a cone and a pyramid. Adapted from [28]; Copyright 2010 American Chemical Society

In Fig. 14.14a are also reported snapshots of the meniscus shape during separation. It shows that the height h of the meniscus remains rather constant. The radius of the contact line r_{tip} therefore decreases linearly as $r_{tip} = (h - z)tan\alpha$ where α is the half-angle of the cone. Since, for large apertures, the total capillary force is nearly equal to the tension term $F_T = -2\pi r_{tip}\gamma \cos \Phi$ (see Eq. 14.2), it is expected to decrease linearly with z.

This model gives a qualitative agreement with the experiments but fails to give good quantitative description of the force curves. Portions of the curves can be fitted but it is difficult to find conditions leading to a good adjustement over the entire curve. Several reasons may explain this discrepancy. Contact angle hysteresis is probably important on the tip, due to possible damage of the gold layer by FIB. In this case, the contact line may be pinned during one part of the retraction curve before receding on the rest of the detachment. Another explanation comes from the fact that, contrary to the hydrophilic tip case (see Fig. 14.8), the precise shape of the tip plays an important role. In Fig. 14.14b are reported the force curves computed for cylindrical and pyramidal tips. In the case of the pyramidal tip, the decrease of force is more pronounced and the bridge rupture occurs for smaller tip-surface separation.

As a result, the motion of the contact line on the hydrophobic tip is a rather complex dynamics which is difficult to describe precisely with our simple method.

14.5 Conclusion

In this chapter, we described the great importance of capillary forces in AFM-based techniques. Although they can be detrimental for imaging resolution or lead to important artefacts in measured height, they can also be used to map surface properties such as hydrophobicity with high resolution. In all cases, a precise understanding of the statics and dynamics of the nanomeniscus is required to extract quantitative information from this technique. AFM also provides a model situation to study the elementary processes involved in tribology, by using the tip as a well-controlled nanocontact. Non conventional effects of relative humidity can be observed provided the tip geometry can be controlled at nanometer scale. The nanomeniscus can also be created on purpose at the tip apex in order to pattern surfaces as in the liquid nanodispensing technique. We have reported a comprehensive study of the capillary force experienced by the AFM tip showing a large variety of behaviors which can be accounted for using numerical simulations based on the Surface Evolver software. The good agreement between experimental and calculated force curves gives a precise description of the process of liquid nanodispensing which helps to optimize tip properties for a desired spot size.

The results presented in this chapter can be interpreted assuming that a meniscus is at equilibrium at each time. In order to get a more complete description, it would be useful to introduce dynamic aspects of the nanomeniscii. Moreover the models used are based on a macroscopic description of the liquid as a continuous medium. However, this hypothesis becomes questionable when meniscus are downscaled to a few nanometer in size. Numerical simulations using lattice gas model and mean-field density functional theories, where used to study the influence of relative humidity on the pull-off force of an AFM tip on a surface [39] or the question of the minimal size of a meniscus [40]. These numerical results suggest that while giving a general evolution in agreement with the experimental data, calculations using macroscopic descriptions may be questionable below contact diameter smaller than 8 nm. Such questions should stimulate experiments at dimensions which can be within reach using AFM-based techniques.

References

1. A.W. Adamson, A.P. Gast, *Physical Chemistry of Surfaces*, 6th edn. (John Wiley and sons, New York, 1997)
2. H. Allouche, M. Monthioux, Chemical vapor deposition of pyrolytic carbon on carbon nanotubes. Part 2. Texture and structure. Carbon **43**(6), 1265–1278 (2005)
3. B. Anczykowski, D. Kruger, H. Fuchs, Cantilever dynamics in quasinoncontact force microscopy: spectroscopic aspects. Phys. Rev. B **53**(23), 15485–15488 (1996)
4. Y. Ando, The effect of relative humidity on friction and pull-off forces measured on submicron-size asperity arrays. Wear **238**(1), 12–19 (2000)
5. J. Arcamone, E. Dujardin, G. Rius, F. Perez-Murano, T. Ondarcuhu, Evaporation of femtoliter sessile droplets monitored with nanomechanical mass sensors. J. Phys. Chem. B **111**(45), 13020–13027 (2007)
6. D.B. Asay, S.H. Kim, Effects of adsorbed water layer structure on adhesion force of silicon oxide nanoasperity contact in humid ambient. J. Chem. Phys. **124**(17), 174712 (2006)
7. A.H. Barber, S.R. Cohen, H.D. Wagner, Static and dynamic wetting measurements of single carbon nanotubes. Phys. Rev. Lett. **92**(18), 186103 (2004)
8. A.H. Barber, S.R. Cohen, H.D. Wagner, External and internal wetting of carbon nanotubes with organic liquids. Phy. Rev. B, **71**(11), 115443 (2005)
9. B. Basnar, I. Willner, Dip-pen-nanolithographic patterning of metallic, semiconductor, and metal oxide nanostructures on surfaces. Small **5**(1), 28–44 (2009)

10. C. Bernard, J.P. Aime, S. Marsaudon, R. Levy, A.M. Bonnot, C. Nguyen, D. Mariolle, F. Bertin, A. Chabli, Drying nano particles solution on an oscillating tip at an air liquid interface: what we can learn, what we can do. Nanoscale Res. Lett. **2**(7), 309–318 (2007)
11. S. Biggs, R.G. Cain, R.R. Dagastine, N.W. Page, Direct measurements of the adhesion between a glass particle and a glass surface in a humid atmosphere. J. Adhes. Sci. Technol. **16**(7), 869–885 (2002)
12. G. Binnig, C.F. Quate, C. Gerber, Atomic force microscope. Phys. Rev. Lett. **56**(9), 930–933 (1986)
13. L. Bocquet, E. Charlaix, S. Ciliberto, J. Crassous, Moisture-induced ageing in granular media and the kinetics of capillary condensation. Nature **396**(6713), 735–737 (1998)
14. D. Huang, C. Sendner, D. Horinek, R. Netz, L. Bocquet, Water slippage versus contact angle: a quasi-universal relationship. PRL **101**(22), 226101 (2008)
15. D. Bonn, J. Eggers, J. Indekeu, J. Meunier, E. Rolley, Wetting and spreading. Rev. Mod. Phys. **81**(2), 739–805 (2009)
16. K.A. Brakke, The surface evolver software. Exp. Math. **1**(2), 141–165 (1992)
17. H.J. Butt, Capillary forces: influence of roughness and heterogeneity. Langmuir **24**(9), 4715–4721 (2008)
18. H.J. Butt, M. Kappl, Normal capillary forces. Adv. Colloid Interface Sci. **146**(1–2), 48–60 (2009)
19. H.J. Butt, B. Cappella, M. Kappl, Force measurements with the atomic force microscope: technique, interpretation and applications. Surf. Sci. Rep. **59**(1–6), 1–152 (2005)
20. S. Cai, B. Bhushan, Meniscus and viscous forces during separation of hydrophilic and hydrophobic surfaces with liquid-mediated contacts. Mater. Sci. Eng. R-Rep. **61**(1–6), 78–106 (2008)
21. I. Casuso, F. Rico, S. Scheuring, High-speed atomic force microscopy: structure and dynamics of single proteins. Curr. Opin. Chem. Biol. **15**(5), 704–709 (2011)
22. J.P. Cleveland, B. Anczykowski, A.E. Schmid, V.B. Elings, Energy dissipation in tapping-mode atomic force microscopy. Appl. Phys. Lett. **72**(20), 2613–2615 (1998)
23. P.G. de Gennes, Wetting—statics and dynamics. Rev. Mod. Phys. **57**(3), 827–863 (1985)
24. A. de Lazzer, M. Dreyer, H.J. Rath, Particle-surface capillary forces. Langmuir **15**(13), 4551–4559 (1999)
25. E.J. De Souza, M. Brinkmann, C. Mohrdieck, A. Crosby, E. Arzt, Capillary forces between chemically different substrates. Langmuir **24**(18), 10161–10168 (2008)
26. E.J. De Souza, L.C. Gao, T.J. McCarthy, E. Arzt, A.J. Crosby, Effect of contact angle hysteresis on the measurement of capillary forces. Langmuir **24**(4), 1391–1396 (2008)
27. M. Delmas, M. Monthioux, Th Ondarçuhu, Contact angle hysteresis at the nanometer scale. Phys. Rev. Lett. **106**(13), 136102 (2011)
28. L. Fabie, H. Durou, T. Ondarçuhu, Capillary forces during liquid nanodispensing. Langmuir **26**(3), 1870–1878 (2010)
29. L. Fabie, T. Ondarcuhu, Writing with liquid using a nanodispenser: spreading dynamis at the sub-micron scale. Soft Matt. 8(18) 4995–5001 (2012)
30. A.P. Fang, E. Dujardin, T. Ondarcuhu, Control of droplet size in liquid nanodispensing. Nano Lett. **6**(10), 2368–2374 (2006)
31. P.L.T.M. Frederix, P.D. Bosshart, A. Engel, Atomic force microscopy of biological membranes. Biophys. J. **96**(2), 329–338 (2009)
32. R. Garcia, R. Perez, Dynamic atomic force microscopy methods. Surf. Sci. Rep. **47**(6–8), 197–301 (2002)
33. D.S. Ginger, H. Zhang, C.A. Mirkin, The evolution of dip-pen nanolithography. Angew. Chem. Int. Ed. **43**(1), 30–45 (2004)
34. W. Gulbinski, D. Pailharey, T. Suszko, Y. Mathey, Study of the influence of adsorbed water on AFM friction measurements on molybdenum trioxide thin films. Surf. Sci. **475**(1–3), 149–158 (2001)
35. J. Hu, X.D. Xiao, D.F. Ogletree, M. Salmeron, Imaging the condensation and evaporation of molecularly thin films of water with nanometer. Science **268**(5208), 267–269 (1995)

36. J.N. Israelachvili, *Intermolecular and Surface Forces* (Academic Press, New York, 1991)
37. C. Jai, J.P. Aimé, R. Boisgard, Dynamical behavior of an evaporating nanomeniscus: a boundary condition problem at the local scale. Europhys. Lett. **81**(3), 34003 (2008)
38. C. Jai, J.P. Aime, D. Mariolle, R. Boisgard, F. Bertin, Wetting an oscillating nanoneedle to image an air-liquid interface at the nanometer scale: dynamical behavior of a nanomeniscus. Nano Lett. **6**(11), 2554–2560 (2006)
39. J.K. Jang, G.C. Schatz, M.A. Ratner, Capillary force in atomic force microscopy. J. Chem. Phys. **120**(3), 1157–1160 (2004)
40. J.Y. Jang, G.C. Schatz, M.A. Ratner, How narrow can a meniscus be? Phys. Rev. Lett. **92**(8), 085504 (2004)
41. M. Jaschke, H.J. Butt, Deposition of organic material by the tip of a scanning force microscope. Langmuir **11**(4), 1061–1064 (1995)
42. J.F. Joanny, P.G. de Gennes, A model for contact-angle hysteresis. J. Chem. Phys. **81**(1), 552–562 (1984)
43. R. Jones, H.M. Pollock, J.A.S. Cleaver, C.S. Hodges, Adhesion forces between glass and silicon surfaces in air studied by -AFM: effects of relative humidity, particle size, roughness, and surface treatment. Langmuir **18**(21), 8045–8055 (2002)
44. M. Kober, E. Sahagun, P. Garcia-Mochales, F. Briones, M. Luna, J.J. Saenz, Nanogeometry matters: unexpected decrease of capillary adhesion forces with increasing relative humidity. Small **6**(23), 2725–2730 (2010)
45. S. Kramer, R.R. Fuierer, C.B. Gorman, Scanning probe lithography using self-assembled monolayers. Chem. Rev. **103**(11), 4367–4418 (2003)
46. P. Lambert, *Capillary Forces in Microassembly* (Springer, New York, 2007)
47. P. Lambert, A. Chau, A. Delchambre, S. Regnier, Comparison between two capillary forces models. Langmuir **24**(7), 3157–3163 (2008)
48. P. Lambert, A. Delchambre, Parameters ruling capillary forces at the submillimetric scale. Langmuir **21**(21), 9537–9543 (2005)
49. D.L. Malotky, M.K. Chaudhury, Investigation of capillary forces using atomic force microscopy. Langmuir **17**(25), 7823–7829 (2001)
50. A. Meister, M. Liley, J. Brugger, R. Pugin, H. Heinzelmann, Nanodispenser for attoliter volume deposition using atomic force microscopy probes modified by focused-ion-beam milling. Appl. Phys. Lett. **85**(25), 6260–6262 (2004)
51. A. Meister, M. Gabi, P. Behr, P. Studer, J. Voeroes, P. Niedermann, J. Bitterli, J. Polesel-Maris, M. Liley, H. Heinzelmann, T. Zambelli, FluidFM: combining atomic force microscopy and nanofluidics in a universal liquid delivery system for single cell applications and beyond. Nano Lett. **9**(6), 2501–2507 (2009)
52. A. Mendez-Vilas, A.B. Jodar-Reyes, M.L. Gonzalez-Martin, Ultrasmall liquid droplets on solid surfaces: production, imaging, and relevance for current wetting research. Small **5**(12), 1366–1390 (2009)
53. P. Nemes-Incze, Z. OsvÃąth, K. KamarÃąs, L.P. BirÃş, Anomalies in thickness measurements of graphene and few layer graphite crystals by tapping mode atomic force microscopy. Carbon **46**(11), 1435–1442 (2008)
54. O. Noel, P.-E. Mazeran, H. Nasrallah, Sliding velocity dependence of adhesion in a nanometer-sized contact. Phys. Rev. Lett. **108**, 015503 (2012)
55. T. Ondarcuhu, L. Nicu, S. Cholet, C. Bergaud, S. Gerdes, C. Joachim, A metallic microcantilever electric contact probe array incorporated in an atomic force microscope. Rev. Sci. Instrum. **71**(5), 2087–2093 (2000)
56. F.M. Orr, L.E. Scriven, A.P. Rivas, Pendular rings between solids— meniscus properties and capillary force. J. Fluid Mech. **67**(FEB25), 723–742 (1975)
57. Q. Ouyang, K. Ishida, K. Okada, Investigation of micro-adhesion by atomic force microscopy. Appl. Surf. Sci. **169**, 644–648 (2001)
58. X. Pepin, D. Rossetti, S.M. Iveson, S.J.R. Simons, Modeling the evolution and rupture of pendular liquid bridges in the presence of large wetting hysteresis. J. Colloid Interface Sci. **232**(2), 289–297 (2000)

59. R.D. Piner, J. Zhu, F. Xu, S.H. Hong, C.A. Mirkin, "Dip-pen" nanolithography. Science **283**(5402), 661–663 (1999)
60. B. Qian, M. Loureiro, D.A. Gagnon, A. Tripathi, K.S. Breuer, Micron-scale droplet deposition on a hydrophobic surface using a retreating syringe. Phys. Rev. Lett. **102**(16), 164502 (2009)
61. Y.I. Rabinovich, M.S. Esayanur, B.M. Moudgil, Capillary forces between two spheres with a fixed volume liquid bridge: theory and experiment. Langmuir **21**(24), 10992–10997 (2005)
62. F. Restagno, L. Bocquet, T. Biben, Metastability and nucleation in capillary condensation. Phys. Rev. Lett. **84**(11), 2433–2436 (2000)
63. S. Rozhok, R. Piner, C.A. Mirkin, Dip-pen nanolithography: âĂL' what controls ink transport? J. Phys. Chem. B **107**(3), 751–757 (2003)
64. E. Sahagun, P. Garcia-Mochales, G.M. Sacha, J.J. Saenz, Energy dissipation due to capillary interactions: hydrophobicity maps in force microscopy. Phys. Rev. Lett. **98**(17), 085504 (2007)
65. K. Salaita, Y.H. Wang, C.A. Mirkin, Applications of dip-pen nanolithography. Nat. Nanotechnol. **2**(3), 145–155 (2007)
66. K. Salaita, Y. Wang, J. Fragala, R.A. Vega, C. Liu, C.A. Mirkin, Massively parallel dip-pen nanolithography with 55000-pen two-dimensional arrays. Angew. Chem. Int. Ed. **45**(43), 7220–7223 (2006)
67. S. Santos, A. Verdaguer, T. Souier, N.H. Thomson, M. Chiesa, Measuring the true height of water films on surfaces. Nanotechnology **22**(46), 465705 (2011)
68. D.L. Sedin, K.L. Rowlen, Adhesion forces measured by atomic force microscopy in humid air. Anal. Chem. **72**(10), 2183–2189 (2000)
69. R. Szoszkiewicz, E. Riedo, Nucleation time of nanoscale water bridges. Phys. Rev. Lett. **95**(13), 135502 (2005)
70. N.R. Tas, P. Mela, T. Kramer, J.W. Berenschot, A. van den Berg, Capillarity induced negative pressure of water plugs in nanochannels. Nano Lett. **3**(11), 1537–1540 (2003)
71. A.A. Tseng, A. Notargiacomo, T.P. Chen, Nanofabrication by scanning probe microscope lithography: a review. J. Vac. Sci. Technol. B **23**(3), 877–894 (2005)
72. S. Uemura, M. Stjernstrom, J. Sjodahl, J. Roeraade, Picoliter droplet formation on thin optical fiber tips. Langmuir **22**(24), 10272–10276 (2006)
73. J.W. van Honschoten, J.W. Berenschot, T. Ondarcuhu, R.G.P. Sanders, J. Sundaram, M. Elwenspoek, N.R. Tas, Elastocapillary fabrication of three-dimensional microstructures. Appl. Phys. Lett. **97**(1), 014103 (2010)
74. C. Viala, C. Coudret, An expeditious route to cis-Ru(bpy)(2)Cl-2 (bpy = 2, 2 '-bipyridine) using carbohydrates as reducers. Inor. Chim. Acta **359**(3), 984–989 (2006)
75. Ch.D Willett, M.J. Adams, S.A. Johnson, J.P.K. Seville, Capillary bridges between two spherical bodies. Langmuir **16**(24), 9396–9405 (2000)
76. C.-C. Wu, D.N. Reinhoudt, C. Otto, V. Subramaniam, A.H. Velders, Strategies for patterning biomolecules with dip-pen nanolithography. Small **7**(8), 989–1002 (2011)
77. X.D. Xiao, L.M. Qian, Investigation of humidity-dependent capillary force. Langmuir **16**(21), 8153–8158 (2000)
78. X.M. Xiong, S.O. Guo, Z.L. Xu, P. Sheng, P. Tong, Development of an atomic-force-microscope-based hanging-fiber rheometer for interfacial microrheology. Phys. Rev. E **80**(6), 061604 (2009)
79. M.M. Yazdanpanah, M. Hosseini, S. Pabba, S.M. Berry, V.V. Dobrokhotov, A. Safir, R.S. Keynton, R.W. Cohn, Micro-wilhelmy and related liquid property measurements using constant-diameter nanoneedle-tipped atomic force microscope probes. Langmuir **24**(23), 13753–13764 (2008)
80. L. Zitzler, S. Herminghaus, F. Mugele, Capillary forces in tapping mode atomic force microscopy. Phys. Rev. B **66**(15), 155436 (2002)

Part V
Conclusions and Perspectives

Chapter 15
Conclusions and Perspectives

Pierre Lambert

These general conclusions fall into three parts: a summary of the contributions of this book, current trends in the field of mechanical applications of surface tension and related perspectives.

15.1 Conclusions

The preface of this book announced the description of surface tensions effects to provide mechanical functions in miniaturized products.

PART I briefly depicted the physical background of the researches presented in this book: scaling laws, surface tension, Laplace's law, interfacial energies and contact angles, as well as control parameters such as temperature or electric field.

PART II devised on static modelling of capillary forces and torques, detailing modelling and computational approaches, boundary conditions, and the coupling with gas compliance, leading to the concepts of **capillary motors** and **bubble robotics**. Note well that so-called immersion forces have not been devised here: let us mention the reference by Kralchevsky [16], which was recently numerically revisited by Cooray et al. [5].

PART III tried to give a possible description of dynamical modelling of liquid bridges, as well for axial elongation of resonating droplets as for lateral shifts towards self-centering applications. Despite a theoretically infinite number of degrees-of-freedom, it could be shown that provided some assumptions, such liquid bridges could be interestingly simplified into second-order systems, well known by mechanical engineers and mass-spring-dashpot systems. It was shown in Fig. 7.17 for axial liquid bridges how the Reynolds and Capillary numbers were good predictors for the

P. Lambert (✉)
BEAMS Department, Université libre de Bruxelles, CP 165/56,
Avenue F.D. Roosevelt, 50, 1050 Brussels, Belgium
e-mail: pierre.lambert@ulb.ac.be

P. Lambert (ed.), *Surface Tension in Microsystems*, Microtechnology and MEMS,
DOI: 10.1007/978-3-642-37552-1_15,

so-called *k*-, *b*- and *m*-states (respectively elastic, viscous and inertial regimes). The results of dynamic lateral forces could be summarized in a non-dimensional map provided in Fig. 8.8.

Finally, **PART IV** presented most case studies, such as the **capillary micromotor** (Chap. 10), **capillary-based sealing** (Chap. 11), **surface-tension driven self-assembly** (Chap. 12), **surface-tension driven actuation** and **energy harvesting** (Chap. 13) and **capillary forces in nanodispensing** (Chap. 14).

15.2 Current Trends

Mechanical applications of surface tension can be further developed using menisci as transducers. Control of surface tension effects will then be required, which can be done with an electric field (electrowetting), an optical actuation [10], a thermal actuation leading to Marangoni-driven micromotors [11], etc. According to Knospe [14], 'capillary force actuation is a new technology holding particular promise for microelectromechanical systems. Relying upon the ability of electrical fields to alter the capillary pressure in a liquid, this technology permits larger actuation forces and strokes than conventional microactuators'. It becomes then necessary to model not only the mechanical behaviour of the liquid bridge (see Chap. 7), but the whole electromechanical behaviour of the actuator, as shown in Fig. 15.1. Combination of mechanical effects of liquids (surface tension and visocity) with their electrical response (capacitive and resistive ones) towards the development of new actuators and sensors [7].

Beside transducers, surface tension can also be applied in pumping devices, as shown by Paust to supply methanol fuel cells [20], by Shabani [22] or Zhang [27].

Another range of possible applications is to use the mechanical models described in this book to develop liquids characterization platforms. Instead of using meniscus viscosity to deduce reponse times such as described in Chap. 8, response time could be used to deduce viscosity. Many other geometries can be envisaged, leading for instance to the development of a miniature capillary breakup extensional rheometer by electrostatically assisted generation of liquid filaments [20].

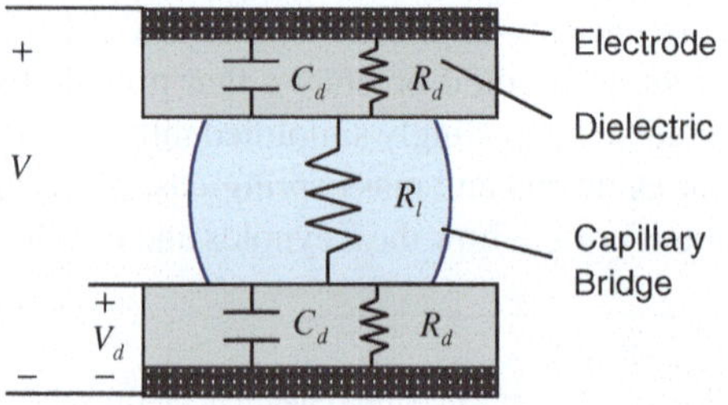

Fig. 15.1 Electrical model of the capillary actuator proposed by Knospe. Reprinted from [15], Copyright (2012), with permission from Elsevier

Deformation of solid surfaces due to capillary forces was already mentioned by Lester in 1961 [17] and later on coupling between capillary effects and materials elasticity was reported by Fortes in 1984 [8]. In 1993, Mastrangelo [18, 19] studied the mechanical stability and adhesion of microstructures under capillary forces. More recently, under Bico's impulse, this domain was founded as the so-called elastocapillary [3]. This domain includes questions such as capillary wrapping, capillary origami and various elasto-capillary systems. Ongoing studies address the dynamics of capillary rise between flexible sheets [4]. These development find applications in the health domain (lungs collapse), in the capillary self-assembly of ultra thin chips (UTC), and in manufacturing of nanostructures (see also hereafter).

Self-centering effects are also extensively studied towards microsystems packaging, let us only mention Berthier's work [2] who generalized the results of Chap. 8: the capillary alignment stiffness is governed by the ratio of the half wetted perimeter $p/2$ over the meniscus gap h, whatever the polygonal shape of the wetted component:

$$k \div \gamma \frac{p}{2h} \tag{15.1}$$

In another contribution, Arutinov [1] published experimental data which could be used to validate the models of Chap. 8 on the case study of mesoscopic polymer foils self-centering.

Sariola and Zhou's studies have also shown a possible alternative to meniscus centering by condensing mist microdroplets instead of depositing on single droplet [21].

Many bio-inspired studies also make use of surface tension effects. Let us cite Ikezoe's work to create a large surface-tension gradient around a metal-organic framework, in order to power a translational motion [12]. Gripping can also be modulated to provide capillary-based switchable adhesion [26], or be inspired by fluid-based adhesion in insects [6].

Surface tension manufacturing is also a fast growing field. Kang et al. [13] could control the shape and size of nanopillar assembly by adhesion-mediated elastocapillary interaction, in order to produce chiral assemblies (Fig. 15.2).

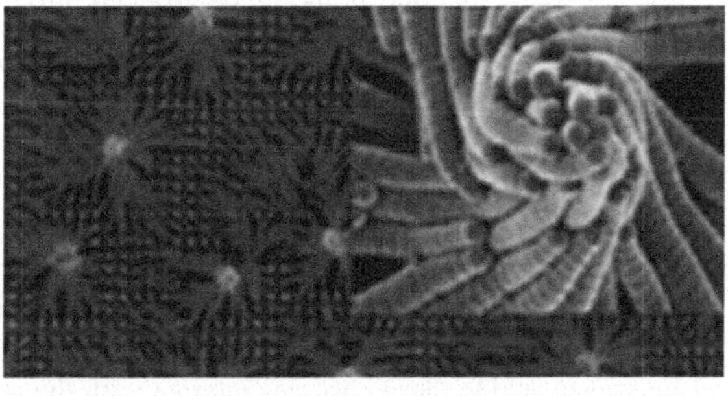

Fig. 15.2 250 nm pillars cluster assembled by capillary forces. Reprinted with permission from [13]. Copyright (2010) American Chemical Society

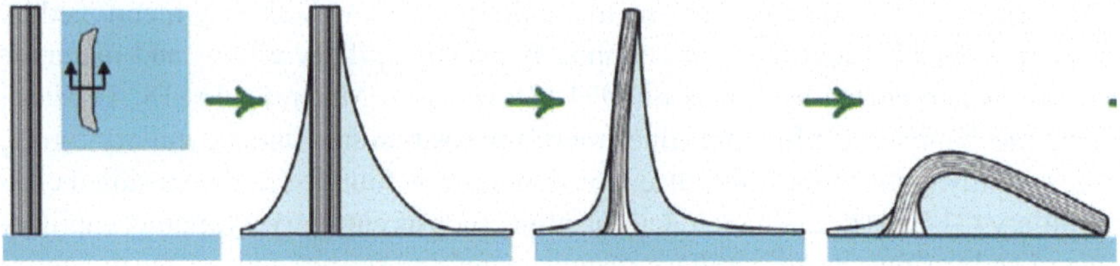

Fig. 15.3 Mechanism of capillary folding described by Tawfick et al: (1) dry carbon nanotube growth; (2) capillary rise; (3) densification and tilt; (4) buckling. Reprinted with permission from [23]. Copyright (2011) American Chemical Society

Tawfick et al. [23] studied the folding conditions of carbon nanotubes under the action of capillary forces (Fig. 15.3).

Tonry et al. recently introduced the concept of electric field assisted capillarity (EFAC, [24]), which is a 'novel process for the single step fabrication of hollow polymer microstructures. This process is claimed to work experimentally on a microscale level using PDMS. This process makes use of both the electrohydrodynamics of polymer at the microscale and the voltage enhanced capillary force exerted on the polymer'.

Finally, surface tension effects can obviously be studied, modelled and used in other fields than microsystems: let us mention for instance the domain of soil mechanics [9, 25].

15.3 Perspectives

The scientific perspectives of the work described in this book are driven by fundamental questions raised in microfluidics and interfacial science. The rational use of surface tension, surface stress and capillary effects could be applied to highly relevant case studies: among many, let us cite capillary gripping, capillary filling, capillary alignment, capillary sealing, capillary self-assembly or droplet manipulation.

These fundamental questions can be grouped into three categories:

1. **Fluid statics and dynamics**: How much force is applied on solids by menisci and micro-flows in a given geometry? What happens if the solid bends when subject to these forces? Are the interfaces stable and what if not? What is the effect of an electric field? How can the microscopic description of wetting be translated into an adequate boundary condition at the macroscopic level?
2. **Surface engineering**: How does a contact line move on a rough surface? Can one pattern the surface microscopically to control this motion? How is the motion affected by evaporation, or by the presence of colloid particles in the liquid or at the interface? Do these particles interact with the micro-patterns on the surface? Can one create highly 3D patterns on the surface by using capillary forces?

3. **Liquid engineering**: How to measure the interfacial properties of complex liquids where apart from surface tension a surface viscoelastic response is present? How to infer macroscopic properties from the dynamics at the molecular scale? And how to engineer liquids and tailor them to the requirements arising from applications? Can one make a liquid that is biocompatible, and has a large surface tension and a low viscosity?

These questions constitute the foundation of a Belgian multidisciplinary project, entitled **MicroMAST**[1]: Micromanipulation and Microfluidics: Multiscale Applications of Surface Tension.

Combining the forefront research in physics, material science, chemistry and engineering, this project will cover topics that range from fundamental theory with atomistic simulations to experiments to investigate the fundamentals and selected more applied case studies. It will address both static and dynamic points of view, and establish the link between the microscopic properties of liquids and surfaces, and the macroscopic performances expected in the case studies.

For these developments to be applied, it will also become necessary to increase the repeatability, which means addressing lacks of repeatability due to wetting and charge trapping.

As a final note, we do hope that reading this book will help researchers, scientists, engineers and students to improve their knowledge on this fascinating and vivid topic.

References

1. G. Arutinov, E.C.P. Smits, M. Mastrangeli, G. van Eck, J. van den Brand, H.F.M. Schoo, A. Dietzel, Capillary self-alignment of mesoscopic foil components for sensor-systems-in-foils. J. Micromech. Microeng. **22**, 115022 (2012)
2. J. Berthier, S. Mermoz, K. Brakke, L. Sanchez, C. Frétigny, L. Di Cioccio, Capillary self-alignment of polygonal chips: a generalization for the shift-restoring force. Microfluid. Nanofluid. **13**, 461–468 (2012)
3. J. Bico, B. Roman, L. Moulin, A. Boudaoud, Elastocapillary coalescence in wet hair. Nature **432**, 690 (2004)
4. T. Cambau, J. Bico, E. Reyssat, Capillary rise between flexible walls. EPL **96**(7), 11047–24001 (2011)
5. H. Cooray, P. Cicuta, D. Vella, The capillary interaction between two vertical cylinders. J. Phys. Condens. Matter **24**, 284104 (2012)
6. J.-H. Dirks, W. Federle, Fluid-based adhesion in insects–principles and challenges. Soft Matter. **7**, 11047–11053 (2011)
7. C. Elbuken, T. Glawdel, D. Chan, C.L. Ren, Detection of microdroplet size and speed using capacitive sensors. Sens. Actuators, A 171(2), 55–62 (2011).
8. M.A. Fortes, Deformation of solid surfaces due to capillary forces. J. Colloid Interface Sci. **100**(1), 17–26 (1984)
9. F. Gabrieli, P. Lambert, S. Cola, F. Calvetti, Micromechanical modelling of erosion due to evaporation in a partially wet granular slope. Int. J. Numer. Anal. Meth. Geomech. **36**, 918–943 (2012)

[1] www.micromast.be, 2012–2017, IAP project funded by Belspo—http://www.belspo.be/IAP/.

10. M. Gaudet, S. Arscott, Optical actuation of microelectromechanical systems using photoelectrowetting. Appl. Phys. Lett. 100(22), 224103–224103-4 (2012).
11. E. Hendarto, Y.B. Gianchandani, Marangoni-driven micromotor in liquid medium, in *Solid-State Sensors, Actuators and Microsystems Conference (TRANSDUCERS), 2011 16th, international*, 2011, pp. 246–249.
12. Y. Ikezoe, G. Washino, T. Uemura, S. Kitagawa, H. Matsui, Autonomous motors of a meta-organic framework powered by reorganization of self-assembled peptides at interfaces. Nat. Mater. **11**, 1081 (2012)
13. S.H. Kang, B. Pokroy, L. Mahadevan, J. Aizenberg, Control of shape and size of nanopillar assembly by adhesion-mediated elastocapillary interaction. ACS Nano **4**(11), 6323–6331 (2010)
14. C.R. Knospe, Capillary force actuation: a mechatronic perspective, ed. by E. Eleftheriou, S.O.R. Moheimani. *Control Technologies for Emerging Micro and Nanoscale Systems*, in Lecture Notes in Control and Information Sciences, vol. 413 (Springer, Berlin, 2011), pp. 201–218.
15. C.R. Knospe, H. Haj-Hariri, Capillary force actuators: modeling, dynamics, and equilibria. Mechatronics **22**(3), 251–256 (2012)
16. P.A. Kralchevsky, K. Nagayama, *Particles et Fluid Interfaces and Membranes* (Elsevier, New York, 2001)
17. G.R. Lester, Contact angles of liquids at deformable solid surfaces. J. Colloid Sci. **16**, 315–326 (1961)
18. C.H. Mastrangelo, C.H. Hsu, Mechanical stability and adhesion of microstructures under capillary forces–Part I: basic theory. J. Microelectromech. Syst. **2**(1), 33–43 (1993)
19. C.H. Mastrangelo, C.H. Hsu, Mechanical stability and adhesion of microstructures under capillary forces–Part II: Experiments. J. Microelectromech. Syst. **2**(1), 44–55 (1993)
20. N. Paust, C. Litterst, T. Metz, M. Eck, C. Ziegler, R. Zengerle, P. Koltay, Capillary-driven pumping for passive degassing and fuel supply in direct methanol fuel cells. Microfluid. Nanofluid. **7**(4), 531–543 (2009)
21. V. Sariola, Droplet self-alignment: high-precision robotic microassembly and self-assembly. Ph.D. dissertation, Aalto University, Jun, 2012
22. R. Shabani, H.J. Cho, Active surface tension driven micropump using droplet/meniscus pressure gradient. Sensors and Actuators, B 1(2), 85–102 (2012 in press).
23. S. Tawfick, M. De Volder, A.J. Hart, Structurally programmed capillary folding of carbon nanotube assemblies. Langmuir **27**(10), 6389–6394 (2011)
24. C. Tonry, M. Patel, C. Bailey, M.P.Y. Desmuliez, S. Cargill, W. Yu, Modelling of the electric field assisted capillarity effect used for the fabrication of hollow polymer microstructures. IEEE, Apr 2012, pp. 1/6-6/6.
25. V.I. Vasil'ev, V.V. Popov, G.G. Tsypkin, Nonlinear problem of unsaturated frozen soil thawing in the presence of capillary forces. Fluid Dyn. **47**(1), 106–113 (2012)
26. M.J. Vogel, P.H. Steen, Capillarity-based switchable adhesion. Proc. Nat. Acad. Sci. **107**(8), 3377–3381 (2010)
27. T. Zhang, T. Cui, High-performance surface-tension-driven capillary pumping based on layer-by-layer self assembly of TiO_2 nanoparticles IEEE, Jun, 2011, pp. 606–609

Appendix A
Navier-Stokes Equations

A.1 Cartesian Coordinates

The deviatoric tensor is:

$$\tau = \mu \begin{pmatrix} 2\frac{\partial u_x}{\partial x} & \frac{\partial u_y}{\partial x} + \frac{\partial u_x}{\partial y} & \frac{\partial u_z}{\partial x} + \frac{\partial u_x}{\partial z} \\ \frac{\partial u_x}{\partial y} + \frac{\partial u_y}{\partial x} & 2\frac{\partial u_y}{\partial y} & \frac{\partial u_z}{\partial y} + \frac{\partial u_y}{\partial z} \\ \frac{\partial u_x}{\partial z} + \frac{\partial u_z}{\partial x} & \frac{\partial u_y}{\partial z} + \frac{\partial u_z}{\partial y} & 2\frac{\partial u_z}{\partial z} \end{pmatrix} \tag{A.1}$$

The flow conservation is:

$$\frac{\partial u_x}{\partial x} + \frac{\partial u_y}{\partial y} + \frac{\partial u_z}{\partial z} = 0 \tag{A.2}$$

The scalar Navier-Stokes equations are:

$$\rho\left(\frac{\partial u_x}{\partial t} + u_x\frac{\partial u_x}{\partial x} + u_y\frac{\partial u_x}{\partial y} + u_z\frac{\partial u_x}{\partial z}\right) = \rho g_x - \frac{\partial p}{\partial x} + \mu\left(\frac{\partial^2 u_x}{\partial x^2} + \frac{\partial^2 u_x}{\partial y^2} + \frac{\partial^2 u_x}{\partial z^2}\right) \tag{A.3}$$

$$\rho\left(\frac{\partial u_y}{\partial t} + u_x\frac{\partial u_y}{\partial x} + u_y\frac{\partial u_y}{\partial y} + u_z\frac{\partial u_y}{\partial z}\right) = \rho g_y - \frac{\partial p}{\partial y} + \mu\left(\frac{\partial^2 u_y}{\partial x^2} + \frac{\partial^2 u_y}{\partial y^2} + \frac{\partial^2 u_y}{\partial z^2}\right) \tag{A.4}$$

$$\rho\left(\frac{\partial u_z}{\partial t} + u_x\frac{\partial u_z}{\partial x} + u_y\frac{\partial u_z}{\partial y} + u_z\frac{\partial u_z}{\partial z}\right) = \rho g_z - \frac{\partial p}{\partial z} + \mu\left(\frac{\partial^2 u_z}{\partial x^2} + \frac{\partial^2 u_z}{\partial y^2} + \frac{\partial^2 u_z}{\partial z^2}\right) \tag{A.5}$$

P. Lambert (ed.), *Surface Tension in Microsystems*, Microtechnology and MEMS,
DOI: 10.1007/978-3-642-37552-1, © Springer-Verlag Berlin Heidelberg 2013

A.2 Cylindrical Coordinates

Cylindrical coordinates are defined by the vectors $\overline{1}_r$, $\overline{1}_\theta$, $\overline{1}_z$. In such a basis, the deviatoric tensor is:

$$\tau = \mu \begin{pmatrix} 2\frac{\partial u_r}{\partial r} r & \frac{\partial u_\theta}{\partial r} + \frac{1}{r}\frac{\partial u_r}{\partial \theta} - \frac{u_\theta}{r} & \frac{\partial u_z}{\partial r} + \frac{\partial u_r}{\partial z} \\ \frac{1}{r}\frac{\partial u_r}{\partial \theta} - \frac{u_\theta}{r} + \frac{\partial u_\theta}{\partial r} & 2\frac{1}{r}\frac{\partial u_\theta}{\partial \theta} + \frac{u_r}{r} & \frac{1}{r}\frac{\partial u_z}{\partial \theta} + \frac{\partial u_\theta}{\partial z} \\ \frac{\partial u_r}{\partial z} + \frac{\partial u_z}{\partial r} & \frac{\partial u_\theta}{\partial z} + \frac{1}{r}\frac{\partial u_z}{\partial \theta} & 2\frac{\partial u_z}{\partial z} \end{pmatrix} \tag{A.6}$$

The flow conservation is:

$$\frac{1}{r}\frac{\partial (r u_r)}{\partial r} + \frac{1}{r}\frac{\partial u_\theta}{\partial \theta} + \frac{\partial u_z}{\partial z} = 0 \tag{A.7}$$

$$\Rightarrow \quad \frac{\partial u_r}{\partial r} + \frac{1}{r}\frac{\partial u_\theta}{\partial \theta} + \frac{\partial u_z}{\partial z} + \frac{u_r}{r} = 0 \tag{A.8}$$

The scalar Navier-Stokes equations are:

$$\rho\left(\frac{\partial u_r}{\partial t} + u_r\frac{\partial u_r}{\partial r} + \frac{u_\theta}{r}\frac{\partial u_r}{\partial \theta} + u_z\frac{\partial u_r}{\partial z} - \frac{u_\theta^2}{r}\right) = \rho g_r - \frac{\partial p}{\partial r} + \mu\left(\frac{\partial^2 u_r}{\partial r^2} + \frac{1}{r^2}\frac{\partial^2 u_r}{\partial \theta^2} + \frac{\partial^2 u_r}{\partial z^2} + \frac{1}{r}\frac{\partial u_r}{\partial r} - \frac{2}{r}\frac{\partial u_\theta}{\partial \theta} - \frac{u_r}{r^2}\right) \tag{A.9}$$

$$\rho\left(\frac{\partial u_\theta}{\partial t} + u_r\frac{\partial u_\theta}{\partial r} + \frac{u_\theta}{r}\frac{\partial u_\theta}{\partial \theta} + u_z\frac{\partial u_\theta}{\partial z} + \frac{u_r u_\theta}{r}\right) = \rho g_\theta - \frac{1}{r}\frac{\partial p}{\partial \theta} + \mu\left(\frac{\partial^2 u_\theta}{\partial r^2} + \frac{1}{r^2}\frac{\partial^2 u_\theta}{\partial \theta^2} + \frac{\partial^2 u_\theta}{\partial z^2} + \frac{1}{r}\frac{\partial u_\theta}{\partial r} + \frac{2}{r^2}\frac{\partial u_r}{\partial \theta} - \frac{u_\theta}{r^2}\right) \tag{A.10}$$

$$\rho\left(\frac{\partial u_z}{\partial t} + u_r\frac{\partial u_z}{\partial r} + \frac{u_\theta}{r}\frac{\partial u_z}{\partial \theta} + u_z\frac{\partial u_z}{\partial z}\right) = \rho g_z - \frac{\partial p}{\partial z} + \mu\left(\frac{\partial^2 u_z}{\partial r^2} + \frac{1}{r^2}\frac{\partial^2 u_z}{\partial \theta^2} + \frac{\partial^2 u_z}{\partial z^2} + \frac{1}{r}\frac{\partial u_z}{\partial r}\right) \tag{A.11}$$

When problems are axisymmetric, the variables do not depend on θ so that relative derivatives with respect to θ are null [1] (of course, $\overline{1}_r$ and $\overline{1}_\theta$ are differentiated with respect to θ). Because we will consider boundary conditions only expressed in along $\overline{1}_r$ and $\overline{1}_z$, we may write down that the angular component of any vector field is null.[1] In this case, suppose the existence of an angular component u_θ of the velocity. This component could be replaced by $-u_\theta$, also matching with boundary conditions since

[1] This is of course not always true since Neustupa and Pokorny [2] interested in the analytical condition for existence of angular component of velocity for axisymmetric problems.

they have a zero θ component. This would gives two different solutions for the same problem, in contradiction with the uniqueness of a solution (for stable solutions). The velocity field is therefore $\overline{u} = u_r \overline{1}_r + u_z \overline{1}_z$. The axial symmetry also imposes the gravity to be along z. The flow conservation and the Navier-Stokes equations can be simplified in:

$$\frac{\partial u_r}{\partial r} + \frac{\partial u_z}{\partial z} + \frac{u_r}{r} = 0 \tag{A.12}$$

$$\rho\left(\frac{\partial u_r}{\partial t} + u_r \frac{\partial u_r}{\partial r} + u_z \frac{\partial u_r}{\partial z}\right) = \rho g_r - \frac{\partial p}{\partial r} + \mu\left(\frac{\partial^2 u_r}{\partial r^2} + \frac{\partial^2 u_r}{\partial z^2} + \frac{1}{r}\frac{\partial u_r}{\partial r} - \frac{u_r}{r^2}\right) \tag{A.13}$$

$$\rho\left(\frac{\partial u_z}{\partial t} + u_r \frac{\partial u_z}{\partial r} + u_z \frac{\partial u_z}{\partial z}\right) = \rho g_z - \frac{\partial p}{\partial z} + \mu\left(\frac{\partial^2 u_z}{\partial r^2} + \frac{\partial^2 u_z}{\partial z^2} + \frac{1}{r}\frac{\partial u_z}{\partial r}\right) \tag{A.14}$$

Appendix B
Vectorial Operators

Divergence of Product $k\overline{u}$

$$\overline{\nabla} \cdot (k\overline{u}) = \overline{\nabla} k \cdot \overline{u} + k\overline{\nabla} \cdot \overline{u} \tag{B.1}$$

Proof The development in cartesian coordinates gives:

$$\begin{aligned}
\overline{\nabla} \cdot (k\overline{u}) &= \partial_x(ku_x) + \partial_y(ku_y) + \partial_z(ku_z) \\
&= (\partial_x k)u_x + k\partial_x u_x + (\partial_y k)u_y + k\partial_y u_y + (\partial_z k)u_z + k\partial_y u_z \\
&= (\partial_x k)u_x + (\partial_y k)u_y + (\partial_z k)u_z + k\partial_x u_x + k\partial_y u_y + k\partial_y u_z \\
&= \overline{\nabla} k \cdot \overline{u} + k\overline{\nabla} \cdot \overline{u}
\end{aligned}$$

This results may be transposed in any coordinate system, thanks to the invariance of the operators.

Divergence of Product $\tau \cdot \overline{u}$

$$\overline{\nabla} \cdot (\tau \cdot \overline{u}) = \left(\overline{\nabla} \cdot \tau\right) \cdot \overline{u} + \tau^{\mathrm{T}} : \overline{\nabla}\,\overline{u} \tag{B.2}$$

Proof The development in Cartesian coordinates gives:

$$\begin{aligned}
\overline{\nabla} \cdot (\tau \cdot \overline{u}) &= \partial_x(\overline{\tau}_{x:} \cdot \overline{u}) + \partial_y(\overline{\tau}_{y:} \cdot \overline{u}) + \partial_z(\overline{\tau}_{z:} \cdot \overline{u}) \\
&= \partial_x(\overline{\tau}_{x:}) \cdot \overline{u} + \overline{\tau}_{x:} \cdot \partial_x\overline{u} + \partial_y(\overline{\tau}_{y:}) \cdot \overline{u} + \overline{\tau}_{y:} \cdot \partial_y\overline{u} + \partial_z(\overline{\tau}_{z:}) \cdot \overline{u} + \overline{\tau}_{z:} \cdot \partial_z\overline{u} \\
&= \partial_x(\overline{\tau}_{x:}) \cdot \overline{u} + \partial_y(\overline{\tau}_{y:}) \cdot \overline{u} + \partial_z(\overline{\tau}_{z:}) \cdot \overline{u} + \overline{\tau}_{x:} \cdot \partial_x\overline{u} + \overline{\tau}_{y:} \cdot \partial_y\overline{u} + \overline{\tau}_{z:} \cdot \partial_z\overline{u} \\
&= (\partial_x\overline{\tau}_{x:} + \partial_y\overline{\tau}_{y:} + \partial_z\overline{\tau}_{z:}) \cdot \overline{u} + \left(\overline{\tau}_{x:}^{\mathrm{T}}\ \overline{\tau}_{y:}^{\mathrm{T}}\ \overline{\tau}_{z:}^{\mathrm{T}}\right) : \left(\partial_x\overline{u}\ \partial_y\overline{u}\ \partial_z\overline{u}\right) \\
&= \left(\overline{\nabla} \cdot \tau\right) \cdot \overline{u} + \tau^{\mathrm{T}} : \overline{\nabla}\,\overline{u}
\end{aligned}$$

P. Lambert (ed.), *Surface Tension in Microsystems*, Microtechnology and MEMS,
DOI: 10.1007/978-3-642-37552-1, © Springer-Verlag Berlin Heidelberg 2013

This results may be transposed in any coordinate system, thanks to the invariance of the operators.

If in addition, the tensor τ is symmetric, we have:

$$\overline{\nabla} \cdot (\tau \cdot \overline{u}) = \left(\overline{\nabla} \cdot \tau\right) \cdot \overline{u} + \tau : \overline{\nabla}\,\overline{u} \tag{B.3}$$

Commutative Matrix Product

$$(\tau \cdot \overline{u}) \cdot \overline{v} = (\tau^{\mathrm{T}} \cdot \overline{v}) \cdot \overline{u} \tag{B.4}$$

Proof

$$\begin{aligned}(\tau \cdot \overline{u}) \cdot \overline{v} &= (\overline{u}^{\mathrm{T}} \cdot \tau^{\mathrm{T}})^{\mathrm{T}} \cdot \overline{v} \\ &= \overline{u}^{\mathrm{T}} \cdot (\tau^{\mathrm{T}} \cdot \overline{v}) \\ &= (\tau^{\mathrm{T}} \cdot \overline{v}) \cdot \overline{u}\end{aligned}$$

The Surface Divergence Operator

The surface divergence operator it a particular case of the volume divergence, considering contribution along direction generating a surface. Consider a parametric surface defined by u and v. The surface divergence operator is defined by:

$$\overline{\nabla}_s \cdot \overline{F} = \frac{1}{I^2}\overline{r_u} \cdot \left(G\frac{\partial \overline{F}}{\partial u} - F\frac{\partial \overline{F}}{\partial v}\right) + \frac{1}{I^2}\overline{r_v} \cdot \left(E\frac{\partial \overline{F}}{\partial v} - F\frac{\partial \overline{F}}{\partial u}\right) \tag{B.5}$$

where

$$\overline{r_u} = \frac{\partial \overline{r}}{\partial u} \tag{B.6}$$

$$\overline{r_v} = \frac{\partial \overline{r}}{\partial v} \tag{B.7}$$

are the vectors tangents to the surface in the direction of increasing u and v respectively, and

$$E = \overline{r_u} \cdot \overline{r_u} \tag{B.8}$$

$$F = \overline{r_u} \cdot \overline{r_v} \tag{B.9}$$

$$G = \overline{r_v} \cdot \overline{r_v} \tag{B.10}$$

$$I^2 = EG - F^2 \tag{B.11}$$

are kind of scaling factor, taking into account the ratio of the increasing of and the arc element with respect to the increasing of u and v in one hand (parameters E and G) and the orthogonality of the tangent vectors (parameter F). Note that usually, the parameter I is denoted by a H that defines here the mean curvature.

As a remark, in the case of principal directions ($F = 0$ since the tangent vector to surface are perpendicular), the surface divergence reads:

$$\overline{\nabla}_s \cdot \overline{F} = \overline{1}_u \cdot \frac{\partial \overline{F}}{\partial u} \frac{1}{||\overline{r}_u||} + \overline{1}_v \cdot \frac{\partial \overline{F}}{\partial v} \frac{1}{||\overline{r}_v||} \tag{B.12}$$

highlighting the elimination of the normal contribution.

The divergence operator is the trace of the gradient matrix, the surface divergence is the trace from which we have removed the normal contribution and we have:

$$\left(\mathsf{I} - \overline{1}_n \otimes \overline{1}_n\right) : \left(\overline{\nabla}\,\overline{u}\right) \tag{B.13}$$

The divergence theorem states that:

$$\iint\limits_{\Gamma} \overline{\nabla}_s \cdot \overline{u}\, \mathrm{dS} = \int\limits_{\partial\Gamma} \overline{u} \cdot \overline{1}_m\, \mathrm{dl} + \iint\limits_{\Gamma} 2H\overline{u} \cdot \overline{1}_n\, \mathrm{dS} \tag{B.14}$$

To understand this theorem, we will illustrate the case for which u and v are the principal directions orthonormed ($E = G = 1$ and $F = 0$). Let's compute the flow balance of the vector field that enters and exits along edges of a small surface $\Delta u \Delta v$ (see Fig. B.1).

The balance along u is:

$$S_u = \left(\overline{F}(u + \Delta u, v) \cdot \overline{1}_u(u + \Delta u, v) - \overline{F}(u, v) \cdot \overline{1}_u(u, v)\right) \Delta v \tag{B.15}$$

and along v:

$$S_v = \left(\overline{F}(u, v + \Delta v) \cdot \overline{1}_v(u, v + \Delta v) - \overline{F}(u, v) \cdot \overline{1}_v(u, v)\right) \Delta u \tag{B.16}$$

The first product is:

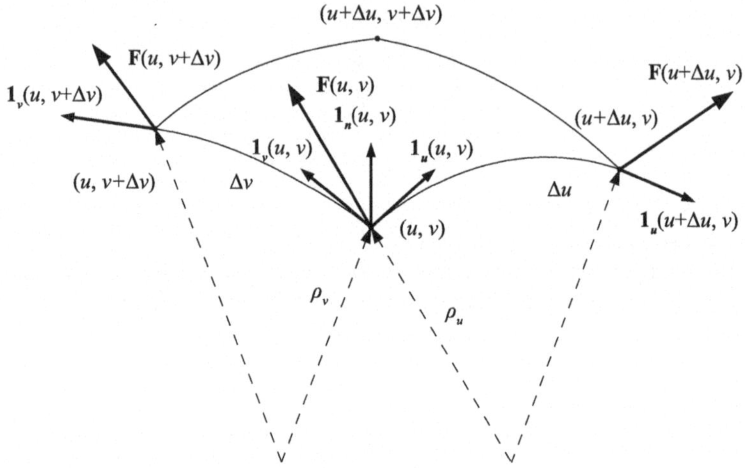

Fig. B.1 Balance of flow on a small surface $\Delta u \Delta v$

$$\overline{F}(u + \Delta u, v) \cdot \overline{1}_u(u + \Delta u, v) = \left(\overline{F}(u, v) + \Delta\overline{F}\right) \cdot \left(\overline{1}_u(u, v) + \Delta\overline{1}_u\right)$$
$$= \overline{F}(u, v) \cdot \overline{1}_u + \Delta\overline{F} \cdot \overline{1}_u + \Delta\overline{F} \cdot \overline{1}_u + \Delta\overline{F} \cdot \Delta\overline{1}_u \quad \text{(B.17)}$$

The latter can be neglected because of the second order of Δ. The balance along u reads now

$$S_u = \left(\Delta\overline{F} \cdot \overline{1}_u + \overline{F} \cdot \Delta\overline{1}_u\right)\Delta v \quad \text{(B.18)}$$

$$= \left(\frac{\Delta\overline{F}}{\Delta u} \cdot \overline{1}_u + \overline{F} \cdot \frac{\Delta\overline{1}_u}{\Delta u}\right)\Delta u \Delta v \quad \text{(B.19)}$$

For $\Delta u \to 0$ and $\Delta v \to 0$

$$S_u = \left(\frac{\partial\overline{F}}{\partial u} \cdot \overline{1}_u + \overline{F} \cdot \frac{\partial\overline{1}_u}{\partial u}\right)\Delta u \Delta v \quad \text{(B.20)}$$

And a similar development for v gives the total balance in $\Delta u \Delta v$

$$S_u + S_v = \left[\frac{\partial\overline{F}}{\partial u} \cdot \overline{1}_u + \frac{\partial\overline{F}}{\partial v} \cdot \overline{1}_v + \overline{F} \cdot \left(\frac{\partial\overline{1}_u}{\partial u} + \frac{\partial\overline{1}_v}{\partial v}\right)\right]\Delta u \Delta v \quad \text{(B.21)}$$

$$= \left(\overline{\nabla}_S \cdot \overline{F} - \overline{F} \cdot 2H\overline{1}_n\right)\Delta u \Delta v \quad \text{(B.22)}$$

The latter equation makes appearing the curvature of the surface. It is true in the form exposed above because the isoline u and v are the principal directions. In this particular case:

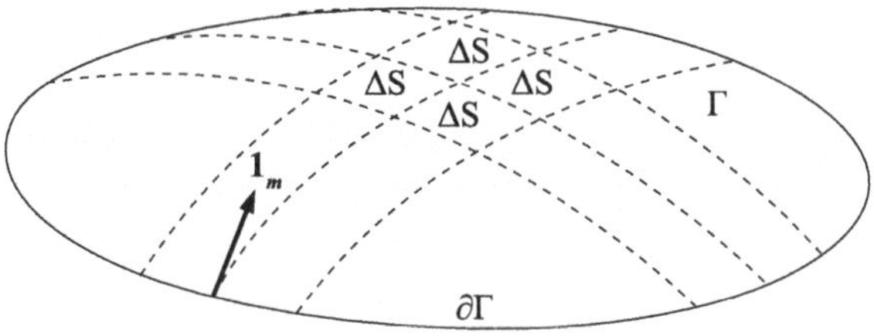

Fig. B.2 Balance of flow on a small surface $\Delta u \Delta v$

$$\frac{\partial \overline{1}_u}{\partial u} + \frac{\partial \overline{1}_v}{\partial v} = -\frac{1}{\rho_u}\overline{1}_n - \frac{1}{\rho_v}\overline{1}_n \tag{B.23}$$

$$= -2H\overline{1}_n \tag{B.24}$$

If we place side by side two inifinitesimal surface ΔS, say in the u direction, the flow exiting the first square is equal to the flow entering in the second one. That is, for a surface Γ delimited by a contour $\partial\Gamma$, the total balance of flow can be deduced by a sum of infinitesimal surfaces where the inflow and outflow can be considered only at $\partial\Gamma$, Fig. B.2. The total balance of flow is hence:

$$\iint_{\Gamma} (S_u + S_v)\,\mathrm{dS} = \int_{\partial\Gamma} \overline{F}\cdot\overline{1}_m\,\mathrm{dl} \tag{B.25}$$

With (B.22), the left hand term is

$$\iint_{\Gamma} (S_u + S_v)\,\mathrm{dS} = \iint_{\Gamma} \overline{\nabla}_S\cdot\overline{F}\,\mathrm{dS} - \iint_{\Gamma} \overline{F}\cdot 2H\overline{1}_n\,\mathrm{dS} \tag{B.26}$$

proving (B.14).

References

1. S. Leonardi, J. Màlek, J. Necas, M. Pokorny, On axially symmetric flows in r3. Zeitschrift für Analysis und ihre Anwendungen **18**(3), 639–649 (1999)
2. J. Neustupa, M. Pokorny, Axisymmetric flow of Navier-Stokes fluid in the whole space with non-zero angular velocity component. Math. Bohem. **126**(2), 469–481 (2001)

Index

A
Actuator
  hydraulic, 212
  micro, 211, 224
  piston-cylinder, 212, 224
Adhesion, 8, 20, 32, 47, 98, 244, 280, 284–286, 311
Atomic Force Microscope (AFM), 279

B
Baxter, x
Biomimetics, 255
Bond number, 5
Boundary conditions, 133
  equilibrium, 27
Brakke, K., 57
Bubbles, 4

C
Capillary, 5
  forces, viii
  length, ix
    energetic method, 20
    exact resolution, 26
    lateral, 46
  length, 52
  number, 5
  pressure, 6, 228, 272
  torque, 46
Capillary forces, 195
Case study, 195
Cassie, x
Circular model, 157
Coating, 7, 235, 237, 244
Cohesion-tension theory, 256, 268
Contact angle, 3, 7–13, 20, 22, 23, 27–34, 36–40, 52, 61, 62, 138, 140, 143, 148, 157, 161, 165, 282, 283, 294, 299–301
  dynamic, 8
  hysteresis, 8
Curvature, 228

D
de Gennes, P.G., 83
Degree-of-freedom, 195
Dispensing, 53
Downscaling, 3
Dip Pen Nanolithography (DPN), 280
Droplet, 4, 5, 31, 71, 72, 74, 76, 77, 82, 83, 86, 88, 119, 181, 280, 281, 289, 290, 293, 299, 311, 312
Dynamic modelling, 195
Dynamical, 33, 129, 133, 136, 139, 179–181, 186, 189, 192, 280, 309
Dynamics, 7, 20, 40, 48, 58, 62, 72, 192, 195, 230–232, 245, 285, 288, 301, 311–313

E
Elastocapillary effects, 229
Electric, 13, 163, 228, 233, 270, 309, 310, 312, 314
Engineering, 3, 13, 62, 211, 224, 227, 228, 255, 312, 313
Euler, 64
Evaporation, 13, 22, 53, 59, 62, 77, 168, 221, 241, 244, 255–257, 260, 265–273, 276, 289–291, 312
Evapotranspiration, 256

P. Lambert (ed.), *Surface Tension in Microsystems*, Microtechnology and MEMS,
DOI: 10.1007/978-3-642-37552-1,

F
Flip-chip, xi
Fluidic assembly, 229
Fluid statics and dynamics, xii

G
Gibbs' inequality, 60, 62

H
Henein, S., 66
Hessian, 57
History, ix
Hooke's law, 262
Hybrid assembly, 229
Hysteresis, 7, 8, 60–62, 166, 211, 213, 287, 301

I
Interfacial energy, 6
Israelachvili, xi, 22, 43

L
Laplace, 96
  equation, 19, 26, 42, 43, 282
  force, 32, 39, 41, 43
  law, 6
  Pierre-Simon, x, 26
  pressure, 228
Lenders number, 6
Liquid Bridge, 129
Liquid engineering, xiii

M
Magnitude, xiii
Manipulation, 19, 72, 115, 243, 245, 312
Mass conservation equation, 132
MEMS, vii, 20, 46, 72, 229, 236, 246, 247, 259, 267
Microactuation, viii
Microfluidic, 135, 212, 229, 270, 271
Microfluidic devices, vii
MicroMAST, xii
Microsystems, 46
Micro-assembly, xi
Microrobotics, xi
Model, x, xi, 8, 29, 33, 36, 38, 54, 56, 99, 104, 259, 260, 271
Modelling, 195, 239, 296, 309
Murray's law, 268

N
NAnoDISpensing Technique (NADIS), 279
Navier–Stokes equation, 132, 133, 135

O
Orders of magnitude, xiii

P
Pèclet number, 5
Parabolic model, 147, 157
Piner, 280
Plateau, J., x
Poiseuille equation, 297

R
Rayleigh, 65

S
Scale, viii
Scaling law, ix, xiii, 3, 4, 19, 90
Seal
  micro, 212–214
  Surface engineering, xiii
  Surface tension, vii, 200, 212–214, 220–225
Self-alignment, 45, 47, 48, 62, 72, 75, 180, 181, 200, 228, 229, 232, 234, 235, 243–245, 247, 248
Self-assembly, 20, 45, 46, 62, 71, 89, 227, 228, 311, 312
Shimoyama, 230
Similitude, 5
STM, 280
Stokes
  theorem, 23
Substrate, 8, 13, 46, 47, 123, 129, 130, 142, 171, 180, 202, 227–229, 264, 282, 284, 286, 288, 291–293, 300, 301
Surface Evolver, 19, 22–26, 35, 36, 42
Surface tension
  Seal, 212, 213
Surface tension, 4, 6, 7, 9

T
Tanner, 10, 11

U
Ultra thin chip (UTC), 229

V
Van der Waals, 5

W
Washburn, x
Weber number, 5
Wenzel, x
Wettability, 7
Wetting, 7, 21, 33, 42, 46, 228–230, 232, 235, 248, 258, 312, 313
Whitesides, 46, 230

Y
Young
  Duprè equation, 7
  modulus

Z
Zisman model, 8

The manufacturer's authorised representative in the EU is Springer Nature Customer Service Centre GmbH, Europaplatz 3, 69115 Heidelberg, Germany. If you have any concerns regarding our products, please contact ProductSafety@springernature.com

Printed and bound by CPI Group (UK) Ltd, Croydon, CR0 4YY
01/10/2026
02225973-0010